Industrial Ecology and Spaces of Innovation

Industrial Ecology and Spaces of Innovation

Edited by

Ken Green

Professor of Environmental Innovation Management, Manchester Business School, University of Manchester, UK

Sally Randles

Research Fellow, ESRC Centre for Research on Innovation and Competition (CRIC), Manchester Business School, University of Manchester, UK

Edward Elgar

Cheltenham, UK • Northampton, MA, USA

Published by
Edward Elgar Publishing Limited
Glensanda House
Montpellier Parade
Cheltenham
Glos GL50 1UA
UK

Edward Elgar Publishing, Inc.
136 West Street
Suite 202
Northampton
Massachusetts 01060
USA

A catalogue record for this book
is available from the British Library

Library of Congress Cataloguing in Publication Data
Industrial ecology and spaces of innovation / edited by Ken Green,
Sally Randles.
p. cm.
Includes bibliographical references and index.
1. Industrial ecology. I. Green, Kenneth. II. Randles, Sally.
TS161.I543 2006
658.5–dc22

ISBN-13: 978 1 84542 097 0
ISBN-10: 1 84542 097 7

Printed and bound in Great Britain by MPG Books Ltd, Bodmin, Cornwall

Contents

Figures

Tables

Contributors

Frans Berkhout is Professor of Innovation and Sustainability and Director of the Institute for Environmental Studies (IVM) at the Vrije Universiteit in Amsterdam, The Netherlands. His research has been concerned with technology, policy and sustainability.

Yuhong Cen is a Lecturer at the College of Economics, Zhejiang Gong Shang University, China. Currently she is studying for a Ph.D. at the Institute of Innovation Research, the University of Manchester. She works on innovation management, international business and issues relevant to sustainable development.

Paul Dewick is a Lecturer in Technology Management at Manchester Business School, University of Manchester. His research interests lie in the area of innovation studies and sustainability.

Suren Erkman is Professor at the Faculty of Geosciences and Environment, Lausanne University, founder of the Institute for Communication and Analysis of Science and Technology (ICAST) in Geneva, Switzerland, co-founder of the Resource Optimization Institute in Bangalore, India and a Board member of the *Journal of Industrial Ecology* (MIT Press). His research focus is the application of industrial ecology in particular to developing country contexts.

Kieron Flanagan is a Lecturer in Science and Technology Policy and Management at Manchester Business School, and a member of PREST, the science and technology policy institute of the University of Manchester. His research and teaching interests range from science and technology policy to studies of technological change and innovation in manufacturing and services sectors.

Chris Foster runs EuGeos (an environmental consultancy) and is a part-time Research Fellow at the Institute of Innovation Research at the University of Manchester. He works on the relationship between technological change and sustainable development.

Ken Green is Professor of Environmental Innovation Management and Academic Dean of Manchester Business School, University of Manchester. He specializes in Life-Cycle Analysis and related systemic environmental evaluations.

Stuart B. Hill is Foundation Chair and Professor of Social Ecology, University of Western Sydney, Australia. He works with practitioners in the redesign/design of sustainable systems and on personal, institutional and cultural transformation.

Jeremy Howells is Professor of the Centre for Research on Innovation and Competition (CRIC) and director of CRIC and PREST at the Institute of Innovation, Manchester Business School, University of Manchester. His research focuses on innovation, knowledge, services, industry-academic links and geographies of innovation.

Xiaodong Li is Professor of the Institute for Thermal Power Engineering, Zhejiang University of China. He specializes in waste thermal treatment and related environmental protection issues.

Simon Marvin is Professor and Director of the Centre for Sustainable Urban and Regional Futures. He specializes in the changing relations between cities, regions and infrastructure networks and developing prospective approaches to understanding urban and regional change.

Will Medd is Research Associate, Department of Sociology and Centre for Sustainable Water Management, Lancaster University. He specializes in the socio-technical organization of water, domestic water consumption and interdisciplinary methodology.

Ian Miles is Professor of Technological Innovation and Social Change at the University of Manchester, where he is a director of PREST and cofounder of CRIC. Research interests include innovation in services; the roles of Knowledge Intensive Business Services (KIBS); Information Society studies; Foresight.

Murat Mirata has a Ph.D. that focuses on sustainability at the regional level and the role of industrial symbiosis in this context. He works as a project manager at DeLabs foundation, Sweden, which aims to contribute to sustainability efforts through establishing innovative partnerships and new business models.

Marcela Miozzo is Reader in Innovation Studies at the Manchester Business School, University of Manchester. Her research interests include the institutional factors that facilitate and inhibit innovation (especially sustainable technologies) in construction.

Richard Pearce is a director of Quantum Strategy & Technology, a management consultancy specializing in resource efficiency and sustainable energy. He is a director of Sustainability Northwest and Envirolink Northwest and has particular interests in innovation, technology transfer and the development of links between industry and academia.

Ramesh Ramaswamy is Director of the Resource Optimization Initiative (ROI), India (www.roi-online.org). His focus is on the applications of Industrial Ecology as a planning platform in developing countries.

Sally Randles is Research Fellow at CRIC, Manchester Business School. Her research interests are the economic sociology of markets; developing the work of Karl Polanyi on institutions and instituted process; the political economy of cities in multi-scalar perspective; and exploring the interface between industrial ecology and innovation studies.

Joe Ravetz is Deputy Director at the Centre for Urban and Regional Ecology, University of Manchester, where he runs a multi-disciplinary programme on sustainable cities and regions, combining technical models with policy analysis.

Alan Warde is Professor of Sociology and Co-Director of CRIC, University of Manchester. His research concerns explanations of consumption.

Matthias Weber is the head of the Department of Technology Policy at ARC systems research in Vienna. His research covers science and technology policy, foresight and futures studies, innovation and sustainable development, and innovation in ICT and transport.

PART 1

Introduction

1. At the interface of innovation studies and industrial ecology

Ken Green and Sally Randles

TRANS-DISCIPLINARITY IN ACTION

Industrial ecology seeks to understand how we can minimize the ecological impacts of materials flows. It has developed and is still developing a unique set of concepts and techniques for modelling socio-economic systems metaphorically as ecological systems. Industrial ecology as a theoretical 'movement' derives partly from a desire to see societies endogenize environmental impacts through new forms of economic development which tip existing systems towards 'better' configurations and practices from the point of view of efficient resource use and reduced environmental impact. When groups of firms/institutions operate collectively, this has the potential (it is argued) to produce new, radical forms of industrial/manufacturing organization.

Innovation studies is concerned with the nature and dynamics of innovation processes and pays particular attention to the capacities (and limits) of innovation to bring about socio-economic transformation. For those innovation scholars also interested in the impact of anthropogenic activities on the environment, concerns turn to whether innovation processes, originating from whatever source (producers, users, households, regulators) and involving whatever form of innovative activity (product, process, institutional, financial, organizational, regulatory) can or should be manoeuvred to bring about positive outcomes for the natural environment. At this normative interface, clearly, innovation studies and industrial ecology have much in common.

To date, however, social scientists in Innovation Studies have not engaged systematically with the Industrial Ecology community, to see what can be gained from introducing understandings about innovation processes to the conceptual and technical armoury of Industrial Ecology. Yet there is clear scope for some fruitful debate between the two communities. What we need to do is to re-think the link between the flow of materials, a flow which Industrial Ecology is especially skilled at analysing, and the technological, social, economic and organizational features and structures which cause physical flows to be patterned, concentrated or dispersed, in particular ways. We can refer to this social and economic structuring of flows as instituting processes (Polanyi 1957; Harvey and Randles 2002; Randles 2003) and key to the idea is that its outcomes are not universal but exhibit contingent spatial and historical variety. We can therefore proceed to excavate empirically the location(s) of realized innovative change within those structures and we can further identify potential sites for innovation together with, importantly, social, economic and technological constraints and limits to change.

We start with a brief discussion of the fundamentals of Industrial Ecology, bearing in mind that the editors and many of the contributors to this book locate their disciplinary and institutional home in Innovation Studies, with its own myriad of contributory disciplines from technology and management sciences, sociology, social ecology and geography. We then summarize the 'basics' of Innovation Studies, as interpreted broadly from the 'Manchester School' in the UK. We identify a series of themes and commonalities as well as conceptual tensions between innovation studies and industrial ecology when they are turned to face each other.

This edited collection has its origins in special session on Industrial Ecology organized for the ASEAT[1] conference at UMIST in April 2003.[2] Session themes were taken up and further developed at a two day workshop organized by the ESRC Centre for Research on Innovation and Competition (CRIC) at the University of Manchester in June 2003 which brought together experts in both Industrial Ecology and Innovation Studies. In discussing the translation of the exciting debates we had at those sessions into an interesting rationale for a book collection, the editors briefed all the contributors to reflect upon the world from the others' point of view. So, all were asked: What are the implications for industrial ecology of your findings or research studying processes of innovation and change? Likewise, how might your work in Industrial Ecology look if you take account of theory and processes of innovation? Ultimately, the question that frames this book is: How can ideas generated by industrial ecologists be of help to innovation scholars and *vice-versa*. In this introductory chapter we highlight and distil those themes and insights that can potentially take this interesting bi-disciplinary alliance forward theoretically and empirically, set out the structure of the book and provide a brief summary of each chapter.

PERSPECTIVES IN INDUSTRIAL ECOLOGY

As Lifset and Graedel (2002) point out, Industrial Ecology is 'industrial' in its interest in product design and processes of manufacture and distribution; it is 'ecological' both in its use of ecosystem analogies as models for

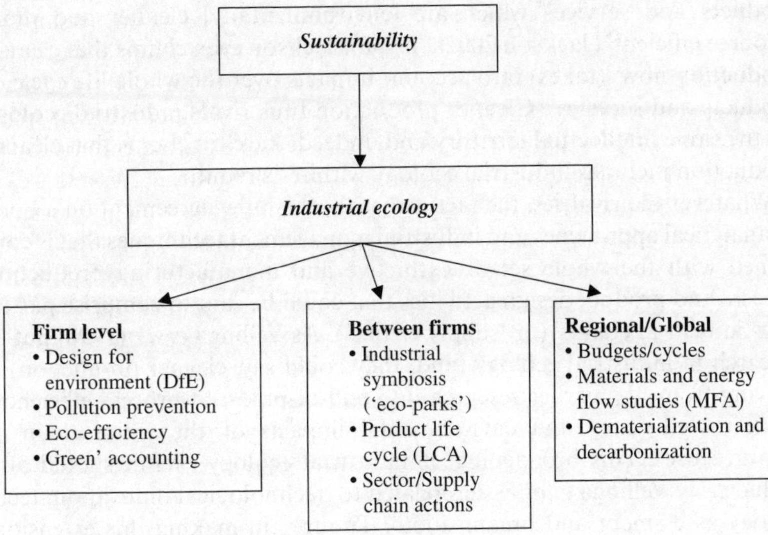

Source: Lifset and Graedel (2002).

Figure 1.1 The elements of industrial ecology seen as operating at different levels

environmental-friendly productive activity (or 'human technological activity') and in its placing of that activity within the larger supporting ecosystems. Lifset and Graedel include under the term 'Industrial Ecology' a large quantity of analytical and policy work that has accumulated over the last 15 years, as Figure 1.1 displays. Industrial ecology now incorporates a number of methods of analysis ('green' accounting, materials flow analysis, life cycle analysis) with a number of practical techniques for product and process redesign (Design for Environment, Eco-efficiency) and a number of broader frameworks for re-design of industrial collectives ('eco-parks', supply chain initiatives) and technological programmes (dematerialisation/de-carbonisation). As the Figure shows, industrial ecology then becomes a vital part of a mission to achieve 'sustainability'.

Interestingly, there is another approach to understanding environmental impacts which, it is claimed, has been converging with IE over the last five years. Thus, Jackson (2002a) argues that the 1980s approach of 'Cleaner production' – which seeks to redesign (with substantial new technological development) production processes to design out the generation of pollution and waste – has been expanding its remit so that it is now 'an approach to environmental management which aims to encourage new processes, products and services which are [environmentally] cleaner and more resource efficient' (Jackson 2002a, p. 36). Jackson even claims that cleaner production now '(takes) into account impacts over the whole life cycle of products and services'. Cleaner production thus rivals industrial ecology for the same intellectual territory and, indeed, Jackson claims that cleaner production includes industrial ecology within its remit.

Whatever the rivalries, the fact is that there is now agreement on a body of analytical approaches and industrial management techniques that is concerned with the whole set of extractive and manufacturing production process and product design activities that could be said to comprise particular 'industrial chains' (or 'supply chains'). As Vellinga *et al.* (1998) put it, research in industrial ecology (and, they could say, cleaner production) is moving from 'the earlier research into end-of-pipe and process efficiency' work to 'efficiency and environmental impacts of the entire chain of resource use'. This broadening of industrial ecology's remit should also include, say Vellinga *et al.* issues related to 'technological innovation, technology assessment and organization'. Though, in making this extension, they bring in yet another term – 'Industrial Transformation' – which extends the notion of industrial ecology towards policy action rather than just analysis. (The concept of 'industrial transformation' has a wider international currency as a subject of one of the Programmes of the International Human Dimensions of Global Environmental Change Programme see IHDP-IT Science Plan (1999).) In particular, this broader setting of Industrial Transformation leads to a set of research questions such as:

- How is the production and innovation process organized and managed?
- What is the nature of the interaction between producers within and across sectors? What are the promising technological developments and development trajectories, products and promising configurations/arrangements of production?
- What are the most efficient ways of organizing the production process across the different sectors?

They conclude that, '(Further) research should identify and analyse various possible technological and organizational trajectories that production units and entire sectors could go through, moving from one dominant way or producing goods and services to another way of doing things.' This programme of research for industrial ecology/industrial transformation/cleaner production brings its concerns close to those of researchers in innovation studies; though, as we will now argue, Innovation Studies researchers add further dimensions for consideration.

PERSPECTIVES FROM INNOVATION STUDIES

We can take for granted that 'innovation', by which we mean technological innovation and the changes in supporting economic and social structures that come with it, as well as innovation originating from non-firm institutions, consumers and users must, in some form or other, be central to the achievement of sustainable production and consumption in all areas of human activity. As one of us has argued elsewhere (Green and Miles 1996), if current systems of production and consumption are unsustainable in terms of their resource usage, ecological impact and long-term environmental effects, then new systems of provision are needed and these will entail new processes, new products, new services and new management practices; if these do not exist, they will have to be invented and launched into social and economic use. Conversely, new forms of social relationships that are innovated with environmental improvement as their goal will inevitably use products and processes in new ways. There is thus a strong relationship between innovation in socio-economic arrangements and innovation in the material products and processes in which they are entwined – 'socio-technical systems of provision' as they could be called. The development of new products and services, new manufacturing and distribution processes, new recycling and disposal methods, based on new technologies or adaptations of existing ones, will strongly influence the sustainability of future systems of provision. Consequently, understanding the processes that are likely to underpin these developments is crucial for policy intervention to achieve desirable forms of sustainability: in short we consider the processes of technological and social innovation and the means of guiding them into sustainable directions.

Understandings of the mechanisms of innovation have changed over the last 40 years. In the 1960s, when innovation was usually seen as the private actions of individual firms carrying out 'R&D' by exploiting scientific 'discoveries' that emerged from public investment. Nowadays, though individual products and services may appear to emerge from individual firms, the process of innovation is seen as involving many social actors (Rothwell 1994). Indeed, some innovations (though not necessarily inventions) are seen as the result of 'social shaping' by actors outside firms (Williams 2000). This happens as a result of interactive processes that involve the exchange of information and knowledge. It can be taken as obvious that these interactions are:

- between firms in the same sector, in partnerships and alliances;
- between firms within a particular supply chain, who act as suppliers and customers to each other;

- between firms and other organizations that regulate them and lobby them.

In short, understanding the dynamics of innovation – where and how it takes place, how it links to sources of scientific and market knowledge – has taken on a systemic dimension. This might be seen as the Innovation Studies equivalent to Industrial Ecology understandings of the flow of materials through chains. However, crucial to an understanding of these systems is another element which until recently has tended to be ignored by both IE and IS: namely, patterns of consumption behaviour. Understandings of consumption of and demand for new products and ways of providing and using them, as opposed to just design, production and supply of those products, is something that innovation studies is becoming increasingly aware of, especially in work done in Manchester (see McMeekin *et al.* 2002; Coombs *et al.* 2001). So, to the three interactions listed above we can add a fourth:

- between firms and their customers, their consumers, their 'markets'.

We call this set of interactions that are required for the introduction of new products and systems, 'distributed innovation' (Coombs *et al.* 2003). We see this as, what we have called elsewhere, a 'meso-level' process that looks at how the 'bottom-up' networks of heterogeneous actors that develop the new products and shape new markets and consumer behaviour are set within macro-structural shifts (Green *et al.* 1999). One of the central aspects of distributed innovation processes is the importance of interactions between innovating firms (or sets of firms) and between those firms and purchasers and users and others like regulators and intermediaries (Medd and Marvin, this volume Chapter 11). With more intense international competition, and rising world incomes, firms have become increasingly sensitive to shifts in consumption behaviour, and many of them attempt to combine this knowledge about consumer demand and markets with knowledge of potential innovation opportunities. We would want to expand the notion of distributed innovation processes from a concentration on economic actors only, to include a wider range of social and political actors: regulatory and standard-setting bodies, lobby groups, professional associations, and publicly-funded science institutes.

The processes through which demand for innovations is identified by firms and articulated by users/consumers and socio-political actors can be summarized thus:

1. A multitude of actors is involved, making 'steering' apparently more complex pathways but actually providing more opportunities for

intervention, given that the process of innovation is both prolonged and wide.

2. Radical innovation is as much about creating markets as about creating things (involves creating firms as well for new technologies; see Green 1991 on biotech).
3. There are system limitations to major transformations ('lock-in').
4. There are opportunities nevertheless for 'niche' exploration of new products ('Strategic Niche Management' or 'Social Niche Management').
5. Societal and political mobilization against industrial regimes can disrupt markets, opening up new 'spaces' for innovation: 'destructive creation' (McMeekin 2001).
6. 'Consumers' should not be restricted to end-consumers, this is especially true for infrastructures – given large energy and water consumption of processing firms (Green *et al.* 2000; Howells, this volume Chapter 9; Medd and Marvin, this volume Chapter 11).
7. Public procurement policies are especially significant (New *et al.* 1999).
8. State sponsored regulation mediated by policy guidance or legislation remain of crucial importance in inducing, re-directing or suppressing innovation (Dewick and Miozzo, this volume Chapter 7; Cen et al., this volume Chapter 8). These regulatory effects may be either direct or indirect in that they operate via their effects on changing demand and consumption practices.
9. Some organized groups of labour are able to carve out a particular occupational niche associated with a corpus of knowledge, competences, and status. Through the strategic endeavour of 'professionalization projects' these groups are able to exert influence on market regulatory processes, the legislative process, processes of technological development and innovation, and processes of opening and growing markets.[3] Environmental consultants are a key group of intermediaries involved in these processes.
10. Innovation and change occurring at one geographical scale has consequences for, or simultaneous impacts on, other scales (for example Beauregard 1995). Further up-scaling and down-scaling are strategic options adopted by agents for exerting control over – 'taming' – resource flows and disciplining boundaries (Roberts 1994). Multi-scalar perspectives are therefore an essential part of a more enlarged understanding of the socio-economic and political consequences[4] of innovation and change but are not typically or traditionally captured by Industrial Ecology models (see Randles and Berkhout, this volume Chapter 14).

Our contention is that these ten summary points capture some of the processes and classes of agent which research in Innovation Studies has already shown to occupy 'spaces of innovation'. Our second contention is that such an analysis might usefully complement existing industrial ecology perspectives providing scope for an enlarged industrial ecology–innovation studies research agenda. Of course, many of these dimensions are already captured in many industrial ecology models and applied case studies, but to our knowledge they are not formalized as such. On the other hand, the systemic holistic dimension integral to industrial ecology analysis is a powerful idea to complement much work in Innovation Studies which is frequently more atomistic or partial in its analysis. A further observation is that the ten dimensions are interlinked. Each influences at least some, if not all, of the others. A methodological approach is therefore warranted (and indeed is already evident in several of the chapters presented in the book) which captures not only the salience of these dimensions separately, but also and importantly outcomes arising from their inter-connectedness, which potentially gives rise to creativity, novelty and change. Of course, we do not suggest that any one case study can or even should try to capture a cross-tabulation of all of these dimensions against all of the others. That is clearly subject to methodological and indeed epistemological limits. However, what we find particularly interesting is that several of the chapters already capture a number of (different) examples of these inter-linked dimensions, regardless of the context of their application, whether they are 'sector' studies, spatially organized studies, or insights into particular interfaces such as consumption-production or regulation-production.

PERSPECTIVES AT THE INTERFACE OF INDUSTRIAL ECOLOGY AND INNOVATION STUDIES

Industrial Ecology and Ecologies of Industries

The term 'industrial ecology' can have different meanings in different (disciplinary) usages. For example, in the sense of 'the ecologies of industries' it can be used to refer to the way industrial structures themselves comprise a variety of differentiated firm-types, with different combinations of firm-size and inter-firm relations privileged through history (Granovetter 1994) and characterizing different cultural and place contexts. Though independent, these differentiated firms none-the-less co-exist in interdependent 'bundles' mediated through various forms of (market and non-market) exchange (Harvey and Randles 2002) occurring across firm boundaries, to form recognizable multiplexes or 'ecologies' of firms.

This meaning is quite different to the resource and material-flows analysis usually and traditionally associated with the term industrial ecology. Of course an understanding of both the 'ecologies of industries' and 'industrial ecology' is important to the study of both. The organization of resources, materials and components flows determines, to an extent, 'the ecology of industry' on the one hand. On the other, the ways that firms come to orientate their activities *vis-à-vis* other firms, including the sources of economic power to control or dominate the operation and shape of the whole system, determine not only the shifting patterns and structures of industries, but also determines, limits or constrains scope for more 'sustainable' material flows (Green and Foster, this volume Chapter 6; Dewick and Miozzo, this volume Chapter 7). Ramaswamy and Erkman (this volume Chapter 5) emphasize that the predominance of independent micro-manufacturers and traders in the industrial structure of India, makes a crucial difference to the sorts of policy interventions that are likely to be successful. Industrial Ecology 'solutions' are thus variegated and must be customized to local conditions rather than assumed transferable from any ubiquitous case (the Kalundborg case is the obvious example).

Complex Systems

Systems thinking lies at the heart of Industrial Ecology and enormous contributions have been made through the conceptualization of resource flows in a systemic way, thus marking a crucial advance on 'linear' representations of the industrial process and its counterpart 'end of pipe' management and policy solutions (Ramaswamy and Erkman, this volume Chapter 2). Equally Industrial Ecology made strides in visualizing the interactivity of different parts of the system, taking a holistic rather than atomized view and mapping complexes of industrial inter-linkages, rather than assuming independent discrete units of production. Further, systems advances attributable to Industrial Ecology derive from its 'circuits and feedbacks' perspective from which follow the 'cradle-to-cradle' and 'waste=food' ideas which take the ecological 'food web' as a key conceptual metaphor.

Of course, as Innovation Studies would crucially add, we are not dealing with static systems. In a primarily capitalist/market organized society we are talking about 'restless' ones (Metcalfe 2001), caused in part by the disruptive 'creative destruction' feature of innovation (Schumpeter 1934 [1911]). In fact we are dealing with open, non-linear, and *ex-ante* indeterminate complex systems (Anderson *et al.* 2000; Randles 2002). These are systems where existing inter-systemic and intra-systemic boundaries cannot be taken for granted but are continually contested and re-instituted by different classes of agent to obtain commercial gain, or new powers, or both (Harvey 2002; Harvey and Randles 2002; Randles 2003). Here we must also include the influence of mobilized non-firm interest groups (McMeekin 2001). Each part or interface of the system can therefore be conceptualized as instituted as an outcome of struggle between different interest groups mobilized (to a greater or lesser extent) to a position of recognizable pattern. Furthermore, these interests are institutionally captured in a range of possible and actual organizational forms such as firms, non-firms for example charities, and governments from where policy making and the legislative processes potentially exert a powerful influence. Analysis must therefore focus on the instituted construction of the separate parts of the system, but it must also pay attention to the relational interdependency of the system. It must identify what and which individual interests and logics press heavily on the whole system. It must also identify how pressures for change in some parts of the system confront pressures for stability and stasis in others, and how these tensions are resolved or accommodated, giving rise to variously stable or unstable outcomes.

Using this conceptual framework, we hope to gain an explanatory handle on the total interdependent logic of the system, as well as appreciating how particular interfaces of connecting (market and non-market) exchange come into being. We can identify sites and agents responsible for innovative change in the past, and put forward scenarios for how particular industrial structures, and therefore material flows, might be reconstituted in the future (Green and Foster, this volume Chapter 6).

Structuring Structures and the Instituted Organization of Socio-economic Life

To repeat, the key question which we wish to introduce is how we can rethink the link between the flow of materials, with the social, economic, and organizational structures which cause physical flows to be and become 'patterned' in particular ways. To elaborate, we can conceive four structural domains which together provide organizational logic to the system. They are: the structuring of materials flow; the structuring and organization of economic activity together with the pecuniary redistributions which arise from the processing of those materials; the social structures and structuring of relations (including power relations) which demarcate classes of agent and finally, the production of structures and meanings of knowledge including how that knowledge (and its associated symbolic significance, the ways meanings are produced and interpreted) is generated and applied. Thus, as noted by some geographers, inspired by Lefebvre, flows of the economy, whether flows of materials, goods, money, people or ideas cannot be considered frictionless. On the contrary, the direction and form that these flows take is materially influenced by social and economic structures and structuring processes which sit astride, refract, and shape flows of energy, commodities and capital (Brenner 1998, 1999, 2000; Randles and Dicken 2004).

There is clear scope for some fruitful debate between the industrial ecology and innovation studies communities. Robert White, then President of the US National Academy of Engineering defined industrial ecology in 1994 as 'the study of the flows of material and energy in industrial and consumer activities, of the effects of these flows on the environment and of the *influences* of economic, political, regulatory and social factors on the flow, use and transformation of resources' (emphasis added). The direction of flow between the 'physical'/'material' world and the 'social/economic/political' world is, in this definition, one in which the social 'influences' the physical. But – as work in innovation studies continues to show – it is possible to see the physical-social relation in a different way, with the process of innovation being 'embedded'[5] in institutionalized structures of social relations (Weber 1978 [1922]; Granovetter 1985; Hamilton 1994), including consumption practices, industrial relations, gender relations, human-to-technology relations, capital/investment relations, and facilities infrastructures such as transport, water and energy. How this can be related to the perspectives already well developed by Industrial Ecology scholars is one of the main themes of this book.

Sustainable Production-consumption and Intermediation

Until recently, consumption was an underdeveloped research area in both industrial ecology and innovation studies. Traditionally both have been production-centred and failed to take account of the proactive role that consumption plays in shaping processes of innovation. For industrial ecology the corollary has been to view consumption as little more than a 'black box' which the industrial system presses upon or alternatively takes its pressures from. Over the last five years however, the research agenda in Innovation Studies has become much more balanced. There is now accepted recognition of the role that consumers and users play as 'active agents' in market formation processes (Green *et al.* 2000; McMeekin *et al.* (eds) 2002). Also, as creatures of habit consumers resist attempts by producers and regulators alike to nudge production systems on to more sustainable trajectories. Indeed more recently there has been a far greater understanding of social processes behind the 'ratcheting upwards' of resource and energy use by domestic or 'end' users and the intimate intertwining and co-dependencies between consumption and infrastructures of provision (see in particular the European Science Foundation funded research programme reported in Southerton *et al.* 2004).

In industrial ecology there has been a similar renaissance of interest in consumption though there is still little evidence yet of either a systematic research programme or collaborative network of researchers focusing their combined attention on consumption. There has nevertheless been a recent high profile awakening of recognition and interest in the importance of the topic (see the Journal of Industrial Ecology special edition on consumption, Hertwich (ed.) 2005) and the emergence of key researchers building a research profile in the area (see Jacobs and Røpke 1999; Princen *et al.* 2002; Hertwich (ed.) 2005; Jackson 2002b, 2005).

More recently still, a new impetus for research has come from the question of how production and consumption articulate. We know that both are mutually constructed by the other, but further than the idea of 'feedbacks' we know little about the agents, knowledge bases, technologies, techniques, media or methods that 'sit between' production and consumption, in a sense 'facing both ways'. Put another way, what is the nature of the feedbacks and how do they operate? This, the topic of intermediation incorporates the need to identify, classify and understand processes of intermediation and the people involved in it. The topic is interesting both to redress the limited attention consumption has received to date, including its total absence from sustainability research, and for the possibilities for policy intervention that may be revealed from a better understanding of the role, activities and influence that intermediaries and intermediation have in the mutual shaping of production and consumption.

Values and Governance

The question first of how systems of inter-related ideas and values are formed and constituted, and second how this relates to innovation processes on the one hand and the organization and ecologies of industries and resource flows on the other, receives scant attention in either traditional innovation studies or industrial ecology. But clearly, understanding how ideas and values are formed is key to understanding creativity, and to appreciating where a capacity to be creative originates, and perhaps more importantly, the normative direction into which that creativity is channelled as societies strive to define and pursue 'progress'.

These questions are pertinent to both innovation, and to attempts to 'shift' economies and industrial organization on to more environmentally sustainable trajectories. This is the case whether we believe creativity is sourced and formed at the unit of the individual, identifying individuals who 'see the world differently' and creatively construct a novel response to a situation or problem (Hill, this volume Chapter 12); or whether we are interested in studying how ideas manifest at a more aggregated societal level and then trace how they play out across arrangements of multi-level governance (Flanagan *et al.* in this volume, Chapter 13). Having a view on how the ideas and values of individuals form and change (Hill proposes a framework for personal values transformation) and/or how societies, at different levels of construction and aggregation (for example contrasting regimes of corporate governance, with how a dominant ideology sweeps, or is imposed, on a nation) is crucial for understanding the opportunities (and limits) involved in bringing about change. The theoretical link therefore between the formation of ideas, values (ethics) and ideologies and their translation into governance regimes on the one hand; and change – to industrial organization, to the organization of resource flows, to energy use, to exchanges of money, goods and services – on the other, is a task well overdue and one which would enrich both innovation studies and industrial ecology, possibly coming from the direction of new cognate disciplines namely social ecology and political ecology.

Newly Industrializing Countries

For both innovation studies and industrial ecology the imperative to undertake in-depth case study research in, and about, newly industrializing countries is real and urgent. As China and India, themselves major economic power houses, take up the additional strains of their new roles as global manufacturer and processors of the world's exported waste, the consequential environmental damage, if left unaddressed, will be immense. Add to this environmental regulation and standards which lag those of the West and populations eager to embrace the material consumption levels of highly-developed market economies, and it is plain to see that the result will be an urgent build up of dilemmas, debates and tensions of international political-economy in the coming years.

Both innovation studies and industrial ecology, separately and together, have an important role to play in contributing to these debates, providing international and national policy-makers with insights into the dilemmas and tensions they face and offering assistance during the sensitive years, indeed decades, of transition. However the challenges and risks of 'getting it wrong' are high. Central to these is the assumption of universality. As the two chapters in this collection (Ramaswamy and Erkman on India, and Cen *et al.* on China) persuasively demonstrate, the context of unique histories, the specifics of contemporary industrial structures, and the diversity of governance and political-economy described mitigates against the insensitive transfer of solutions developed in the West to unique and context-specific situations of countries in transition. Contingency must be the guiding principle. And for contingency to be taken seriously, deep and thoroughly researched case-studies involving researchers knowledgeable about specific national and local situations and their path-dependent histories must inform policy and management/consultancy practice.

BOOK STRUCTURE AND CHAPTER SUMMARIES

Immediately following our scene-setting chapter, Erkman and Ramaswamy (Chapter 2) provide an introduction to Industrial Ecology. They offer a brief canter through the cognitive and conceptual base, the origins and history of the discipline, and the key techniques written for the non-specialist reader.

This is followed by three chapters written from the 'heartland' of industrial ecology, but with each author asked to consider the implications of an explicit or implicit 'injection' of perspectives from innovation studies into their account. Ravetz (Chapter 3) provides a conceptual framework interlinking Regional Development (RD), Industrial Ecology (IE), and Business Environment (BE) and he uses this RD:IE:BE model to structure his analysis and identify innovation opportunities. He reports on a range of resource productivity methodologies which underpin a range of diagnostic quantitative modelling tools for industrial ecologists, particularly relevant to analysis at the sub-national regional scale. He distinguishes between production-centred mass-balance techniques and consumption-centred 'tracing back' of embodied resources through the supply chain. Several of these analytical tools are still at the exploratory/testing stages and themselves represent innovations in terms of how we visualize and model the flow and utilization of resources. Finally he reflects on the role that product and process enabling innovations, such as information and computer technology, has on our understanding of the structuring and monitoring of resource flows.

Staying with the sub-national region, Mirata and Pearce report on three large but embryonic industrial symbiosis projects in the UK. The authors provide an assessment of the critical success factors of these initiatives. They note that the systematic attempt to identify and implement industrial symbiosis arrangements represents innovative activity in its own right. However such initiatives require, as do all innovations, considerable investments of time and resources in the face of inertia, perceived risk, or both. The significant reorganization of industrial 'complexes' which extensive industrial symbiosis projects require need to affect mutual learning across collaborating groups if the ambitious industrial symbiosis projects are to be operationalized. The authors suggest that setting up and maintaining the complex relationships of industrial symbiosis depends less on hard-wiring activities, such as data collection, and more on the social dimensions of trust, social networks, and the pro-active involvement of a committed project champion or advocate. Nevertheless the low success rate within the reported cases suggests that, as with all innovations, failure is a normal and to-be-expected part of the innovation process.

Finally in this group, Erkman and Ramaswamy report on their work undertaking resource flow models of sub-sectors of the Indian economy spatially located at the regional scale. They again find that a crucial first step for the actors involved (the consultants, firms, and regulatory agencies) is to 'see the problem differently'. In this case there is a collective realization of both the scale and location of 'hot spots' of resource depletion or inefficiency (for example captured in the rate and scale of water use) when data are aggregated to a 'newly visible' scale above that of the individual (small) firm. An interesting outcome in one of the cases was the appearance of a local entrepreneur who on recognizing the nature of the problem introduced, through the market, desalination machines for sale to local manufacturers who were then able to make cost savings. The point is that seeing 'the problem' differently, or indeed seeing it at all, enabled the emergence of a creative response from local business.

The next group of three chapters are located in innovation systems theory. They each describe the multiple actors, including service providers such as designers, retailers and the like involved in the emergence (or resistance) to new products, processes, regulatory mechanisms, university and scientific institutional affiliations and so on, what might be called 'distributed innovation processes' (Coombs *et al.* 2003). Each of the chapters takes a historical perspective to capture the dynamic and transformatory nature of the systems under investigation. In addition, Green and Foster (Chapter 6) isolate the powerful players in pea production–consumption capable of shaping the entire socio-technical system. Using scenario methodologies, they offer two alternative – more sustainable – models of interdependent system contrasting with the current industrialized/modern one. These are an 'organic' model and a new-industrialized model. Both would require significant change (involving 'distributed' innovation) played out as the comprehensive reconfiguration of existing socio-economic arrangements.

Dewick and Miozzo (Chapter 7) similarly describe the actors and characteristics of the domestic construction sector as being highly fragmented, conservative, mature and offering low profit margins. They are interested in the insertion and diffusion of energy-saving technologies of thermal insulation and solar heating into this specific context and find that there are interesting national differences. Their chapter illuminates therefore, both the need to understand the general features of the industry in question which may exhibit high levels of resistance to innovation, and the specifics of national difference which may, by contrast, witness very different experiences in terms of the development and uptake of environment-friendly technologies.

Finally the piece by Cen *et al.* (Chapter 8) uses a similar methodology to describe the historical emergence of a 'problem' of municipal solid waste in China resulting in the appearance of a range of regulatory responses and a variety of technological solutions. The chapter focuses on one of these, the suite of 'waste to energy' incineration technologies. A key contribution of this chapter is the backdrop of transition in China from the Mao period to open-market liberalism entailing rising domestic consumption, reduced domestic recycling, rapid economic growth, urbanization, and the emergence of technological/industrial responses to the very recent municipal solid waste problem.

The remainder of the book is dedicated to probing, in a very preliminary way, some of the missing or under-researched themes of both industrial ecology and innovation studies. First, we turn to the question of consumption and intermediation. As we might expect from such a new field, the scope of these topics is broad and has not yet settled into either a coherent body, or competing strands of literature. The three chapters therefore represent simply a taste of different perspectives.

Howells (Chapter 9) asserts that industrial consumption is an under-researched area of innovation studies. He uses the term 'intermediate consumption' to refer to consumption within industrial supply chains or supply webs rather than domestic or individual consumption. Howells draws attention to two important points. First the investment of time and resources that organizations must make in order that relevant personnel are able to 'learn to consume', especially in the context of the consumption of new products or investment in new processes. He sees this as a hidden cost to the organization which importantly constrains the development and uptake of technical or indeed organizational innovation. Second, he draws attention to the service dimension of the interface between seller, buyer, and user of new products and processes. This service dimension, which ranges from informal advice to formal training programmes and market research is crucial to processes of knowledge acquisition on the part of both the seller (as products are adapted to the needs of the buyer/user) and the user who must 'learn' how to consume them. Indeed buyers/users need to contribute to the development of new products, services and processes if the latter are to be integrated into the buying organization's business effectively. A key implication for industrial ecology of the Howells chapter is that intermediate consumption is both sticky and messy yet is rendered invisible in the simplistic flow diagrams beloved by industrial ecologists.

The 'stickiness' of consumption is a theme taken up also by Randles and Warde (Chapter 10). They draw on Bourdieu-inspired practice theory to suggest that much 'ordinary' consumption – the everyday consumption of water, electricity, and dealing with domestic waste for example – is unreflexive and habitual. It also follows, to an extent, position within social strata and affiliations with a range of social groups. Taking two examples of practice, namely the daily routines of managing domestic rubbish and showering, the authors note that both are overlain with historically changing understandings of 'what is waste?' and 'why and when take a shower?' The authors also note the co-dependency of practice with institutional and physical infrastructures (the size of bins, the space in the kitchen dedicated to pre-sorting, the municipal provision of collection facilities and so on). This observation requires that both consumption practices and physical and institutional infrastructures be nudged towards more sustainable arrangements together, not separately or without consideration each of the other. They note also that the stickiness of consumption combined with opposite pressures to increase resource use (taking more showers, installing energy and water guzzling power showers) may make this process very difficult to bring about at all. Further, because practice theory rejects rational-choice agency models it implies that intervention which relies upon rational-choice reflection and decision making, such as educational campaigns, are unlikely to be successful in isolation. Rather, change involves incremental (but sometimes disruptive) shifts in practice compatible with, adjusted to, or responding to, changes in associated suites of interdependent technologies and institutional infrastructures.

Medd and Marvin (Chapter 11) take up the question of intermediation in response to the problem of the production-consumption dichotomy. They see intermediaries as organizations which stand in-between, and mutually shape, material flows, technologies, social practices and social organization. They suggest that intermediaries play an important role in bringing together and mediating different interests. The authors illustrate this by looking at the operation and influence of intermediaries in the water sector. In contrast to the whole-system approach of Industrial Ecology, research on intermediaries must focus on very specific practices and groups to reveal the influence that derives from their in-between position. Equally, the policy intervention possibilities that derive from understanding the role, powers, and influence of intermediaries are highlighted.

Our final two substantive chapters explore issues of values, ideas, power, and socio-political systems. Hill (Chapter 12) is optimistic about individuals' capacities to become self-aware and learn our way forwards to better, more environmentally and socially sensitive futures. Agents, in Hill's view are both reflexive and creative but these capabilities are suppressed by the dominant world views and the pressures under which modern-industrialized societies live. He advocates 'deep industrial ecology' encompassing radical shifts in mindset and lifestyle and jettisoning many of the incremental – in particular industrial or technological – pathways to alternative futures. However his approach is neither confrontational nor polemical. He illustrates his ideas with reference to his own transformation from industrial chemist. He also describes the work of the innovative agriculturalist P.A. Yeomans who, in the 1940s, rejected modern scientific methods for controlling and manipulating landscapes and instead, through observation, creativity and determination, designed radically different methods of land management oriented towards the guardianship of nature and optimization of water-use.

The concern of Flanagan *et al.* (Chapter 13) is the multi-dimensional examination of governance issues in a 'knowledge-based economy'. Considering the themes of transition and industrial transformation, as part of a wider brief to facilitate the reflection of senior policy makers and industrialists upon alternative futures of manufacturing in Europe, they discuss the often invisible dimensions of shifts in cultural assumptions and social values looked at through concepts such as learning organizations, service economy, information society, risk-society and post-industrial structures. Their chapter could therefore perhaps be viewed as a whole-society counterpart to the piece by Hill.

Finally in their brief endnote, Randles and Berkhout return to the opportunities, problems, and missing links in the central book theme of bridging perspectives from Industrial Ecology and Innovation Studies. They note that, in some respects, such an aim may be both possible and commendable but there are also major differences in the ontology and epistemology of the two disciplines which militate against the achievement of such an objective. They explore this via a brief consideration of four theoretical areas relevant to both disciplines, but highlighting deep problems in the attempted incorporation or glossing over of such theoretical disharmonies when translating conceptual work into a research programme situated at the interface of industrial ecology and innovation studies. The areas they explore are: (a) the validity and compatibility of underpinning conceptual metaphors; (b) the question of scale and multi-scalarity; (c) conceptualizing knowledge and understanding 'information failure'; and (d) assumptions about agency and the role of the agent.

A RESEARCH NETWORK AND RESEARCH AGENDA AT THE INTERFACE OF INDUSTRIAL ECOLOGY AND INNOVATION STUDIES

An enthusiastic international group of multi-disciplinary academics, industrialists and environmental consultants interested in working across the disciplines of industrial ecology and innovation studies has formed an informal collaborative network. A number of questions continue to prompt debate within the group and help to orientate its focus and activities. These formed the agenda for the earlier workshops and still orientate a proposed research agenda at the interface of industrial ecology and innovation studies, though naturally not all of these questions or themes have been covered in the present collection. Pertinent research questions can nevertheless be organized around seven subheadings:

Industrial Ecology and Innovation

(a) What scientific base of techniques, methods and models are emerging from academia, consultancy and industry to strengthen and expand the jurisdiction of industrial ecology and how is diffusion of this epistemology occurring?
(b) What technological innovations (including information technology, and 'nano' technology) are emerging to facilitate, measure, monitor and manage pollution remediation, waste-to-food chemical transformation and resource and information flows?
(c) What new services are/could emerge to aggregate/disaggregate or re-scale and intermediate resource and waste streams to re-package waste into 'right-size' units for market/non-market exchange?
(d) What new markets are emerging or being intentionally created for re-usable materials, and to what uses are they being put?
(e) What new forms of economic/non-economic organization of exchange and intermediation are evident in applied Industrial Ecology case studies? How have exchange and intermediation changed over time, and what evidence is there that 'missing intermediaries' are preventing the establishment of more desirable material-money exchanges and material flows.
(f) What evidence is there of innovation as new forms of industrial organization, new relationships, new classes of economic/non-economic agent, new 'business models' and new economic/non-economic roles and activities?
(g) In inherently dynamic innovation-enabled capitalist economies, is the objective of 'Closing the Loop' either feasible or desirable?

Governance, Institutions and Geo-politics

(a) Can we better understand and integrate the role of the State, of self-regulation and 'systemic' governance issues in industrial ecology models.
(b) Are ideas from industrial ecology compatible with existing local area planning processes? What implications, opportunities and constraints do local planning processes pose for Industrial Ecology?
(c) Why do (for example) industrial symbiosis arrangements 'appear' (self-organize) in some places and not others?
(d) What real societal-institutional constraints and limits are there to the practical application of industrial ecology models?
(e) How can questions of scale and multi-scalarity be integrally captured or taken account of in industrial ecology models and analyses?
(f) What role and degree of influence do financial systems (the availability and access to investment capital, credit, shareholder pressures and so on) have on encouraging environmental and social responsibility on the part of individuals and corporations?

Industrial Ecology and Consumption

(a) How can we move beyond a 'black-box' representation of consumption in Industrial Ecology models and analyses?
(b) Can we better understand consumption practices of industrial ecology systems including recycling and re-use? What new consumption patterns and practices are emerging to take up 'recyclable' industrial materials? Who is the discerning user of recycled materials (for example domestic re-use in gardens, public sector re-use in municipal parks and play areas, the collection, transformation and re-use of 'waste' materials in households, in the construction sector, in art, in design)?
(c) Can we better understand inter-organizational buying behaviours and consumption?

Industrial Ecology in Action

(a) What opportunities and barriers exist for translating theory into practice?
(b) How can we contribute to existing wide portfolios of comparative case studies in industrial ecology – of products, materials, companies, territories, collections of firms/institutions, in order to combine methods and insights from industrial ecology and innovation studies?

Industrial Ecology and Policy

(a) What are the implications for such questions and perspectives for the evaluation of existing legislation and the development of 'new' policy?
(b) How can foresight and other futures methodologies assist transitions to more sustainable production-consumption configurations and what innovations would be needed to 'back-cast' them into the ideas and imaginations of potential innovators?

Ethics, Values and 'Deep' Approaches to Industry Ecology

(a) What role does learning, the formation of ideas, and systems of ethics and values play in propensities to move further towards, or further away from socio-economic arrangements deemed by advocates to be more 'sustainable' than their antecedents.

This chapter has provided schematic and simplified representations of perspectives in industrial ecology and innovation studies. It has suggested that an interesting research agenda bubbles-up at the interface of the two approaches and has described an initiative to bring together scholars from both these communities to investigate, debate, and research further topics which occupy the 'spaces of innovation' at this interface.[6] Our contention is that these themes and research questions can be fruitfully explored and further developed through an extended programme of theoretical reflection and empirical (especially applied and comparative case study) research. We hope that others with an academic interest, specialist expertise or practical/industrial experience in these areas will be interested to join us.

NOTES

1. 'Advances in the Economic and Social Analysis of Technology', the 6th ASEAT Conference of 7-9 April 2003 marked the launch of the Institute of Innovation Research, then a joint venture between UMIST and the Victoria University of Manchester.
2. The objectives, scope and content of the ASEAT session and workshop were informed by findings from a small number of exploratory interviews conducted initially to assist CRIC's sister research centre PREST in their analysis for the EU project FUTMAN, commissioned by EC-DG 'Competitive and Sustainable Growth', see Flanagan *et al.* this collection Chapter 13.
3. See Randles and Tether (2002, 2003) on the emergence of a new 'profession' of practitioners in environmental services and technologies.
4. Smith 1996; Swyngedow 1997; Brenner 2000; Randles and Dicken 2004.
5. Though see Sayer (2002) who critiques the embeddedness concept, among others.
6. This idea is complementary to the agenda recently put forward advocating closer links between the dominant natural science and engineering aspects of Industrial Ecology and

management and policy studies (Korhonen 2004, p. 289). Although to be fair, as we well know, industry was a prime mover in the construction and importantly the diffusion of the key principle that industrial processes are analogous to natural eco-systems and could be 'managed' as such (Frosch and Gallopoulos 1989; Erkman and Ramaswamy, this volume Chapter 2).

REFERENCES

Andersen, B., J.S. Metcalfe and B. Tether (2000), 'Distributed innovations systems and instituted economic processes', in J.S. Metcalfe and I. Miles (eds), *Innovation Systems in the Service Economy: Measurement and Case Study Analysis*, Boston: Kluwer.

Ayres, R.U. (1996), 'Creating industrial ecosystems: a viable management strategy', *International Journal of Technology Management*, **12**(5/6) special issue, 608–24.

Ayres, R. and L. Ayres (eds) (2002), *A Handbook of Industrial Ecology*, Cheltenham,UK and Northampton, MA, USA: Edward Elgar.

Baas, L.W. (1998), 'Cleaner production and industrial ecosystems, a Dutch experience', *Journal of Cleaner Production*, **6**, 189–97.

Beauregard, R. (1995), 'Theorising the global-local connection', in P.L. Knox and P.J. Taylor (eds), *World Cities in a World System*, Cambridge: Cambridge University Press.

Brenner, N. (1998), 'Between fixity and motion: accumulation, territorial organisation and the historical geography of spatial scales', *Environment and Planning D: Society and Space*, **16**(4), 459–81.

Brenner, N. (1999), 'Beyond state-centrism? Space, territoriality, and geographical scale in globalization studies', *Theory and Society*, **28**, 39–78.

Brenner, N. (2000), 'The urban question as a scale question: reflections on Henri Lefebvre, urban theory and the politics of scale', *International Journal of Urban and Regional Research*, **24**(2), 361–78.

Chertow, M.R. (2000), 'Industrial symbiosis: literature and taxonomy', *Annual Review of Energy and Environment*, **25**, 313–37.

Côté, R.P. and E. Cohen-Rosenthal (1998), 'Designing eco-industrial parks: a synthesis of some experiences', *Journal of Cleaner Production*, **6**, 181–8.

Coombs, R. and A. McMeekin (1996), *The Use of Demand Analysis in R&D Projects*, Manchester: CROMTEC/Technology Strategy Forum.

Coombs, R., K. Green, A. Richards and V. Walsh (eds) (2001), *Technology and the Market: Demand Users and Markets*, Cheltenham, UK and Northampton, MA, USA: Edward Elgar.

Coombs, R., M. Harvey and B. Tether (2003), 'Distributed processes of provision and innovation', *Industrial and Corporate Change*, **12**(5), 1051–81.

Desrochers, P. (2000), 'Market processes and the closing of industrial loops historical reappraisal', *Journal of Industrial Ecology*, **4**(1), 29–43.

Desrochers, P. (2001), 'Eco-industrial parks: the case for private planning', *The Independent Review*, 5(3), 345–71.

Dewick, P. and M. Miozzo (2003), 'Sustainable technologies and the construction industry: an international assessment of regulation, governance and firm networks', paper presented to the Industrial Ecology and Spaces of Innovation meeting, CRIC Workshop, 17-18 June, University of Manchester, Manchester.

Erkman, S. (1998), *Vers une Ecologie Industrielle*, Paris: Editions Charles Léopold Mayer.

Erkman, S. (2002), 'The recent history of industrial ecology', in R. Ayres and L. Ayres (eds), *A Handbook of Industrial Ecology*, Cheltenham, UK and Northampton, MA, USA: Edward Elgar.

Erkman, S. and R. Ramaswamy (2000), 'Cleaner production at the system level: industrial ecology as a tool for development planning (case studies in India)', paper presented at UNEPS 6th International seminar on cleaner production, 16-17 October, Montreal, Canada.

Erkman, S. and R. Ramaswamy (2003), *Applied Industrial Ecology: A New Platform for Planning Sustainable Societies*, India: AICRA.

Frosch R.A. and N.E. Gallopoulos (1989), 'Strategies for manufacturing', *Scientific American*, **261**(3), 94–102.

Gertler, N. and J.R. Ehrenfeld (1996), 'A down-to-earth approach to clean production', *Technology Review*, **99**(2), 48–54.

Granovetter, M. (1985), 'Economic action and social structure: the problem of embeddedness', *American Journal of Sociology*, **91**(3), 481–510.

Granovetter, M. (1994), 'Business groups', in N. Smelser and R. Swedberg (eds), *The Handbook of Economic Sociology*, Princeton, NJ: Princeton University Press, Chapter 18.

Green, K. (1991), 'Shaping technology and shaping markets: creating demand for biotechnology', *Technology Assessment and Strategic Management*, **3**(1), 57–76.

Green, K., R. Hull, V. Walsh and A. McMeekin (1999), 'The construction of the techno-economic: networks vs. paradigms, *Research Policy*, **28**, 777–92.

Green, K. and I. Miles (1996), 'A clean break? From corporate research and development to sustainable technological regimes', in R. Welford and R. Starkey (eds), *Reader on Business and the Environment*, London: Earthscan, pp. 129–43.

Green, K., B. Morton and S. New (2000), 'Greening organisations: purchasing, consumption and innovation', *Organisations and Environment*, **13**(2), 206–25.

Hamilton, G. (1994), 'Civilisations and the organisation of economies', in N. Smelser and R. Swedberg (eds), *The Handbook of Economic Sociology*, Princeton, NJ: Princeton University Press, Chapter 8.

Harvey, M. (2002), 'Markets, supermarkets and the macro-social shaping of demand: an instituted economic process approach', in A. McMeekin, K. Green, M. Tomlinson and V. Walsh (eds), *Innovation by Demand*, Manchester: Manchester University Press.

Harvey, M. and S. Randles (2002), 'Market exchanges and "instituted economic process" an analytical perspective', *Revue d'Economie Industrielle*, special issue, December.

Hertwich, E. (ed.) (2005), 'Special edition on consumption and industrial ecology', *Journal of Industrial Ecology*, 9(1/2).

den Hond, F. (2000), 'Industrial ecology: a review', *Regional Environmental Change*, **1**(2), 60–69.

IHDP-Industrial Transformation Project (1999), 'Science plan', IHDP report no 12, Bonn: IHDP.

Jacobs, M. and I. Røpke (1999), 'Special issue on consumption', *Ecological Economics*, **28**(3).

Jackson, T. (2002a), 'Industrial ecology and cleaner production', in R.U. Ayres and L.W. Ayres (eds), *A Handbook of Industrial Ecology*, Cheltenham, UK and Northampton, MA, USA: Edward Elgar.

Jackson, T. (2002b), 'Evolutionary psychology in ecological economics: consilience, consumption and contentment', *Ecological Economics*, 41(2), 289–303.
Jackson, T. (2005), 'Live better by consuming less: is there a "double dividend" in sustainable consumption?', in E. Hertwich (ed), special edition on consumption and industrial ecology, *Journal of Industrial Ecology*, 9(1/2), 19–36.
Korhonen, J. (2004), 'Industrial ecology in the strategic sustainable development model: strategic applications of industrial ecology', *Journal of Cleaner Production*, special issue 'Applications of Industrial Ecology'.
Lifset, R. and T.E. Graedel (2002), 'Industrial ecology: goals and definitions', in R.U. Ayres and L.W. Ayres (eds), *A Handbook of Industrial Ecology*, Cheltenham, UK and Northampton, MA, USA: Edward Elgar.
McGlade, J., R. Murray, J. Baldwin, B. Winder and K. Ridgway (2003), 'A co-evolutionary framework for understanding transformation and resilience of socio-economic systems: the example of South Yorkshire mining communities', paper presented to the Industrial Ecology and Spaces of Innovation CRIC Workshop, 17–18 June, University of Manchester.
McMeekin, A. (2001), 'Innovation, demand and environmental sustainability', thesis submitted to UMIST for PhD degree, Manchester School of Management, UMIST, Manchester.
McMeekin, A., K. Green, M. Tomlinson and V. Walsh (2002) (eds), *Innovation by Demand*, Manchester: Manchester University Press.
Metcalfe, J.S. (2001), 'Restless capitalism: increasing returns and growth in enterprise economics', in A. Bartzokas (ed.), *Industrial Structure and Innovation Dynamics*, Cheltenham, UK and Northampton, MA, USA: Edward Elgar.
Metcalfe, S. and A. Warde (eds), *Market Relations and the Competitive Process*, Manchester: Manchester University Press.
Miles, I., M. Weber and K. Flanagan (2003), 'The future of manufacturing in Europe 2015–2020, the challenge for sustainability: governance, social attitudes and politics', (WP7) draft paper from grant ref G1MA-CT-2001-00010/DGRES presented at the CRIC/IIR workshop 'Industrial Ecology and Spaces of Innovation', 17–18 June, Manchester: University of Manchester/UMIST.
Mirata, M. and R. Pearce (2003), 'Industrial symbiosis in the UK', paper presented at the CRIC/IIR workshop 'Industrial Ecology and Spaces of Innovation', 17-18 June, Manchester: University of Manchester/UMIST.
New, S., B. Morton and K. Green (1999), 'Deconstructing green supply and demand: PVC healthcare products and the environment', *Risk Decision and Policy*, **4**(1), 221–54.
Polanyi, K. (1957), 'The economy as instituted process', in *Trade and Market in the Early Empires*, New York: The Free Press, Chapter 13.
Princen, T., M. Maniates and K. Conca (2002), *Confronting Consumption*, Cambridge, MA: MIT Press.
Quist, J., M. Knot, W. Young, K. Green and P. Vergragt (2001), 'Strategies towards sustainable households: using stakeholder workshops and scenarios', *International Journal of Sustainable Development*, **4**(1), 75–89.
Randles, S. (2002), 'Complex systems applied: the merger that made GlaxoSmithKline', *Technology Analysis and Strategic Management*, **14**(3), 331–54.
Randles, S. (2003), 'Issues for a Neo-Polanyian research agenda in economic sociology', *International Review of Sociology*, in M. Harvey and R. Ramlogan (eds), special section, 'Polanyian perspectives on instituted economic processes, development and transformation', **13**(2), 409–34.

Randles, S. and P. Dicken (2004), 'Scale and the instituted construction of the urban: comparing the cases of Manchester and Lyon', *Environment and Planning A*, **36**(11), 2011–32.
Randles, S. and B. Tether (2002), 'Services, scale and structures of internationalisation: Northwest England's environmental technologies firms', in M. Miozzo and I. Miles (eds), *Internationalisation, Technology and Services*, Cheltenham, UK and Northampton, MA, USA: Edward Elgar.
Randles, S. and B. Tether (2003), 'North West England's environmental technologies and services (ETS) firms', paper for the environment 2003 on-line conference, October.
Ravetz, J. (2003), 'Regional industrial ecology and resource productivity – new approaches to analysis and communication', paper presented at the CRIC/IIR workshop 'Industrial Ecology and Spaces of Innovation', 17-18 June, Manchester: University of Manchester/UMIST.
Roberts, S. (1994), 'Scalar dynamics', in *Disciplining Boundaries*, an edited volume arising from the University of Kentucky Committee on Social Theory, Spring.
Rothwell, R. (1994), 'Industrial innovation: success, strategy, trends', in M. Dodgson and R. Rothwell (eds), *The Handbook of Industrial Innovation*, Cheltenham, UK and Northampton, MA, USA: Edward Elgar.
Sayer, A. (2002), 'Markets, embeddedness and trust: problems of polysemy and idealism', in J.S. Metcalfe and A. Warde (eds), *Market Relations and the Competitive Process*, Manchester: Manchester University Press.
Schumpeter, J. (1934), *The Theory of Economic Development: An Enquiry into Profits, Capital, Credit, Interest and the Business Cycle*, first published in 1911, Cambridge, MA: Harvard University Press
Smith, N. (1996), 'Spaces of vulnerability: the space of flows and the politics of scale', *Critique of Anthropology*, **16**, 63–77.
Southerton, D., H. Chappells and B. Van Vliet (2004), *Sustainable Consumption: The Implications of Changing Infrastructures of Provision*, Cheltenham, UK and Northampton, MA, USA: Edward Elgar.
Swyngedow, E. (1997), 'Excluding the other: the production of scale and scaled politics', in R. Lee and J. Wills (eds), *Geographics of Economies*, London: Arnold, Chapter 13.
Vellinga, P., J. Gupta and F. Berkhout (1998), 'Towards industrial innovation: the way ahead', in P. Vellinga, J. Gupta and F. Berkhout (eds), *Managing a Material World: Perspectives in Industrial Ecology*, Amsterdam: Kluwer.
Weber, M. (1978), *Economy and Society*, first printed in 1922, Berkeley, CA: University of California Press.
White, R. (1994), 'Preface', in B.R. Allenby and D.J. Richards (eds), *The Greening of Industrial Ecosystems*, Washington, DC: National Academy Press.
Williams, R. (ed) (2000), *Concepts, Spaces and Tools: Recent Developments in Social Shaping Research*, final report, Edinburgh: RCSS.

2. Industrial ecology: an introduction

Suren Erkman and Ramesh Ramaswamy

INTRODUCTION

Industrial ecology? A surprising, intriguing expression that immediately draws our attention. The spontaneous reaction is that 'industrial ecology' is a seeming contradiction in terms, the general perception being that industries cause ecological damage.

We are used to considering the industrial system as isolated from the natural ecological system or biosphere, with factories and cities on one side and nature on the other, the problem being perceived as one of minimizing the impact of the industrial system on what is 'outside' of it: its surroundings, the 'environment'. Since long, studies by ecologists have focused on the consequences of the various forms of pollution on nature. As early as the 1950s, strategies were conceived in order to diminish the impact of pollution, which essentially consisted in building filters to ensure that the waste from industries did not 'leak' into the environment. This is illustrated in the classical end-of-pipe approach for the treatment of pollution, which has proved to be quite useful, but not entirely adequate in the long run.

Analysis showed that better strategies were required because the process of building filters was often just transferring the pollutant from one medium to another (for example, from water to land). Second, the process of building filters was not very economical as there were no savings accruing from the process. This approach did not also pay adequate attention to the issue of resources. Considering the increasing population, the rising aspirations of the people and the earth's limited resources, the issue of a more efficient use of resources certainly needed to be addressed.

Cleaner production, pollution prevention and eco-efficiency strategies were then evolved, which looked at possible changes in the process or parts of the process, to minimize waste. With this, the economies of production were very often better as lower waste meant better material utilization. By addressing the issues in a preventive way, they definitely represent important progress. However, today, these strategies remain mainly targeted towards specific manufacturing processes and business strategies within individual companies.

But in all these perspectives, the industrial system was not fully seen as part of the biosphere. A broader view was needed. One needed to think of going even further, and trying to apply strategies like cleaner production at the level of a cluster of companies, or at the level of an industrial zone, or even for a whole region – in other words, to apply cleaner production and similar approaches at the level of a system. This idea stems from the recognition that substantial additional gains, both economic and environmental, can be achieved by addressing issues at the level of a system (a cluster of companies, an industrial zone, a region, and so on), as compared to individual and isolated approaches.

Industrial ecology explores the assumption that industrial activities should not be considered in isolation from the wider world but rather in terms of an industrial ecosystem functioning within the natural ecological system or biosphere. The industrial system, in a similar way to a natural ecosystem, essentially consists of flows of materials, energy and information, and furthermore relies on resources and services provided by the biosphere. It is important to stress at the outset that the word industrial, in the context of industrial ecology, refers to all human activities occurring within the modern technological society. Thus tourism, housing, medical services, transportation and agriculture are all a part of the industrial system. And the word ecology, here, refers to the science of ecosystems.

Industrial ecology can also be seen as a practical approach to sustainability. It is an attempt to address the question: How can the concept of sustainable development be made operational in an economically feasible way?

INDUSTRIAL ECOLOGY IN A NUTSHELL

So far, there is no standard definition of industrial ecology. Whatever the definitions may be, all authors more or less agree on at least three key elements of the Industrial Ecology perspective:

(a) Industrial ecology is a systemic, comprehensive, integrated view of all the components of the industrial economy and their relations with the biosphere.
(b) It emphasizes the biophysical substratum of human activities, for example, the complex patterns of material flows within and outside the industrial system, in contrast with current approaches which mostly consider the economy in terms of abstract monetary units, or energy flows.
(c) It considers technological dynamics, for example the long term evolution (technological trajectories) of clusters of key technologies as

a crucial (but not exclusive) element for the transition from the actual unsustainable industrial system to a viable industrial ecosystem.

Industrial ecology does not merely address issues of pollution and environment, but considers as equally important, technologies, process economics, interrelationships of businesses, financing, overall government policy and the entire spectrum of issues that are involved in a socio-economic system.

Two terms are often used while talking about industrial ecology. These are 'industrial ecology' and 'industrial metabolism' and it may be useful to clarify what we mean by these expressions.

'Industrial metabolism' is the whole of materials and energy flows through an industrial system. It is studied through an essentially analytical and descriptive approach, mainly Material Flow Analysis (MFA), based on the principle of conservation of mass. MFA is aimed at understanding the circulation of the materials linked to human activity, from their initial extraction to their inevitable reintegration, sooner or later, into the overall biogeochemical cycles. The expression metabolism of economic activities (or sometimes socio-industrial metabolism) is also in use and can be considered as synonymous.

Industrial ecology goes further: the idea is first, on the basis of industrial metabolism, to understand how the industrial system works, how it is regulated, and how it interacts with the biosphere; then, on the basis of our scientific understanding of ecosystems, we try to determine how the industrial system could be restructured to make it compatible with the way natural ecosystems function.

INDUSTRIAL ECOLOGY: A BRIEF HISTORY

There is little doubt that the concept of industrial ecology existed well before the expression, which began to appear sporadically in the literature of the 1970s. In fact, and not surprisingly, specialists of scientific ecology had all along the intuition of the industrial system as a subsystem of the biosphere. But this line of thought had never been actively investigated. The concepts of industrial ecology have been discussed on and off from the 1960s.

The expression re-emerged in the early 1990s, at first, among a number of industrial engineers connected with the National Academy of Engineering in the USA. Every September, the popular scientific monthly *Scientific American* publishes an issue on a single topic. The September 1989 special issue was on 'Managing Planet Earth'. It featured an article, 'Strategies for manufacturing', by Robert Frosch and Nicholas Gallopoulos, both then at General Motors (Frosch and Gallopoulos 1989).

In their article, the two authors offered the idea that it should be possible to develop industrial production methods that would have considerably less impact on the environment. This hypothesis led them to introduce the notion of industrial ecology. Projections regarding resources and population trends lead to the recognition that the traditional model of industrial activity – in which individual manufacturing processes take in raw materials and generate products to be sold plus waste to be disposed of – should be transformed into a more integrated model: an industrial ecosystem. The industrial ecosystem would function as an analogue of biological ecosystems. (Plants synthesize nutrients that feed herbivores, which in turn feed a chain of carnivores whose wastes and bodies eventually feed further generations of plants.) An ideal industrial ecosystem may never be attained in practice, but both manufacturers and consumers must change their habits to approach it more closely if the industrialized world is to maintain its standard of living – and the developing nations are to raise theirs to a similar level – without adversely affecting the environment.

However, as Robert Frosch indicated during his lecture, 'Towards an industrial ecology', presented before the United Kingdom Royal Society in 1990: 'The analogy between the industrial ecosystem concept and the biological ecosystem is not perfect, but much could be gained if the industrial system were to mimic the best features of the biological analogue' (Frosch and Gallopoulos 1992).

In contrast to preceding attempts, Frosch and Gallopoulos's article sparked off strong interest. There are many reasons for this: the prestige and wide audience of *Scientific American*, Frosch's reputation in governmental, engineering and business circles, the weight carried by the authors because of their affiliation with General Motors, and the general context, which had become favorable to environment issues, with, among other features, discussions around the Brundtland Commission report on sustainable development. The article manifestly played a catalytic role, as if it had crystallized a latent intuition in many people, especially in circles associated with industrial production, who were increasingly seeking new strategies to adopt, to deal with environmental issues.

Although the ideas presented in Frosch and Gallopoulos's article were not, strictly speaking, original, the *Scientific American* article can be seen as the source of the current development of industrial ecology. Ideas on industrial ecology were also disseminated among business circles on the basis of the *Scientific American* article, but indirectly. Hardin Tibbs, a British consultant who was working in Boston in 1989 for the company

Arthur D. Little, says that reading Frosch and Gallopoulos's article inspired him to write a 20 page brochure called *Industrial Ecology: An Environmental Agenda for Industry*. Arthur D. Little published the text in 1991. It was published again in 1993 by the Global Business Network, a consulting company near San Francisco, joined by Hardin Tibbs, which develops prospective scenarios for its member companies (Tibbs 1993). The Hardin Tibbs brochure quickly sold out, then thousands of photocopies of it were circulated, spreading Frosch and Gallopoulos's ideas throughout the business world. Other authors, also inspired by the Frosch and Gallopoulos's article, began to write papers disseminating the idea in various academic and business circles.

Today, industrial ecology is being pursued with unprecedented vigor. It is gaining recognition not only in business communities, but in academic and government circles as well. In 1997, the *Journal of Industrial Ecology* was launched (MIT Press, http://mitpress.mit.edu/JIE) and the International Society for Industrial Ecology (ISIE) (http://www.is 4ie.org) was founded in 2000.

THE INDUSTRIAL ECOLOGY AGENDA: RESTRUCTURING THE INDUSTRIAL SYSTEM

The principal objective of Industrial Ecology is to reorganize the industrial system (including all aspects of human activity) so that it evolves towards a mode of operation that is compatible with the biosphere and is sustainable over the long-term. The strategy for implementing the concepts of Industrial Ecology is often referred to as eco-restructuring and can be described in terms of four main elements:

1. Optimizing the use of resources;
2. Closing material loops and minimizing emissions;
3. Dematerializing activities;
4. Reducing and eliminating the dependence on non-renewable sources of energy (Ayres and Simonis (eds) 1994; Erkman *et al.* 2001).

Optimizing the Use of Resources

Optimizing the use of materials and energy in any industrial activity starts with an analysis of production processes in order to eliminate unnecessary losses. This is a step that is carried out by individual companies on their own activities and is called pollution prevention or cleaner production.

While there have been considerable efforts in this area during the past 10–15 years, there is still room for further improvement, particularly in the newly-industrializing countries that will represent the principal manufacturing base in the future.

Once we begin to consider the biological analogy underlying industrial ecology, we realize that additional aspects of resource optimization are not covered by the approaches mentioned above. In natural ecosystems, certain species feed on the waste of other species and thereby contribute to the creation of a food chain. Industrial ecology therefore suggests the idea of an 'industrial food chain' in which companies are linked in some form of network in order to exploit unutilized resources or by-products and thereby increase resource utilization.

Thus, the concept of Eco-Industrial Park (EIP) was born in the early 1990s. EIPs are areas in which companies cooperate to optimize resource use, namely, by mutually recovering the waste they generate (the waste produced by one enterprise is used as raw material by another) (Côté 1997; Côté and Rosenthal 1998; Francis and Erkman 2001; Lowe 2001).

The notion of 'park' should not be considered in the sense of a geographically confined area: an eco-industrial park can very well encompass a neighboring city, even a remote enterprise, especially if the latter is the only one around capable of recovering a rare type of waste impossible to process at other factory sites. Hence the new term, 'eco-industrial networks', where parks represent a particular case, is appropriate. The notion of eco-industrial parks (or networks) is quite different from traditional waste exchange programs. Indeed, it involves a systematic recovery process of overall resources in a given region, within the conceptual framework of scientific ecology. It does not merely recycle waste on an ad hoc basis.

One idea that fits in with the notion of eco-industrial parks is that of industrial biocoenoses. In biology, the concept of biocoenosis refers to the fact that, in ecosystems, various species of organisms always meet according to characteristic patterns of association. Just as in natural ecosystems, there are key species in industrial biocoenoses. Power plants, for instance, are an obvious key species. All kinds of different eco-industrial complexes could develop around thermal power plants (coal, oil, gas, nuclear), given the extent of matter flows involved and the enormous quantity of energy wasted as heat.

Once the best possible associations are determined, including the most appropriate combinations of various industrial activities, the concept can then be extended to industrial complexes. For example, instead of building an isolated sugarcane production unit, one should attempt, from the outset, to plan an integrated complex whose objective is to use all the flows of matter and energy linked to sugarcane processing in the best possible way. In this instance, a number of units could be attached – a paper mill, a distillery, a thermal power station – in order to recover all the different by-products of sugarcane. A variety of possibilities come to mind: pulp–paper complexes, fertilizer–cement ventures, steelworks–fertilizer–concrete partnerships, and so on. Granted, there are examples of partial and spontaneous complexes that have been around for a long time. However, the main focus now should be on developing these complexes in a more explicit and systematic way (Nemerow 1995).

The Eco-Industrial Park (EIP) is proving to be an important tool within the industrial ecology approach and at present there are around 50 EIP projects under way, particularly in North America, Western Europe and Asia (Chiu 2002).

Closing Material Loops and Minimizing Emissions

In natural ecosystems all materials flow cyclically in the form of a quasi-closed loop. For example, bacteria, fungi and small invertebrates break down dead matter or waste products from plants into simpler chemical compounds that can once again be used by plants. Companies that carry out this function of recycling wastes in the industrial ecosystem are usually referred to as 'recyclers'. Unfortunately, while natural ecosystems are very effective at closing the material loops, the industrial ecosystem is still far from optimal. Only a small fraction of the waste is returned to the system; the majority is 'lost from the industrial system' (a) through the creation of waste during the manufacturing of products, (b) as waste that is formed by a product when it is considered to be of no further useful value, and (c) in the form of products that are designed to be completely or partially dispersed during their use. At present, the losses of materials due to consumption patterns (for example, types (b) and (c) above) greatly exceed those during the manufacturing process.

Closing material loops within the industrial ecosystem, therefore, means addressing the complete life cycle of the product. One way is to make the recycling industry more effective, both with respect to technological solutions as well as logistics. However, energy is required to close the material loop in a natural or an industrial ecosystem. As long as we continue to use fossil fuels as our main source of energy in the industrial ecosystem, recycling will also contribute to the creation of waste from the combustion process. The energy associated with recovery of a material must therefore be considered when deciding on a strategy for closing the loop. In the case of the recovery of aluminum from scrap, for example, the energy requirement for recycling is much lower than that for extraction and purification of aluminum from bauxite. The environmental impact due to recycling is only one-tenth of that to produce virgin aluminum.

Although it is possible to envisage closing material loops for consumption patterns (a) and (b) above, there are some materials that are designed to be completely or partially dispersed during their use. Some examples are pharmaceuticals, fertilizers, pesticides, detergents, solvents, and so on. Such materials clearly cannot be recycled after use and will always represent a loss of resources. Minimizing dissipation of this type of product is a difficult challenge and may be addressed (in some cases) by rethinking about the service demanded.

One area where open material flows can no longer be accepted is when such materials are toxic/hazardous and, in particular, when they are persistent and bio-accumulate (accumulate in living organisms). Whether the material is lost due to inefficient recycling or through dissipative use, sustainability arguments imply that its future use must be seriously questioned and alternative solutions provided.

Dematerializing Activities

An important objective of industrial ecology is not only to create cyclic flows of materials but also to minimize the total flow of matter and energy used to provide equivalent services. Technical progress often makes it possible to obtain more service from a smaller amount of matter, such as by producing lighter objects or by replacing one material by another (for example, a few kilograms of optical fiber allows for more telecommunications throughput than one ton of copper cable). However, dematerialization is not as simple as it may seem – less massive products may have shorter life spans and will therefore ultimately consume more resources and generate more waste. Furthermore, dematerialization does not apply only to consumer goods, but also to the heavy infrastructure of the industrial system, such as buildings, roads, transportation networks, and so on (Herman *et al.* 1989, see also the examples of increased resource productivity exposed in von Weizsacker *et al.* 1997).

At present, two strategies are being debated: *relative dematerialization* so as to obtain more services and goods from a given quantity of matter, and *absolute dematerialization*, which strives to reduce the total amount of matter circulating within the industrial system. There has been a recent surge of interest in dematerialization in the context of the so-called 'new economy', or 'knowledge based economy', and there have been many claims that the emerging information technologies will contribute to the dematerialization of the economy. However, this is far from proven, and at this stage we must acknowledge our ignorance about the real impact of new information technologies on resource consumption. An introduction to the debates about the impact of the knowledge based economy on sustainability can be found in GeSI 2002, http://gesi.org). For a preliminary assessment see Berkhout and Hertin 2002.

Probably one of the best ways to dematerialize the economy is to emphasize the service rendered, or the function, and so on to market the use of the product rather than the product itself. For many years our economic system has been organized to maximize production. Within the context of industrial ecology, the objective is to prioritize use in order to evolve towards a genuine service-oriented society, also referred to as functional economy. This involves strategies such as durability (extending the useful life of a product), renting rather than owning, and selling use rather than the actual product. To illustrate the point, a photocopy machine manufacturer who sells the photocopy service rather than the machine itself, will run a more profitable operation if the photocopy machine, of which he retains ownership, requires as little matter inputs as possible, has the longest possible useful life, is easily recyclable, and so on (Mont 2002; Stahel; 2003).

Reducing and Eliminating the Dependence on Non-renewable Sources of Energy

Energy is an extremely important factor in the eco-restructuring of the industrial system. All efforts have to be made to increase energy efficiency through developments such as co-generation and energy cascading.

However, fossil fuels (coal, oil or natural gas) are a crucial factor in powering the engines of industrial economies. Combustion of fossil fuels is fundamentally dissipative and lies at the root of many environmental problems, including the enhanced greenhouse effect, smog, oil spills, acid rain, and so on. Eco-restructuring, therefore, must involve a change in the way that we obtain energy so as to make it more compatible with the goals of industrial ecology. In the first phase we can try to make fossil fuel consumption less harmful – for example, by recovering carbon dioxide gas or by decarbonizing the energy supply via a change from coal and oil to natural gas (and eventually perhaps hydrogen). However, it is clear that this is only a temporary solution and the move from fossil fuels to alternative renewable energies must be made quickly (Nakicenovic 1997; Socolow (ed.) 1997).

INDUSTRIAL SYMBIOSIS IN KALUNDBORG

As a matter of fact, industrial ecology is already more than a nice theoretical idea: the industrial symbiosis, which has evolved during the last three decades in the small city of Kalundborg, in Denmark, offers the best evidence that such an approach can be very practical and economically viable. Kalundborg, located 130 km west of Copenhagen, can be seen as a successful example of an industrial complex minimizing pollution and optimizing the use of various resources. Before addressing the specific issues of developing countries, a short discussion of the Kalundborg symbiosis would be useful.

The history of Kalundborg really began in 1961, with a project to use surface water from Lake Tissø for a new oil refinery in order to save the limited supplies of groundwater. The city of Kalundborg took the responsibility for building the pipeline while the refinery financed it. Starting from this initial collaboration, a number of other collaborative projects were subsequently introduced and the number of partners gradually increased. By the end of the 1980s, the partners realized that they had effectively 'self-organized' into what is probably the best-known example of a working industrial ecosystem, or to use their term – an industrial symbiosis (Christensen 1999; Ehrenfeld and Chertow 2002, http://www. symbiosis.dk).

The Kalunborg Industrial Symbiosis today consists of six main partners:

- Asnæs power station, part of SK Power Company and the largest coal-fired plant producing electricity in Denmark;
- Statoil, an oil refinery belonging to the Norwegian Statoil company;
- Novo Nordisk, a multinational biotechnology company that is a leading producer of insulin and industrial enzymes;
- Gyproc, a Swedish company producing plasterboard for the building industry;
- The town of Kalundborg, which receives excess heat from Asnæs for its residential district heating system; and
- Bioteknisk Jordrens, a soil remediation company that joined the Symbiosis in 1998.

In addition, several other companies participate as recipients of materials or energy. The status of the industrial symbiosis in 1999 is shown in Figure 2.1.

Thanks to the symbiosis, the reduction in the use of groundwater has been estimated at close to two million cubic meters per year. However, in order to reduce overall water consumption by the partners, the Statoil refinery supplies its purified wastewater as well as its used cooling water to Asnæs power station, thereby allowing this water to be used twice and saving additionally one million cubic meters of water per year.

Asnæs power station supplies steam both to Statoil and Novo Nordisk for heating in their processes and, since it is therefore functioning in a co-generation mode, it is able to increase its efficiency.

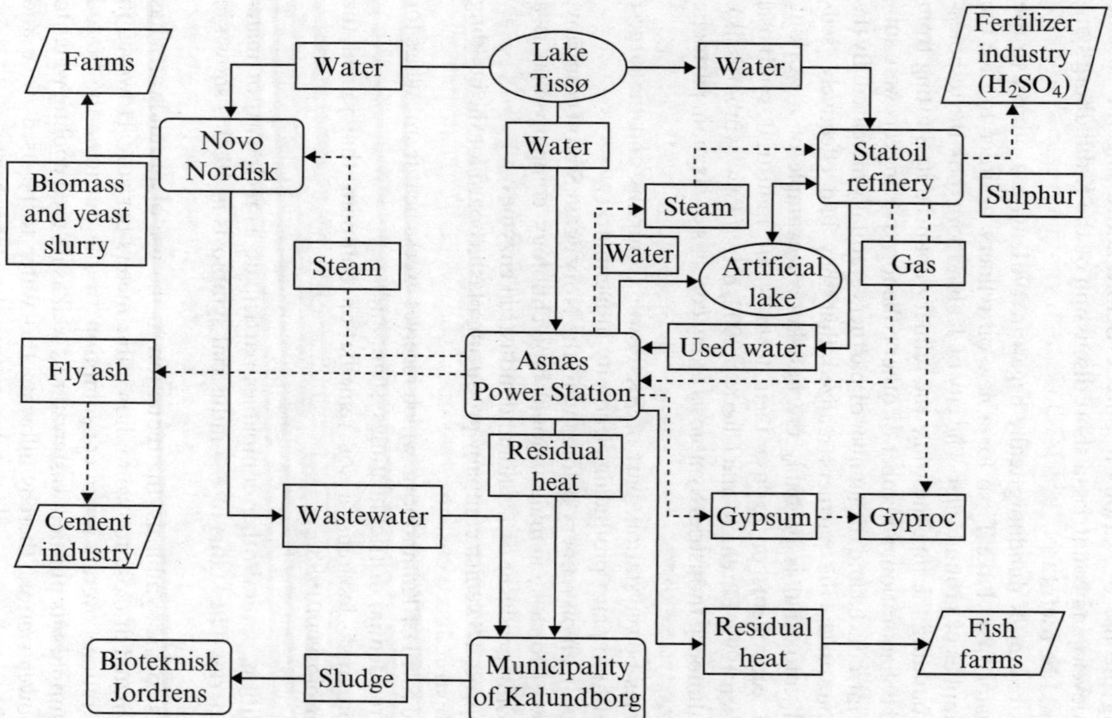

Source: C. Francis, adapted from Christensen, 1999.

Figure 2.1 Resource flows in the Kalundborg industrial ecosystem (status 1999)

Excess gas from the operations at the Statoil refinery is treated to remove sulfur, which is sold as a raw material for the manufacture of sulfuric acid, and the clean gas is then supplied to Asnæs power station and to Gyproc as an energy source.

In 1993, Asnæs power station installed a desulfurization unit to remove sulfur from its flue gases, which allows it to produce calcium sulfate (gypsum). This is the main raw material in the manufacture of plasterboard at Gyproc. By purchasing synthetic 'waste' gypsum from Asnæs power station, Gyproc has been able to replace the natural gypsum that it used to buy from Spain. In 1998, approximately 190 000 tonnes per year of synthetic gypsum were available from the power station.

Novo Nordisk creates a large quantity of used bio-mass coming from its synthetic processes and the company realized that this could be used as a fertilizer since it contains nitrogen, phosphorus and potassium. The local farming communities use more than 800 000 cubic meters of this liquid fertilizer each year as well as over 60 000 tonnes of a solid form of the fertilizer.

Finally, residual heat is also provided by Asnæs power station to the district heating system of the town. The system functions via heat exchangers so that the industrial water and the district heating system are kept separate.

The investment made to put the different material and energy exchanges in place has been estimated at $75 million. This is the cost of the 18 projects established up to and including 1998. Keeping in mind that each exchange is based on a separate contract between the two partners involved, revenues can be estimated as coming from selling the waste material and from reduced costs for resources. The partners estimate that they have saved $160 million so far. The average payback time of a project is less than five years. Therefore the clear lesson is that a more rational utilization of resources is not only good for the environment, but also saves money.

In any discussion of industrial ecology, the Kalundborg Symbiosis has tended to take center stage as the model to emulate. The importance of the Kalundborg example is not just how a few companies can share their waste for improved profit and societal gain, but more importantly, how local communities and societies can find strategies that can improve their sustainability by using their resources more efficiently. The Kalundborg example is more important from the point of view that it successfully exemplifies a development strategy that is different from the conventional wisdom of the time.

There is no doubt that the Kalundborg model has fruitfully inspired the recent thinking on environmental management of industrial estates and eco-industrial networks. Yet there is also a growing recognition that we need to look beyond Kalundborg. This is especially true regarding the implementation of Industrial Ecology in developing countries, where the industrial pattern is very unlike Kalundborg.

THE SCOPE OF INDUSTRIAL ECOLOGY

The term industrial ecology appears to suggest that it has something to do with just industry or ecology. But its scope goes far beyond that. In its very essence, in its broader definition, industrial ecology aims to study the 'flow' of all resources (material, energy, land, forest, human resources, or any other) through an entire identified socioeconomic system (a town, region, state) with a view to strategically optimizing their use. The 'flow' refers to the consumption of the resource (both the quantity and method of use) by different entities in the socioeconomic system. By this definition, industrial ecology lays emphasis on not just 'production' but on 'consumption' as well, either by individuals or by commercial entities.

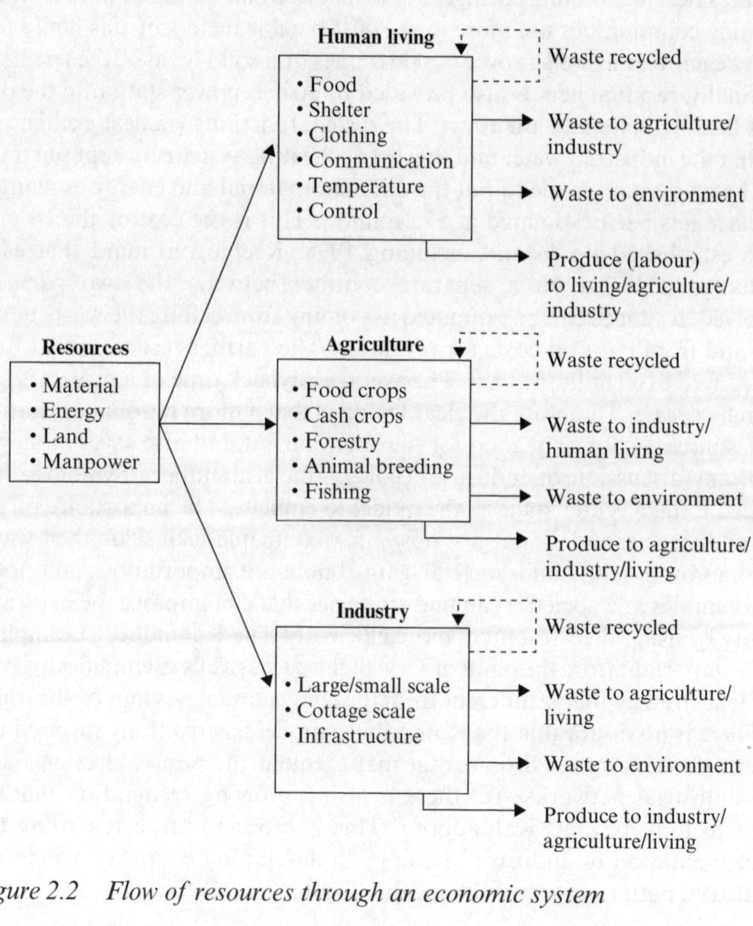

Figure 2.2 Flow of resources through an economic system

The scope of industrial ecology at a regional level could be as depicted in Figure 2.2. The utility of such an understanding is obvious. To develop strategies for optimizing the use of resources, it is essential to make a detailed analysis of the quantitative data about their consumption by different entities in society. Such knowledge about the flows and patterns of use of resources, besides contributing to the sustainability of a region in a broad sense, also offers specific advantages. New business and employment opportunities can emerge from creating value from certain resources previously considered as wastes, or from detection of possible innovative linkages between companies. It also allows the anticipation of potential environmental problems, an invaluable asset for planners and public authorities.

ACKNOWLEDGEMENTS

This chapter is reproduced from Erkman, S. and R. Ramaswamy (2003), *Applied Industrial Ecology: A New Platform for Planning Sustainable Societies*, Bangalore, India: Aicra Publishers, Chapter 1.

REFERENCES

Ayres, R.U. and U.E. Simonis (eds) (1994), *Industrial Metabolism. Restructuring for Sustainable Development*, Tokyo and New York: United Nations University Press.

Berkhout, F. and J. Hertin (2001), 'Impacts of information and communication technologies on environmental sustainability: speculations and evidence', report to the OECD, SPRU-Science and Technology Policy Research Unit: University of Sussex, Brighton, UK, 25 May.

Chiu, A. (2002), 'Ecology, systems and networking: walking the talk in Asia', *Journal of Industrial Ecology*, **5**(2), 5–8.

Christensen, J. (1999), 'Industrial symbiosis: a profitable potential for environmental benefits', in P. Pangotra, S. Erkman and H. Singh (eds), *Proceedings of the Workshop on Industry and Environment*, Ahmedabad, India: Indian Institute of Management, pp. 56–77.

Côté, R.P. (1997), 'The environmental management of industrial estates', technical report no 39, compiled by, UNEP Industry and Environment, Paris (United Nations Publication: 92-807-1652-2).

Côté, R.P. and C.E. Rosenthal (1998), 'Designing eco-industrial parks: a synthesis of some experiences', *Journal of Cleaner Production*, **6**(3/4), 181–8.

Ehrenfeld, J. and M. Chertow (2002), 'Industrial symbiosis: the legacy of Kalundborg', in R.U. Ayres and L.W. Ayres (eds), *A Handbook of Industrial Ecology*, Cheltenham, UK and Northampton, MA, USA: Edward Elgar, pp. 334–48.

Erkman, S., C. Francis and R. Ramaswamy (2001), '*Industrial ecology: an agenda for the long-term evolution of the industrial system*', Alliance for a Responsible,

Plural and United World. Sections 1.3 and 1.4 of this chapter are based on a contribution by Dr C. Francis to this last reference.
Francis, C. and S. Erkman (2001), *Environmental Management for Industrial Estates. Information and Training Resources*, prepared for UNEP–DTIE by ICAST, Paris: United Nations Publication.
Frosch, R.A. and N.E. Gallopoulos (1989), 'Strategies for manufacturing', *Scientific American*, **261**(3), 94–102.
Frosch, R.A. and N.E. Gallopoulos (1992), 'Towards an industrial ecology', in A.D. Bradshow et al. (eds), *The Treatment and Handling of Wastes*, London: Chapman and Hall, pp. 269–92.
Global e-Sustainability Initiative (2002), 'Information and communications technology', GeSI and United Nations Environment Programme, Division of Technology, Industry and Economics.
Herman, R., S.A. Ardekani and J.H. Ausubel (1989), 'Dematerialization', in J.H. Ausubel and H.E. Sladovich (eds), *Technology and Environment*, Washington, DC: National Academy Press, pp. 50–69.
Hileman, B. (1995), 'Eco-industrial parks offer economic and environmental advantages', *Chemical & Engineering News*, p. 34.
Lowe, E.A. (2001), 'Eco-industrial park handbook for Asian developing countries', prepared for Asian Development Bank.
Mont, O. (2002), 'Functional thinking – the role of functional sales and product service systems for a function-based society', Swedish Environmental Protection Agency report 5233, Stockholm, July.
Nakicenovic, N. (1997), 'Freeing energy from carbon', in J.H. Ausubel and H.D. Langford (eds), *Technological Trajectories and the Human Environment*, Washington, DC: National Academy Press, pp. 74–88,
Nemerow, N. (1995), *Zero Pollution for Industry. Waste Minimization Through Industrial Complexes*, New York: John Wiley & Sons.
Socolow, R. (ed.) (1997), 'Fuels decarbonization and carbon 18 industrial ecology: an introduction, sequestration: report of a workshop', The Center for Energy and Environmental Studies PU/CEES report no 302, Princeton University, Princeton, NJ.
Stahel, W.R. (2003), 'The functional society: the service economy', in D. Bourg and S. Erkman (eds), *Perspectives on Industrial Ecology*, Sheffield, UK: Greenleaf Publishing, pp. 264–82.
Tibbs, H. (1993), *Industrial Ecology. an Environmental Agenda for Industry*, Emeryville, CA: Global Business Network.
Weizsäcker, E.V., A.B. Lovins and L.H. Lovins (1997), *Factor Four. Doubling Wealth, Halving Resource Use*, London: Earthscan Publications Ltd.

PART 2

Industrial ecology: techniques and cases

3. Regional industrial ecology and resource productivity: new approaches to modelling and benchmarking

Joe Ravetz

INTRODUCTION

There is a topical debate on how far industrial innovation can further the goals of industrial ecology, in terms of physical resource productivity, waste minimization and closed-loop material flows. One way to approach this is through analytic modelling and benchmarking of the interactions of economic flows with environmental flows. Such modelling can potentially work well at the regional scale, where physical flows often show a close fit with economic activity, and where there is added potential for recycling, industrial symbiosis and other forms of integrated resource management.

However, experience shows that analytic modelling and benchmarking tools have to be situated within policy and business practices, if they are to be used and useful. It is also clear that few modelling or benchmarking tools are well equipped to deal with the complexities of supply chains, actors and networks, and evolutionary or structural change. However, there are growing aspirations and rapid learning from economic development policy-makers on the issues of regional resource productivity. This is forcing the pace of development and applications of such tools even before they are functional.

This chapter is a brief review of work in progress on the UK regional agenda for resource productivity, including a conceptual structure, modelling tools and policy applications. We first outline a resource productivity 'framework' for mapping the interactions of different forms of flows and capitals, in economic, physical and social forms. This framework then helps to underpin a 'diagnostic toolkit', in the form of two modelling and benchmarking tools currently in development or on trial in the regions of the UK. Based on the REWARD program and the Mass Balance programme, these are drawn at the regional scale, with the potential for extension to the local, sectoral, firm and product levels.

Thirdly, such tools have applications 'before', during and 'after' the innovation process, in benchmarking either the potential or the impact of changes, in either industrial production or consumer demand. One topical application is to explore the interaction of the 'weightless' economy of ICT-based resource management, with the 'weighty' economy of the resources themselves, in a regional industrial symbiosis system which matches waste to material supplies. For this and similar applications, such tools can act as enablers and facilitators to a broader socio-technical innovation process.

CONTEXT

The UK's national strategy recognizes the structural challenge of the goal of sustainable development, and aspires towards a far-reaching policy experiment in 'Sustainable Consumption and Production'. In practice this is seen not so much as a strategy, as a framework for future policy development (HMG 2005, Department of Trade and Industry and DEFRA 2003). Within this agenda, the supply side or production side can be interpreted in terms of 'resource productivity', which promotes the themes of eco-efficiency and dematerialization as a driver for business competitiveness, risk minimization, shareholder value and others (Performance and Innovation Unit 2001; Leadbeater 1998).

This theme of 'resource productivity' follows several strands in industrial ecology ('IE') thinking. One is the inter-dependence of material flows and waste exchanges in industrial clusters and along industrial supply chains (Chertow 2000). Another is the 'Factor Four' approach which focuses on the overall reductions in environmental impact, as a combination of both supply sides and demand sides (von Weizsacker *et al.* 1997). This then points towards the 'frontiers' of technological innovation, and the benchmarking of firms and sectors in relation to best or average practice (Tyteca 1996). A further approach is the 'eco-modernization' of industrial sectors (Ravetz 1999), and the implications for new approaches to environmental regulation (Gouldson and Murphy 1999). There is consensus that these approaches aim to meet both economic and environmental goals, but there is also a realization that the results do not always coincide.

Meanwhile there is an emerging agenda for regional development and governance, across the diverse geography of the UK. This has seen the growth of institutions such as the Regional Development Agencies (RDAs) (University of Dundee *et al.* 2001): the growth of practices such as 'integrated appraisal' and 'Regional Spatial Strategies' (Haughton and Counsell 2004): and the general awareness of the regional or city-region scale in the sustainable development agenda. Key strands of the RDA economic

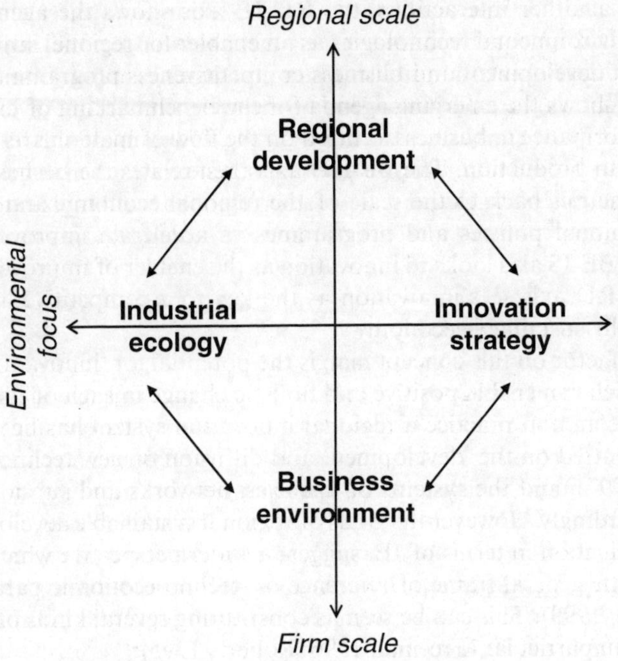

Figure 3.1 Regional industrial ecology agenda

strategies include industrial innovation, the role of clusters, market development and business competitiveness programmes (Cooke *et al.* 2003). More recently, environmental management and pollution control technologies have emerged as a key theme for industrial cluster development (Environment Agency 2003). Underlying this is an emerging agenda for policy integration through a more transparent, accountable and evidence-based practice of evaluation and participation (Ravetz *et al.* 2004).

The theme of this chapter centres on the intersection of these agendas of regional development, industrial ecology, business environment benchmarking, and innovation strategy – with shorthand versions as RD: IE: BE: and IS. The concept mapping below shows the Industrial Ecology and Innovation Strategy axis as discussed by Randles and Green (this volume), somewhere between the firm scale and regional scale, as an environment-economic agenda (Figure 3.1). The Business Environment agenda is more focused on the firm or sector level, and concerns both the 'push factors' of regulation, cost and liability, and the 'pull' factors of improved image and new markets. Meanwhile the Regional Development agenda takes more of an overview of the economy and environment of the region.

Looking at other interactions, the RD-IE axis shows the agenda for emerging environmental technologies as an enabler for regional scale clusters, market development and business competitiveness programmes. The IE-BE axis shows the emerging agenda for the benchmarking of environmental performance in business, centred on the flow of materials as wastes and inputs to production. The BE-RD axis then relates the business performance metrics back to the scale of the regional economy, and points towards regional policies and programmes to accelerate improvements. Finally, the BE-IS axis looks to innovation as the enabler of improvements, and the IS-RD axis sees innovation as the key to a competitive and environmentally sustainable economy.

The new factor on this concept map is the potential for 'innovation' as a catalyst which can enable positive and holistic change in each of its applications. In common practice a regional innovation system has been conceived as centred on the 'development and diffusion of new technologies' (Morgan 1997), and the systems of agencies, networks and subsidies are geared accordingly. However the goals of regional sustainable development and its application in terms of IE, suggest a wider perspective which aims to expand the general frame of reference or 'techno-economic paradigm' (Green *et al.* 1999). This can be seen as constituting several kinds of paradigm shift, in particular (Freeman 1996; Roberts 1995):

- awareness of local and global thresholds and limits for environmental resources and assimilation capacity;
- awareness of economic growth not as an end, but as a means to the goal of social welfare (however that may be defined).

All this suggests the need for improved analytic tools to examine the intersections of the themes of 'RD-IE-BE-IS' above, in the general spirit of evidence-based policy-making. Such tools might take the form of technical diagnosis, comparative benchmarking, quantitative analysis and multi-criteria decision support methods. Examples of such diagnostic systems are currently being developed in the UK as 'databases' and 'models', mainly focused on the quantitative appraisal of alternative scenarios for regional production and consumption. When models, methods and support systems are combined into a 'toolkit' this would aim at applications at the product scale, in terms of environmental impact through the supply chain: at the firm scale, in terms of environmental performance benchmarking: at the sectoral scale, in terms of indices for overall performance and comparison between regions: and at the regional policy scale, where the trends, targets and priorities for regional interventions may be assessed.

At each of these levels, there are baseline applications, in terms of monitoring and reporting: comparison of best and worse cases: and of incentives or barriers to individual firms. There are also applications in future studies, in terms of trends, projections, alternative scenarios, targets and trend-target analysis (Ravetz 2000b). Such approaches can then be used in practice for evaluation and appraisal, in terms of the *ex-ante* and *ex-post* assessment of opportunities, interventions and technological improvements.

At present it is clear that such analytic tools are in their infancy: the sophistication of financial accounting can be contrasted with the data scarcity of most physical models, although even such national economic accounts are relatively recent (Pedersen and de Haan 2006; Vaze and Balchin 1996). It is also clear that the analysts and modellers comprise only one element in a combined learning and innovation process, involving firms, consultants, regulators, academics, consumers and others. The approaches described here are at the start of a long process.

A RESOURCE PRODUCTIVITY FRAMEWORK

At the centre of the relationship of RD-IE-BE is the theme of 'resource productivity'. Resource productivity has many definitions and many incentives – as a driver of business competitiveness and quality management, as cost saving or dependency reduction, and as a measure of industrial innovation. A narrow definition would see resource productivity as a measure of economic output per unit of input, whether these are in terms of finance, labour or physical units (Performance and Innovation Unit 2001). A wider perspective would look beyond the 'output' to that of 'outcome' in terms of human welfare achieved at the end of the supply-demand chain.

To pursue this wider perspective on resource productivity, we demonstrate a general framework for its definition and quantification. This has been extended from the 'Integrated Sustainable Cities Assessment Method' (ISCAM), a package of methods and tools for systems mapping and modelling (Ravetz 2000a). The discussion in this section includes first an overall framework for the interaction of economic, environmental and social flows and capitals. Then we examine how this can be applied to various concepts of resource productivity, at the regional, firm and product level.

A Set of Cycles

A useful mapping to identify the interaction of economic, environmental and social flows and capitals is shown in the diagram below (Figure 3.2). This

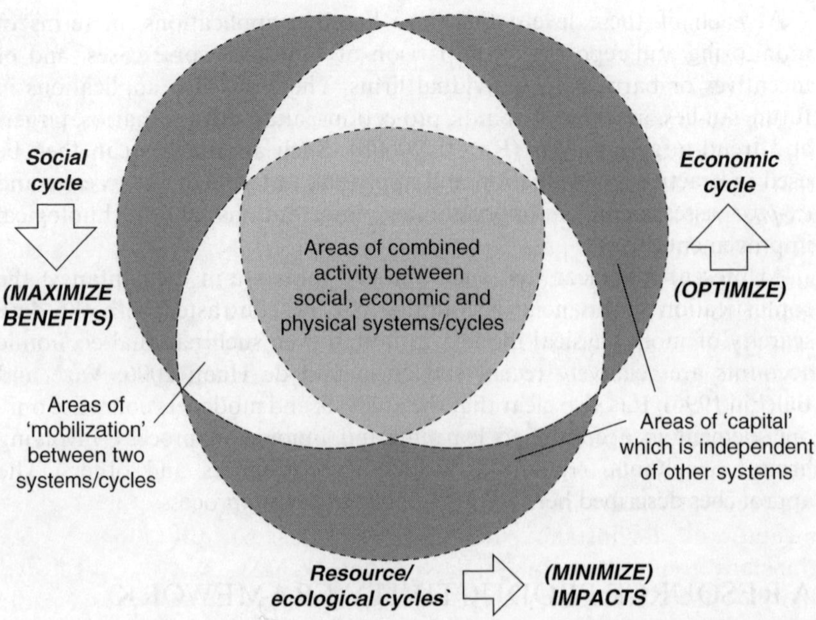

Figure 3.2 Resource productivity framework (a): mapping of systems and cycles

shows several kinds of flow, which can be conceptualized and potentially modelled as cyclic processes:

- *Resource/material flows*, from extraction, manufacture, use, disposal and return to the environment.
- *Economic flows*: shown as the conventional 'economic' circular flow of money/capital.
- *Social flows*, concerning the cycle of labour, production, consumption/'utility'/'outputs', and social welfare 'outcomes'.

It should be noted that this diagram is strictly a heuristic mapping device, not a social theory, and any deterministic modelling on this theme is of course more problematic. The diagram also shows the generalized policy goal or normative direction for each of these, in the light of the concepts of 'dematerialization' (Leadbeater 1998) and 're-socialization' (Robinson and Tinker 1995):

- *Resource flows*: to MINIMIZE impacts, in order to maintain life support systems;

- *Economic flows*: to OPTIMIZE, in the light of the above;
- *Social flows*: to MAXIMIZE social welfare and social capital.

The integrated framework here of course is vastly simplified, in order to identify the fundamental types of interactions between each of the circles, that is physical, economic and social systems/processes. Its use is mainly as a conceptual tool, not necessarily as explanatory theory in itself. In particular it provides a typology of interactions between physical, economic and social systems/processes. We should also note that there is nothing sacred about the division into three circles. One alternative scheme uses four types of capital (natural, human, manufactured and social), in a prototype evaluation framework for sustainable regional development (GHK *et al.* 2003).

The conceptualization of the circular processes or 'cycles' may be conceived as running in either direction, depending on the issue at hand, which is a recurring theme in system dynamics methodology. For the environmental cycle, for instance, there is a fairly clear path in terms of mass transfers, from primary material extraction, to use, to waste, and back to the biosphere. However there is also an opposite kind of causal path, whereby the demand for materials at the point of consumption then motivates or 'causes' their extraction through the pull of market forces. The circular flow of money may be equally bi-directional in the nature of its cause and effect.

Identifying 'Capital'

Where we can identify assets or resources or stored/maintained qualities, whether these are economic, environmental, or social, then this correlates with the concept of 'capital'. In economic terms this is fundamental and quite familiar, subject to the many possible financial contingencies of liquidity, interest, equity, time preference and so on. For environmental capital assets it is less clear: it may be tangible in terms of economic functions (for example, a hectare of commercial forestry), but quite volatile or fuzzy in terms of social functions (for example, a hectare of mixed community woodland). This perspective points to the way that 'capital' is not necessarily a straightforward factor such as money in the bank, but more akin to the potentiality or latent qualities for mobilization, qualities which are effective for each of the circles in relation to the others (economic, environmental or social). In other words, just as economic capital is only 'realized' when the money is drawn from the bank, environmental capital is only 'realized' when brought into social or economic processes.

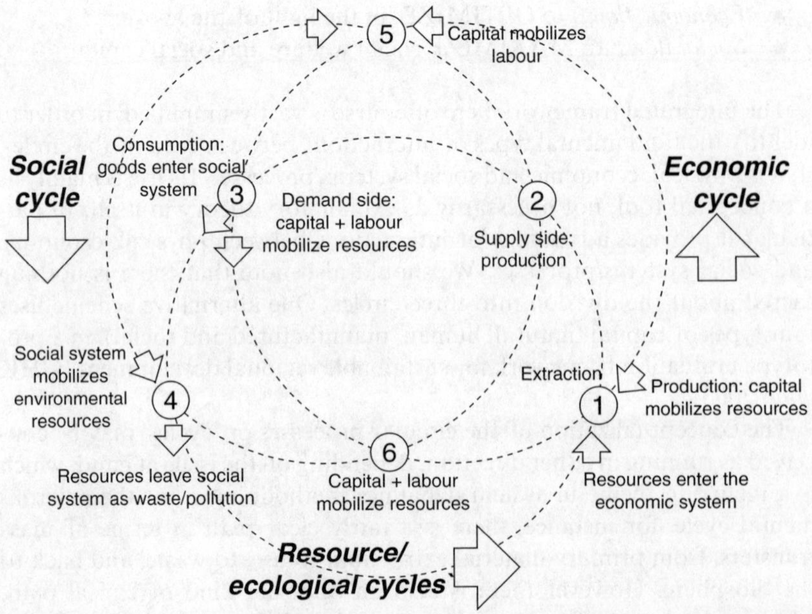

Figure 3.3 Resource productivity framework (b): mapping of interactions

Where we can identify types of 'capital' which are mobilized by combination or transformation with another type of capital, then we have a generic typology of possible interactions between each of the circles. The various points 1–6 on the analytic Figure 3.3 represent the range of possible generic interactions between flows in the economy, environment and society.

This provides an overall template for more detailed modelling and analysis. The flow in each direction is shown as far as this can be visualized. The summary Table 3.1 shows a more complete listing of each interaction or crossing point in each system cycle.

Mapping Resource Productivity

A more evolved perspective on resource productivity looks at the interaction of each part of the cycles of economic, environmental and social systems, and the mobilization or impact which is effected at each point. One analysis of the regional metabolism in construction minerals, for instance, shows the different kinds of capital involved (McEvoy *et al.* 2004):

Table 3.1 Generic interactions in resource productivity

	Physical cycle		
1	Physical resources extracted and enter economic system	4	Physical environment enables/ mobilizes social activities and systems
2	Physical resources are processed by labour in the economic system	3	Physical community resources mobilized by economic system and capital
3	Physical resources are sold from economic system into social consumption	2	Physical resources transformed into independent capitalized commodities
4	Physical resources leave social system as waste, to return to ecological system	1	Physical resources leave economic system as wastes, to return to ecological system
	Economic cycle		
5	Capital mobilizes social system to generate productive labour	1	Economic production: capital mobilizes physical resources
3	Capital uses labour, to act on physical resources in economic production	6	Mobilized resources are processed with added value from labour input
6	Physical goods and commodities, are the result of economic production	3	Economic value mobilizes social system independent of physical resources
1	Economic value/capital is generated, independently of physical/social effects	5	Economic value/surplus/ accumulation disengages from social system
	Social cycle		
4	Social system engages with physical environment in time and space	5	Social system mobilized as labour, by economic capital
6	Social system mobilized, via capital in economic production	2	Mobilized labour in economic production, to process and transform physical resources
2	Social system conditioned by non-material consumption/production	6	Social consumption of physical resources, after economic activity
5	Social experience, independent of economic consumption/production	4	Welfare gained in social system, after consumption/engagement of physical resources

- Economic capital mobilizes industrial plant, to extract minerals to provide construction concrete, which serves the demand for social housing.
- Employment is generated in local quarries to supply distant minerals markets, which promotes local resident spending, while creating

disturbance in a National Park area and reducing the attraction to tourists.

The social system shown here is often left out of conventional resource productivity calculations, as it is often complex, volatile and difficult to quantify. For instance, it would be simple to assume that the 'value' of construction minerals is equal to their commercial value at the retail or wholesale stages, until we consider the longer term sustainable development issues (Hammersley 1996). This valuation may be workable and plausible, up until the point at which the social system and its embedded 'capital'/ 'value' is predominant: for example, the 'value' of the undisturbed landscape: or the 'value' of the finished social housing to the community and neighbourhood. This of course raises a challenging agenda for institutional ecological economics, and the measurement of externalities (Ravetz 2000b, p. 235; Jacobs 1997).

The framework above can be used as a basis for mapping different types of resource productivity, in terms of ratios between the various key stages on the economic, environmental and social cycles (Figure 3.4).

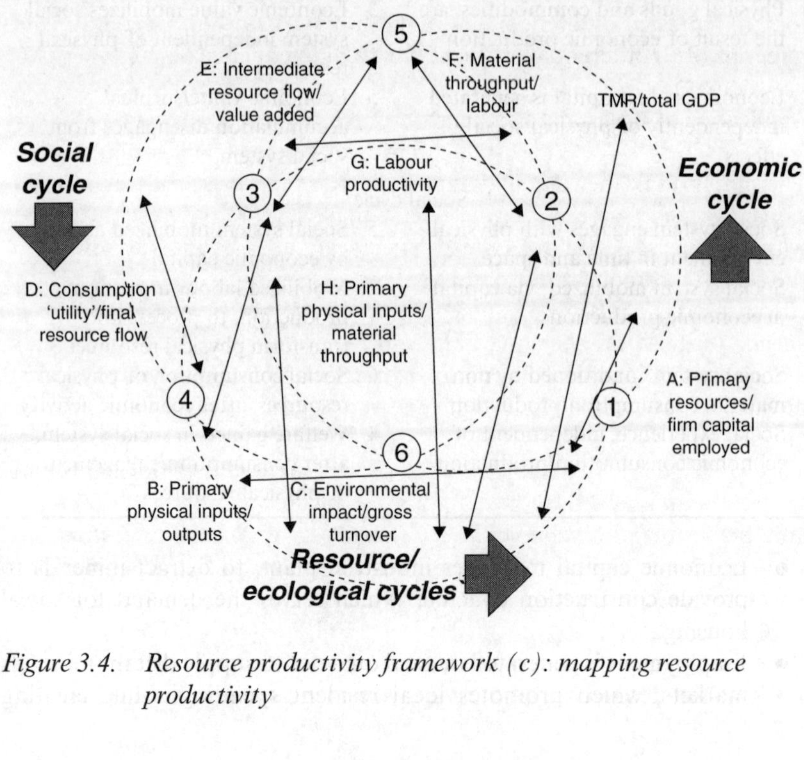

Figure 3.4. Resource productivity framework (c): mapping resource productivity

The mapping shown here is not fixed and final, but shows the possible scope of a larger set of resource productivity indicators:

- A: Primary resources/firm capital employed
- B: Primary physical inputs/outputs
- C: Environmental impact/gross turnover
- D: Consumption 'utility'/final resource flow
- E: Intermediate resource flow/value added
- F: Material throughput/labour
- G: Labour productivity: net output/employee
- H: Primary physical inputs/material throughput
- The aggregate 'resource productivity' may be summed up as Total Material Requirement/total GDP.

Naturally, the selection of a practical set of performance indicators for any real industry, process or product can be difficult. The complexities of supply chains, labour effects, social impacts, inter-generational effects and so on, rapidly exceed the available data and the willingness to collect it. In the minerals case study referred to above, there was lengthy debate from senior industrialists and consultants on the choice of appropriate indicators, to represent different stages in the supply chain, for different types of minerals, with different applications, with different environmental impacts.

Application to Firms and Products

The simplified framework above can be seen to represent an ideal theoretical case, where one firm uses one material to make one product with one worker, with one type of social outcome. In reality of course, each of these domains can be vastly complicated by industrial processes, supply chains, institutional effects, market effects, global/local externalities and so on. This is then the agenda for business benchmarking, which can be applied to environmental management, environmental performance, or resource productivity metrics in various ways. First, we look here at the generic typologies in which any business may engage with resource flows, and hence resource productivity. This works generally within the input-output methodology for inter-industry transactions and the conceptualization of upstream/backward linkages, and downstream/forward linkages (Wiedman *et al.* 2005; Giljum and Hubacek 2003). Broadly, the types of impacts can be classed as direct, indirect and 'induced'.

1. Direct production impacts: direct or on site consumption of energy/ material resources, in the processing and manufacture of physical

products. This applies more to material- or product-intensive businesses in manufacturing sectors, for instance the manufacture of plastic containers.
2. Indirect production impacts: embedded energy/resources in the upstream or downstream stages of the supply chain. This applies more to processes at one stage in a more lengthy and complex supply chain, for instance the impacts of manufacturing above may be outweighed by the production of the plastic itself.
3. Induced production impacts: embedded energy/resources where the material flows are removed at some distance from the supply chain. This applies to producer services or consumer services, for instance an environmental consultancy which advises on the manufacturing process above.

A similar breakdown can be identified on the consumer side, for both households and the public or non-profit sectors, for example, direct, indirect and induced impacts. Such a typology can then be applied to the question of resource productivity, with the added dimension of what is here termed 'resource flow proximity', for example, the distance from the main resource flow path in terms of number of supply chain links (Table 3.2).

Finally there is a significance in the degree of material intensity, which indicates how much of the economic value is identified or represented in the material flow and how much in other forms of capitals or flows. The summary Table 3.3 represents the supply chain issues for different scales of business, from an 'ideal' one product/one firm supply chain to the complexity of a multi-national firm.

To summarize, this resource productivity framework so far has deliberately been simplified in order to map the key features. We have said little here about the consumption side: recent work under the UK Sustainable Consumption and Production programme looks at what might be termed the counterpart agenda of 'resource consumptivity' (Jackson and Michaelis 2003). For the present chapter, the resource productivity framework then serves to underpin a set of regional models and information systems as below.

RESOURCE PRODUCTIVITY MODELS

One response to complex problems – not the only one – is to simplify with a model. This brings its own pitfalls and drawbacks: it also brings into focus the pre-conceptions of the model makers, the model users, the social construction of knowledge which is formalized in the model, and a host of

Table 3.2 *Resource productivity benchmarking, upstream and downstream*

	Externality/ factor inputs	Backward linkages	Activity	Forward linkages	Externality 'goods'	Externality 'bads'
Product/ operation MFA			Product operation	Users of product operation	Welfare from product operation	Operational waste emissions
Direct MFA	Energy transport	Primary materials and components	Product manufacture	Assembly distribution retail	Products and by-products	Production waste emissions
Indirect MFA	Inter-industry energy/transport	Inter-industry demands		Value added services		
Induced MFA	Inter-industry resource demands	Inter-industry services	Inter-industry value added	Consumption	Welfare from consumption	Post consumer waste, etc.

Table 3.3 Material intensities in the business supply chain

	Primary	Secondary	Tertiary	Direct consumption	Indirect consumption	Direct emissions	Indirect impacts
MFA	Materials enter economy	Added value to make products	Added value in services	Materials consumed directly in products and services	Materials consumed indirectly as part of system/ infrastructure		
MFA intensity	Material intensive	Product intensive	Service intensive	Consumption intensive			
Thresholds and limits	Resource/ capacity depletion			Demand side	Infrastructure side	Pollution thresholds	Eco-system change
Ideal supply chain	Material extracted	Single product	Simple sales	One type of consumer		Emissions and waste	Minor outcomes
Average business	Many materials	Range of products	Range of services	Range of customers			
Multinational business		Many complex products	Many logistics of services				

other considerations. (Darier and Shackley 1998; Ravetz 1998). Here we focus on two related examples of models for regional consumption and production, currently in prototype development and testing around the UK.

The REWARD Programme

The conventional approach to regional analysis has at its core an economic or econometric base. This has recently been extended to include environmental and resource issues in a current regional programme in England and Wales. The REWARD programme ('Regional and Welsh Appraisal of Resource Productivity and Development') aims to provide an information base for resource productivity initiatives at the regional level.[1] It was formed by a partnership of Regional Development Agencies and similar bodies between 2002–2004, and now forms the beginning of a longer term programme,[2] with three main objectives:

1. Development of a computer model – the REEIO ('Regional Economy-Environment Input-Output model'). This provides a new level of analysis of the effects of economic trends and economic policies on resource use and environmental pressures.
2. A research programme and database on the resource productivity of the regions of England and Wales, and the implications for policy and business.
3. An applications and capacity building programme in each of the regions of England and Wales – enhancing strategic intelligence through workshops, training, toolkits, information systems, analysis and communications.

The REEIO Model

The REEIO software model provides a relatively detailed quantitative analysis of regional strategy and policy appraisal, providing a solid technical foundation for other analysis, and links to other technical models and databases. The REEIO is based on a detailed econometric input-output model of each regional economy, based on the widely used 'Local Economy Forecasting Model' and its parent the MDM model of the UK economy (Brettell and Gardiner 2003; Barker 1998). This uses a 50 sector economic classification aggregated up from the 123 national SIC classification, and the labour market is shown in six types of employment and 25 types of occupation. The REEIO then links economic and employment changes with key environmental and resource pressures:

- Waste sector: arising from household, industrial/commercial, construction, agriculture and so on, biodegradeable, non-biodegradeable, inert and non-inert compositions: disposal routes to landfill, incineration, recycling/re-use.
- Energy sector: demand from households, transport, industrial/commercial activity: energy supply is by 13 sectors and six fuels.
- Air emissions: including greenhouse gases, SO_x, NO_x, VOCs, PM and eight others.
- Water sector: aggregate water demand metered or non-metered is related to households and to economic activity.

The user inputs are arranged in a series of 'what-if' scenario assumptions, from overall population trends to the details of waste or energy management. These are generally arranged as policy inputs or technological change, but short term interventions, projects and shocks can also be simulated. The outputs can be taken to spreadsheets for charting, and further analysis on policy or business implications. To cover more detailed questions such as economic clusters, transport strategy or environmental technologies, a series of 'off-model' components is being developed in the form of smaller spreadsheets. A key resource is a comprehensive database of economic and environmental indicators, trends, projections, and scenario inputs for each region. One of the main components is the 'Linking-Up' study, which looks in detail at the policy applications of the model, in terms of future studies, strategic planning, evaluation/appraisal, and policy training (CURE 2002). One of the route maps produced by the Linking Up study shows the range of applications of the REEIO and related models (Figure 3.5).

The model so far has been applied in several regions including the North West of England, where it helped to analyse the trends, projections and the potential for commercial/industrial waste minimization (CURE 2004). This project aimed to quantify the opportunities for waste minimization by increasing the scale of activity in business-environment programmes. The study process included a regional workshop, a detailed report on modelling and regional initiatives, and the setting up of a forum to take it forward. This also aimed to link the REEIO system to the material flow analysis method in the next section, although this turned out to be difficult without the new industrial and commercial waste datasets currently being developed by DEFRA.

Mass Balance Approach

An assessment of material and energy flows within a defined boundary is termed a Material Flow Analysis (MFA). This looks at the material inputs

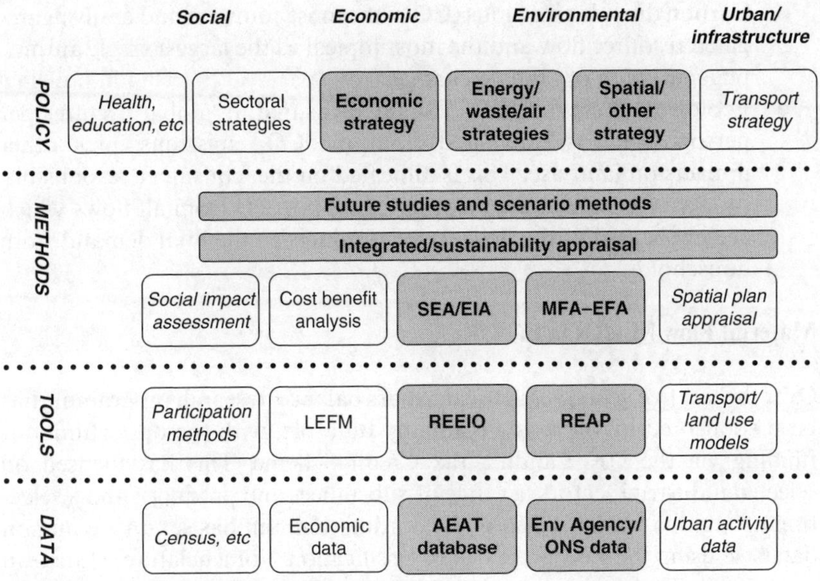

Figure 3.5 Toolkits for regional sustainable development

to a region in terms of raw materials and products, and at outputs in terms of waste and emissions, plus any changes in stocks. The analysis focuses on the consumption of goods and services by households and the commercial sector, including materials directly used and consumed. It may also look at 'hidden' material flows including ores and wastes from extraction or harvesting, energy used for extracting, transporting and producing materials: and greenhouse gas emissions from energy use. This kind of data is generally arranged in terms of 'consumption sectors', for example, the material requirements of the functions generated by final consumer needs, rather than the detailed breakdown of economic 'production sectors' in the REEIO model and most economic accounts. As a result of these two complementary approaches, a number of key physical indicators can be generated (Eurostat 2000; Bringezu and Schutz 2001; Brunner and Rechberger 2004):

- Direct Material Consumption (DMC): the total amount of materials directly used in the regional economy and consumed in the region, for example, excluding exports.
- Total Material Consumption (TMC): the total material use associated with regional consumption, including DMC together with the indirect or 'hidden' material flows generated by that flow. Again, this excludes exports and their associated indirect flows.

- Carbon dioxide emissions (CO_2): the most common and easily aggregated resource flow and the most topical as the largest single anthropogenic cause of climate change.
- Ecological footprint (EF), usually measured in 'global hectares per person'. This is calculated from the CO_2 emissions, plus other impacts on land use. This is allocated on the 'consumer responsibility' basis, an aggregate measure of all impacts from all flows which are implicated in the delivery of products to the final demand from households.

Material Flow Models in the UK

Over the last five years, a large scale 'mass balance' research programme has been sponsored by the waste company Biffa plc, with the opportunity of funding via the UK Landfill Tax Credit Scheme. This has focused on selected industrial sectors: a range of substances and products: and a selection of regions or sub-regions.[3] A coordination unit has set up a common database using the European CN (Classification Nomenclature) (Linstead *et al.* 2003). There are two main approaches:

- Production-centred mass balance: this takes a sectoral approach to raw materials and manufacturing, and includes exports plus regional final demand. This is more compatible with the REEIO model approach.
- Consumption-centred mass balance: this focuses on the products and services delivered to final consumers in private households or government, and traces the direct and indirect material consumption along product supply chains, with their impacts, which could be anywhere in the world. This approach is suited to a LCA method, and its simplified version the Ecological Footprint.

In principle a combined and integrated system should be developed with both production and consumption as part of a whole. However, existing data is generally inadequate for making detailed links between one approach and the other. For instance current UK waste data does not generally contain details of its material content, its industry source, or its location of origin. The consumption data now being assembled from a variety of databases including PRODCOM, COICOP and the IVEM energy database, does not have detailed information on the waste arising from each stage in the supply chain, or its material content, or the inter-industry transfers of materials and waste. However there is enough current data to at least provide an outline five-stage model of the UK and regional economy in material flow terms (Figure 3.6) with explanations as follows:

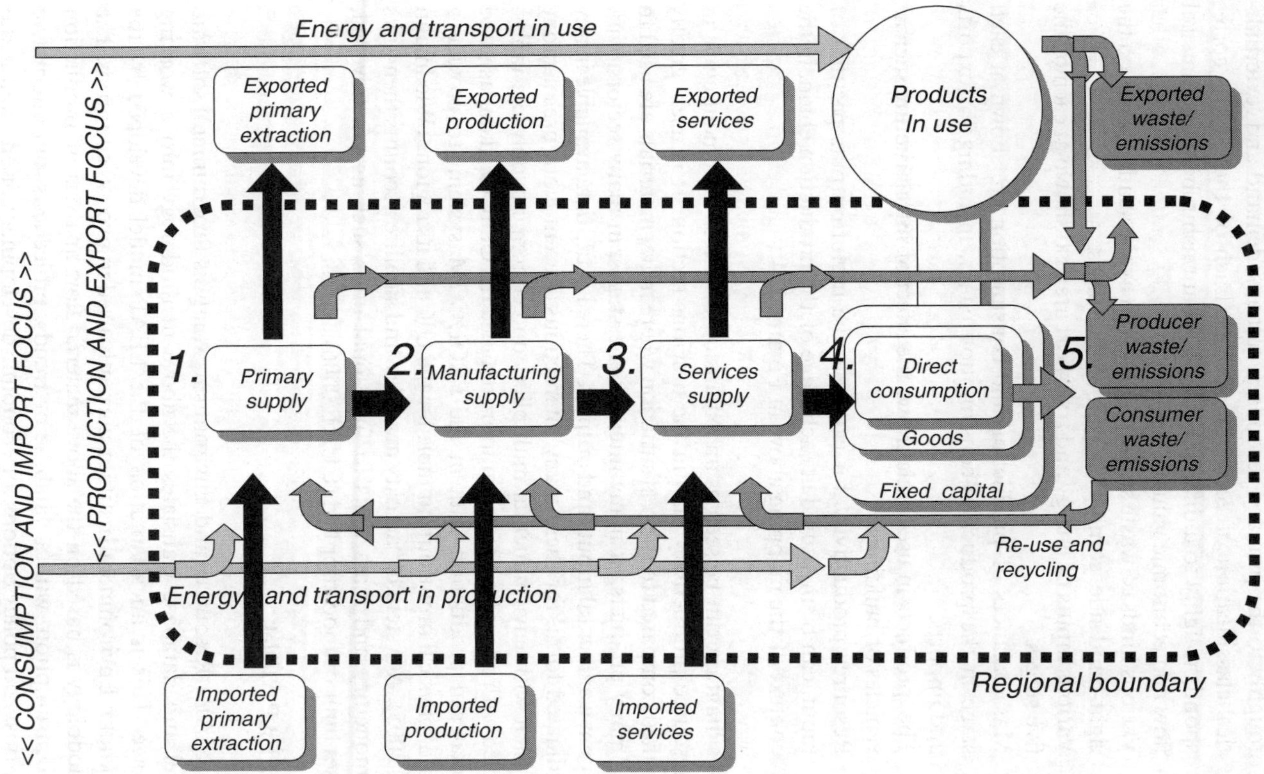

Figure 3.6 Regional material flow analysis

- The framework is organized in a five stage process, corresponding roughly to the primary, secondary, tertiary, demand and 'externalities' classification of economic sectors. Each of these types corresponds roughly to a different kind of relationship between material flow and economic value.
- Various kinds of waste streams are shown by the shaded boxes on the right-hand side, coming off each of the stages.
- Various inputs of energy and transport are also shown at each of the five stages.
- Mass balances of production and consumption are shown at each stage in the production-consumption chain, including for exports and imports.
- The 'products in use' circle shows the effect of infrastructure such as vehicles or buildings.
- Resource productivity, for example, the useful outputs per unit of input, can be measured at each stage of the production chain, in the context of the mapping above at Figures 2–4.

This diagram can be seen as an expansion of the environment circle in the resource productivity map in the previous section. It is itself greatly simplified compared to the real situation where many materials are used to make many products, at many intermediate stages, in many sectors, with many environmental inputs and outputs. There is little data available in any coordinated form, for these many interactions. Recent work uses a proxy approach with environmental multipliers on economic supply-use tables, coupled with allocation of expenditure data via the COICOP database, and the material production data in the PRODCOM system, to provide a 'hybrid' physical input-output table for the UK and its regions (Wiedmann *et al.* 2005). A system of 'activity modules' and satellite accounts then links the monetary and mass units to functional units such as km travelled, houses built or food supplied (CURE 2006).

The REAP Model

The REAP ('Resources and Environment Analysis Programme') software model and database translates the above methodology into a working package. This is an adaptation of the LEAP model developed by the Stockholm Environment Institute, currently used in over 40 countries. The methodology is based on the above material flow analysis of production and consumption, with a database of trends, projections and alternative scenarios, and policy options for economic development or environmental management under a range of alternative assumptions, with proxies for economic development, technology innovation, price and fiscal effects and so on. The REAP system provides several unique features over and above other available tools in the UK:[4]

- Analysis of inter-dependencies between sectors and supply chains, via the hybrid physical input-output table and database for the UK regions.
- Analysis of total impacts of consumption to meet final demand, through a detailed model of international trade and UK imports.
- Analysis of material flow at regional and local authority level; the local level is calculated through applying physical throughput to household expenditure data and then to the local area Acorn classification.

The REAP system is arranged around a 'functional' concept, with four types of components:

- Population and demand: factors that affect the overall size of the economy, labour force and consumption: regional migration, demographic factors, and household incomes/savings.
- Technology and production: factors that affect the share between economic sectors, and the transactions between each of the sectors: for example, the size of the waste management sector, and its use of transport services.
- Productivity and eco-efficiency: the resource intensity or the amount of waste/emissions produced for each £ of turnover in each sector: for example, the waste from construction activity.
- Environmental management: for some topics, there are further choices to be made: for example, waste disposal/recycling methods.

The modelling method starts with the form of a simple environmental accounting model, organized in principle around the five-stage mass balance framework above. However in practice the data at each stage are in different types of units – raw materials at the primary stage, products at secondary stage, composite items and services such as floorspace or transport kilometres at the tertiary and demand stage. Also, to design scenario settings with policy relevance involves a wider set of parameters than a purely material flow centred mass balance model can deliver. The way forward is seen as a loose-coupled modelling framework, where the core mass balance model is linked to a range of other models with compatible formats:

- econometric-based physical input-output model, which provides the environmental multipliers;

- materials, products, components and environmental coefficients model/database, adapted from the PRODCOM system;
- activity sectors and policy issues at the regional and urban level, including urban development, land use, housing, transport, and so on.

This is generally based on the ISCAM modelling approach, which provides a consistent format for the 'off-model' calculations needed to link policy inputs with the mass balance information (Ravetz 2000). In turn, this is based on the 'decomposition' approach to disaggregating compound trends and dynamic relationships into discreet factors (Ekins 2004; Kaya 1990).

Applications to Benchmarking

One of the main applications of the REAP tool and the regional production-consumption framework is a method for benchmarking the actual environmental performance of businesses. In this case the role of the benchmarks is to identify the interaction of economic performance with environmental impact/resource consumption, within the typology above of direct/indirect and induced resource flows. There are several recent approaches to this:

- The ENWORKS on-line data capture tool is focused on the 'opportunities' for better practice which have been discussed in firm-level site visits and other outreach work, as part of a regional programme (Enviros Consulting 2003; www.enworks). This then serves to track progress in the project pipeline of concept, targets, feasibility, implementation, monitoring and evaluation stages. There is a drawback in that the analysis only counts energy, solid waste and water effluent production, but is focused on monetary values rather than physical quantities.
- The ASSESS on-line environmental management package focuses on environmental policy, but also contains a trial application of direct 'mass balance' questions (www.egeneration.co.uk). Experience shows that these are difficult to translate for different business sectors, and difficult for business to find data for without direct pressures or incentives from regulators, customers or supply chains.
- The PERFORM database of firms across the EU in selected industrial sectors is very comprehensive and analytically sound. However difficulty was found both in getting the primary data, and in finding applications and users for the completed work (Hertin and Berkhout 2005; www.perform.ac.uk).

The experience so far is that on-line questionnaire type survey forms work most effectively with human contact as backup, which increases

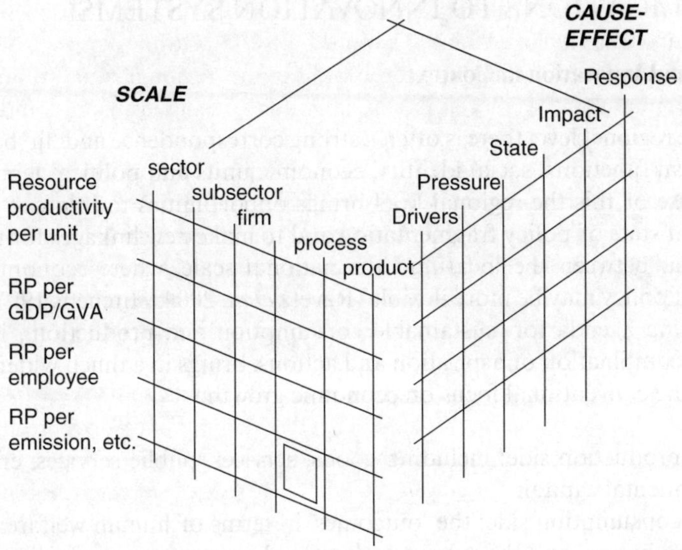

Figure 3.7 Eco-region: benchmark framework

the incentives for data collection. However on-line benchmarking is often technically complex, costly and prone to these failures. Hence a hybrid approach is followed for the REAP prototype business benchmarking scheme, between manual and on-line access: it is also triangulated between production (sectoral) and consumption (product) level analysis. The template under development contains at the time of writing, as per the 'Rubik's cube' visualization in Figure 3.7, the following components:

- environmental factors in waste, materials, transport, energy, water, minerals, toxicity burden if known and so on;
- economic/social factors: GDP/turnover, GVA, employees, capital investment, other EHS/corporate responsibility;
- average/best practice for similar firms/products;
- average/worst/best practice for the sector and sub-sector;
- trends, projections, targets and trend-target distance for the key factors above;
- comparison wherever possible with regional thresholds, pressure points, limits, goals and targets; and
- more qualitative information on opportunities and threats, specific to each business type in each sector.

APPLICATIONS TO INNOVATION SYSTEMS

Regional Innovation in Context

At the regional level there is often a strong correspondence and 'fit' between physical functions, social identity, economic units and political territories. Because of this the regional level brings opportunities to improve on the current state of policy fragmentation and to make new linkages for the SD agenda, between the local and the national scale, where economic and spatial policy may be more flexible (Ravetz *et al.* 2004; Hitchens 1997). The emerging agenda for 'sustainable consumption and production', itself a fuzzy combination of aspiration and actions, brings in a much wider scope than the conventional focus on economic growth:

- production side: including goods, services, public services, environmental capital;
- consumption side: the 'outcomes' in terms of human welfare, social cohesion, and the culture and psychology of consumers, clients, citizens, institutions; and
- quality of life/added value: a wider view of the interactions of social, economic and environmental capitals and flows, as represented in the RP mapping in the second section of this chapter.

Each of these represents a challenging agenda for institutional change or shift in 'techno-economic paradigm', via a process of innovation, either indigenous or catalysed by public interventions (Freeman 1996). Such a shift can be observed at the regional scale in the UK and EU, in terms of strategies for regional sustainable development, and emerging concepts for integrated planning and management for energy, waste, physical resources and so on. In order to facilitate such physical systems, integrated concepts are also emerging for finance and investment, governance and accountability, planning and management, monitoring and mapping, technological diffusion and so on. The point here for the RD-IE-BE-IS agenda, is to highlight the many dimensions of innovation which may be involved in such a programme, beyond the conventional boundaries of bringing technology to the market-place:

- innovation in institutions to handle such networks and partnerships;
- innovation in financial models, trading schemes and market developments;
- innovation in consumer and public services on the demand and consumption side;

- innovation in social enterprise and citizen responsibility to enhance social capital and cohesion; and
- innovation in logistics and supply chain networks to enable integrated resource management.

Such a broader picture has been linked back to a practical policy agenda for innovation in RP in a UK government initiative (Performance and Innovation Unit 2001). This identified market failures and institutional barriers to innovation, and then proposed a combination of market development, fiscal subsidies, capacity building measures, regulatory improvements and a strategic research programme – not so much a new agenda as a consolidation of current thinking.

Structural Change and Resource Productivity

The analytic models in the fourth section of this chapter are focused on more on a quantitative and technical version of RP, and the wider regional agendas sketched here tend to cut across the formal boundaries of the technical models, with little relevance to the input–output tables and similar structures. The question here is how far such technical models can help to inform such institutional innovation, or whether there are other more useful approaches?

One of the precursors to the REAP modelling system was a case study project on the regional metabolism of construction minerals, aggregates and inert wastes (McEvoy *et al.* 2004). Based on a detailed analysis of resource flows and impacts, this pointed to the emergence of resource management thinking at the regional level and firm level, and the potential for more integrated 'resource management enterprises'. These were proposed as using advanced ICT to closely match supply and demand for re-used/recycled material in time and space, along similar lines to the 'Industrial Symbiosis' programme (Murata and Pearce, this volume). While there are prototype 'waste exchanges' now operating, it is clear that integrated resource management across the whole of the construction industry is dependent on innovation, not only in the logistics of matching supply and demand, but in institutions, management practices, regulatory and accounting procedures, design and specification constraints and so on.

This makes for an interesting contrast between the flow of materials represented by the resource models, and the dematerialized flow of digital information which is apparent in e-commerce. To pursue this we look at the theme of e-commerce as one powerful catalyst of such innovation in markets, technologies, institutions and so on (Wilsdon 2001), where e-commerce is beginning to provide functions such as:

- tracking of resource – waste demand, with specific in space, time, ownership;
- matching demand with supply of new, re-used, recycled products and materials;
- interactive markets/shadow markets which enable trading between public, private, third and consumer sectors;
- lean design specifications to minimize waste and maximize effectiveness;
- integrated and participative assessments of impacts, costs, values and benefits to different social and economic groups.

For the implications of this we could look at current analysis of the impacts of e-commerce, which tends to assume that markets, production processes, societies and so on will remain the same, except for the e-commerce effect on speed and the globalizing scale of activity (OECD 2001). In contrast it can be argued that e-commerce is already more instrumental in shaping much more fundamental and qualitative innovation and structural change, even while it is now used actively by a minority of consumers and businesses (Castells 2001):

- change in economic and market structures: for example, instant/virtual markets, virtual distributed corporations, virtual stakeholder networks, consumer agglomeration markets, reverse auctions, consumer-to-consumer markets;
- change in institutional structures, for example, relations between governments and markets, transparency and accountability of corporations;
- change in social and cultural norms: for example, global media and styles: mobile telephony as a generator of social interactions;
- change in industrial and technological processes: for example, just-in-time production, outsourcing, multi-agent contracting;
- changes in the logistics of retail and distribution are difficult to predict: but examples such as the e-Bay internet trading system points to the possibilities.

Clearly e-commerce has the potential for rapid restructuring of markets, production and trading interactions, in new configurations at local, regional and global level. This can be characterized with the concept of 'intermediation' – in other words the agencies and actions involved at each step in a supply chain or distribution chain (Pakko 2002). 'Dis-intermediation' represents the process of removing intermediaries, suppliers, brokers, distributors and other middlemen, who are rendered obsolete by the more rapid and cost-effective access of e-commerce. In contrast, 're-intermediation' is the process of establishing new agencies which act as brokers of information, access and capital in new patterns of trading and exchange. It is interesting to compare such an ICT-based supply chain perspective with current regional development perspectives such as the 'learning region' (Morgan 1997), or social perspectives such as the 'richness of cities' and their capacity for creativity and cohesion (Christie and Levett 1999).

This discussion of industrial ecology, innovation and inter-mediation might then continue in various directions beyond the scope of this chapter. One is the re-configuration of global, regional and local supply chains, as catalysed by the internet. Another is an institutional dimension on structural change and the facilitation of innovation. A third direction is the ethical internet agenda, where the technological risks and impacts on vulnerable social groups and economies are seen as mitigated by public policy. Each of these can contribute to this chapter's theme of resource productivity and the contribution of analytic tools to innovation strategy.

Linking Analytic Models to Regional Innovation Systems

This section has raised a very challenging agenda, and it is clear that the examples of analytic models described here are only at the start of a development process for their technical applications, let alone the wider frame of structural and institutional change. However institutional learning is rapid, with the current UK programmes of regional workshops on strategies, scenarios, and sustainable consumption and production. While such learning is not necessarily in a straightforward path, with parallel levels from awareness raising to target setting, various applications to regional innovation strategies are beginning to emerge.

A useful way to identify the application potential is through a typology of regional innovation systems (Cooke *et al.* 2003, Braczyck *et al.* 1998). This characterizes the institutional style and context in terms of two axes: a business innovation axis ranges from 'globalist–interactive–localist': and a public governance axis ranges from 'grassroots–networked–centralized'. The applications of the resource productivity models can then be identified in terms of their functions in monitoring, benchmarking, forecasting and evaluation.

For the globalist business model, there are long term regulatory pressures and technological potentials which may be represented by the models; a centralist mode of governance would aim to relate these to firm policy targets and public interventions. Meanwhile for a localist business model, a more entrepreneurial approach may be more concerned with performance benchmarking and the related apparatus of incentives, risk management, training, market development and so on. A networked business model then focuses the resource productivity models on promoting regional scale industrial ecology, through supply chain analysis, waste exchanges and possible business-consumer trading. Each of these may work differently with a centralist style of government, where analytic models may be used to set targets and monitor progress: or a grassroots style of government, where the analytic data is used in a more entrepreneurial way. In each of these there are different approaches to linking the sectoral and firm agenda to that of regional resource productivity, and in linking the pattern of production through the market, to the pattern of consumption through a wider view on society and the environment.

Implications and Future Research

This chapter has reviewed work in progress on the UK regional agenda for resource productivity. It has shown an outline of two types of analytic models, and a review of the potential applications to regional innovation systems.

At present the agenda for 'resource productivity' in the UK is being expanded to that of 'sustainable consumption and production', and this is finding new possibilities at the regional scale of policy. Likewise there is growing demand from the corporate social responsibility agenda, for firms and sectors to monitor and benchmark their performance on a wider frame. Generally there is much mutual learning between the regional governance and the level of sectors and firms: there is also rapid learning between the environmental management and economic development professions: and between developers and users of the analytic models.

Each of these models cannot directly represent structural change and the innovation process, but provides valuable functions in monitoring and benchmarking, scenario modelling, and appraisal and evaluation, all of which promote and facilitate the process of innovation for improved resource productivity. Likewise they point the way towards new business and market configurations for the idealized 'resource management enterprise' with more sustainable inter-mediation and logistic systems. Such enterprises may be oriented around a simple 'resource productivity' agenda, for example, 'doing more with less': or a wider frame which includes social, economic and environmental interactions at each stage in their supply chain and product life-cycles.

All this points to several directions for future research. One concerns the technical dimension of model development, data management, and communicating to different parties for different applications. Another concerns the business dimension of integrating technical information with management and market information. A third concerns the public governance agenda in environmental management and economic development, and the task of integrating public policy information systems with those of the market. Finally, it is clear that there are new techno-economic paradigms emerging, which demand the integration of technical knowledge with a wider framework on sustainable consumption and production.

NOTES

1. Full details, documents and databases are available as of end 2005 on www.scpnet.org.uk.
2. As of end 2005, the partnership includes: Environment Agency; North West Development Agency (NWDA); North East Regional Assembly (NERA); South East England Development Agency (SEEDA); East Midlands Development Agency (EMDA); East of England Development Agency (EEDA); Advantage West Midlands (AWM); and the National Assembly of Wales (NAW). Other contributors include Cambridge Econometrics, AEA Technology, Caleb Management Services Ltd, and the Centre for Urban and Regional Ecology at Manchester University.
3. Details available as of end 2004 on www.biffaward.org.uk: www.massbalance.org.
4. Details on www.ecologicalbudge.org.uk.

REFERENCES

Barker, T. (1998), 'Large-scale energy-environment-economy modelling of the European Union', in I. Begg and B. Henry (eds), *Applied Economics and Public Policy*, Cambridge: Cambridge University Press.

Braczyck, H., P. Cooke and M. Heidenreich (eds) (1998), *Regional Innovation Systems*, London: UCL Press.

Brettell, S. and B. Gardiner (2003), 'Application of Econometric Models to Evaluation', in: *Methods for Evaluation of Structural Programmes, Luxembourg, EC:* (MEANS) Handbook.

Bringezu, S. and H. Schütz (2001), 'Total material resource flows for the United Kingdom', Department of the Environment, Transport and Regions contract EPG 1/8/62. DETR, London.

Brunner, P. and Rechberger (2004), *A Practical Handbook of Material Flow Analysis*, London: Lewis Publishers.

Castells, M. (2001), *The Internet Galaxy: Reflections on the Internet, Business and Society*, London: Oxford University Press.

Chertow, M.R. (2000), 'Industrial symbiosis: literature and taxonomy', *Annual Review of Energy and Environment*, **25**, 313–37.

Christie, I. and R. Levett (1999), 'Towards the Ecopolis: sustainable development and urban governance', report no. 12.

Cooke, P., S. Roper and P. Wylie (2003), 'The golden thread of innovation and Northern Ireland's evolving regional innovation system', *Regional Studies*, **37**(4), 365–80.

CURE (2002), 'The linking-up report: linking environmental-economic modelling into sustainable planning in Wales and the regions', report to Environment Agency, accessed at www.reward-uk.net.

CURE (2004), 'REWARD NW – Building the evidence to inform regional commercial and industrial waste policy', report to NWDA and Environment Agency, accessed at www.reward-uk.net.

CURE (2006), 'REAP technical report 5: the REAP activity model specification', accessed at www.ecologicalbudget.org.uk.

Darier, E. and S. Shackley (1998), 'The seduction of the sirens: global climate change and modelling', in F. Fisher and M. Hajer (eds), *Living with Nature*, Oxford: Oxford University Press.

Department of Trade and Industry (DTI) and DEFRA (2003), Changing Patterns: UK Government Framework for Sustainable Consumption and Production, London: DEFRA publications, accessed at www.defra.gov.uk/environment/business/scp.

Ekins, P. (2004), 'Step changes for decarbonising the energy system: research needs for renewables, energy efficiency and nuclear power', *Energy Policy*, 32, 1891–904.

Enviros Consulting Limited (2003), 'Enworks: a baseline review', accessed at www.enworks.com.

Eurostat (2000), *Economy-wide Material Flow Accounts and Derived Indicators. A Methodological Guide*, Luxembourg: Office for Official Publications of the European Community.

Freeman, C. (1996), 'The two-edged nature of technological change: employment and unemployment', in W.H. Dutton (ed), *Information and Communication Technologies: Visions and Realities*, Oxford: Oxford University Press.

GHK, Policy Studies Institute, Institute for European Environmental Policy, Cambridge Econometrics (2003), *The Contribution of the Structural Funds to Sustainable Development: A Synthesis Report*, vol 1 to DG Regio, EC. Available as of March 2006 http://europa.cu.int/comm/regional_policy/sources/docgener/evaluation/ rado_cn.htm.

Giljum, S. and K. Hubacek (2004), 'Alternative approaches of physical input-output analysis to estimate primary material inputs of production and consumption activities', *Economic Systems Research*, **16**, 301–10.

Gouldson, A. and J. Murphy (1999), *Regulatory Realities: The Implementation and Impact of Industrial Environmental Regulation*, London: Earthscan.

Hammersley, R. (1995), 'An inquiry into prospects for sustainability in minerals planning', *Sustainable Development*, **3/2**, ERP Environment, Chichester: Wiley.

Haughton, G. and C. Counsell (2004), *Regions, Spatial Strategies and Sustainable Development*, London: Routledge.

Hertin, J. and F. Berkhout (2005), 'Environmental policy integration for sustainable technologies – Rationale and practical experiences at EU level, in J. Tait and C. Lyall (eds), *New Modes of Governance: Developing an Integrated Policy Approach to Science, Technology. Risk and the Environment*, Aldershot, Ashgate Publishers.

Her Majesty's Government (HMA) (2005), *Securing the Future: the UK Sustainable Development Strategy*, CM6467, London: TSO.

HMG (2005), *Securing the Future: A National Strategy for Sustainable Development*, London: TSO.

Hitchens, D. (1997), 'Environmental policy and implications for competitiveness in the regions of the EU', *Regional Studies*, **31**(8), 813–19.

Jackson, T. and L. Michaelis (2003), 'Policies for sustainable consumption', London: Sustainable Development Commission, accessed at www.sd-commission.gov.uk.

Jackson, T. and P. Roberts (1997), 'Greening the Fife economy: ecological modernization as a pathway for local economic development', *Environmental Planning and Management*, **40**(5), 615–30.

Jacobs, M. (1997), 'Environmental valuation, deliberative democracy and public decision-making institutions', in J. Foster (ed.), *Valuing Nature: Economics, Ethics and Environment*, London: Routledge.

Kaya, Y. (1990), 'Impact of carbon dioxide emission control on GNP growth: interpretation of proposed scenarios', paper presented to the IPCC Energy and Industry Subgroup, Response Strategies Working Group, Paris.

Leadbeater, C. (1998), 'Welcome to the knowledge economy' in A. Hargreaves and I. Christie (eds), *Politics of the Future: The Third Way and Beyond*, London: Demos.

Linstead, C., C. Gervais and P. Ekins (2003), *Mass Balance UK: An Essential Tool for Understanding Resource Flows*, Newark: Royal Society of Wildlife Trusts, also accessible at www.massbalance.org.

McEvoy, D., J. Ravetz and J. Handley (2004), 'Managing the flow of construction minerals in the North West region of England – a mass balance approach', *Journal of Industrial Ecology*, **8**(3), 121–40.

Morgan, K. (1997), 'The learning region: institutions, innovation and regional renewal', *Regional Studies*, **31**(5), 491–504.

OECD (2001), *Science Technology and Industry Scoreboard. Towards a Knowledge-Based Economy*, accessed at www1.oecd.org/publications/e-book/92-2001-04-1-2987/

Pakko, M. (2002), 'What happens when the technology growth trend changes? transition dynamics, capital growth and the "New Economy" ', *Review of Economic Dynamics*, **5**(2), 376–407.

Pedersen, O.G. and M. de Haan (2006), 'The system of environmental and economic accounts–2003 and the economic relevance of physical flow accounting', *Journal of Industrial Ecology*, **10**(1–2), 19–42.

Performance and Innovation Unit (2001), *Resource Productivity: Making More With Less*, London: TSO.

Ravetz, J. (1998a), 'Integrated assessment models: from global to local', *Impact Assessment and Project Appraisal*, **16**(2), 147–54.

Ravetz, J. (1999), 'Economy, Environment and the Sustainable City: Notes from Greater Manchester', in P. Roberts and A. Gouldson (eds), *Integrating Environment and Economy: Local and Regional Strategies*, London: Routledge.

Ravetz, J. (2000a), 'Integrated assessment for sustainability appraisal in cities and regions', *Environmental Impact Assessment Review*, **20**, 31–64

Ravetz, J. (2000b), *City-Region 2020: Integrated Planning for a Sustainable Environment*, (with a foreword by the UK Secretary of State for the Environment), London: Earthscan, in association with the TCPA.

Ravetz, J. (2002), *Local Sustainable Development Indicators: a Review and Evaluation*, Web-based Report from DG Environment / Sustainable Towns & Cities Campaign.

Ravetz, J., H. Coccossis, R. Schleicher-Tappeser and P. Steele (2004), 'Evaluation of regional sustainable development – transitions and prospects', *Journal of Environmental Assessment Planning & Management*, **6**(4), 585–619.

Roberts, P. (1995), *Environmentally Sustainable Business: a Local and Regional Perspective*, London: Paul Chapman Publishing.

Robinson, J. and J. Tinker (1995), *Reconciling Ecological, Economic and Social Imperatives: Towards an Analytical Framework*, British Columbia: Sustainable Development Research Institute.

Tyteca, D. (1996), 'On the measurement of the environmental performance of firms – a literature review and a productive efficiency perspective', *Journal of Environmental Management*, **46**, 281–308.

University of Dundee, Aston Business School, Arup Economics and Planning (2001), 'Evaluation of RDA strategies and action plans', final report to the Department for Transport, Local Government and the Regions, accessed at www.odpm.gov.uk.

Vaze, P. and S. Balchin (1996), 'The pilot UK environment accounts', *Econ. Trends*, **514**, 41–67.

von Weizsacker, E., A. Lovins and L.H. Lovins (1997), *Factor Four: Doubling Wealth, Halving Resource Use*, London: Earthscan.

Wiedmann, T., J. Minx, J. Barrett and M. Wackernagel (2005), 'Allocating ecological footprints to household consumption activities by using input-output analysis', *Ecological Economics*.

Wilsdon, J. (2001), *Digital Futures: Living in a Networked World*, London: Earthscan.

4. Industrial symbiosis in the UK

Murat Mirata and Richard Pearce

INTRODUCTION

Industrial Symbiosis (IS) networks, or synonymous concepts of eco-industrial parks and industrial ecosystems containing such networks, are regarded to manifest the regional application of the main principles of the emerging Industrial Ecology (IE) field (Ayres 1996; Ehrenfeld and Gertler 1997; Chertow 2000; den Hond 2000). In essence, IS networks harvest improvement potentials present at the inter-organizational interfaces via collaborative interactions among anthropogenic activities, mostly located within physical proximity to each other. Webs of synergistic linkages within IS networks allows improvements in the efficiency and effectiveness by which different resources are utilized, in addition to that which can be achieved by fragmented pursuit of improvements in individual units (Mirata 2004). Consequently, IS networks offer a potential to improve the sustainability profile of regional economic activities. More specifically, IS networks offer potential for:

- Environmental benefits linked to reductions in resource use, dependence on non-renewable resources, pollutant emissions and waste handling needs;
- Economic benefits emerging from reductions in the costs of resource inputs, production, and waste management and from generation of additional income due to higher value of by-product and waste streams;
- Business benefits due to improved relationships with external parties, development of a green image, new products and their markets; and
- Social benefits by generating new employment and raising the quality of existing jobs, and by creating a cleaner, safer natural and working environment.

One of the best known examples of IS networks that provide such benefits is located in Kalundborg, Denmark. In Kalundborg, the first inter-firm exchange of resources took place in the early 1970s and included the delivery of excess steam from Asneas power plant to the neighbouring statoil refinery, soon after followed with another delivering the refinery's surplus gases to the power plant. Over a period of more than 30 years more synergistic linkages developed forming a network of resource exchanges mainly among the power plant, the refinery, a plasterboard manufacturer, a pharmaceutical plant, a sulphuric acid manufacturer, a cement company, local farmers, fish farms and the municipality.

Ravalorization of resources embedded in waste or by-product streams through inter-organizational synergistic interactions forms the main motivation for developing IS networks. However, as shown by numerous examples from industrial history (Desrochers 2000, 2002), this is neither a new idea nor is it new to businesses or policy makers. Nevertheless, there are a limited number of comprehensive, operational IS networks in modern times. What is more novel and innovative about the concept is perhaps something else. First, there is conscious inquiry into the factors that influence the successful development and operation of IS networks. Second, using this emerging understanding, and departing from the assumption that there are many regions where potential for gains through inter-organizational synergies are either present or can be created, but remain unexploited due to the lack of necessary production processes or organizational settings, change agents are trying to catalyse the development of IS networks in different parts of the world in a systematic way (for examples of such efforts the reader can refer to Baas (1998); Côté and Cohen-Rosenthal (1998); Chertow (2000)). Most of these efforts, however, can be considered in their early stages and none have thus far succeeded in creating a system that is comparable to Kalundborg, or any other similar network. Thus, sharing experiences from such efforts are of great importance for the advancement of the IS related work that has a large learning need and potential.

As two such change agents,[1] the authors of this chapter have been involved in the efforts to catalyse the development of IS networks in different UK regions and that of a national IS programme in the UK. Here, we first present IS networks as an innovative approach with potential to contribute to the more sustainable development of regional economies. We then provide an overview of the interplay between the technical, political, informational, economic, and organizational factors and the take-up of IS by regional parties as an innovative way of organizing their activities and elaborate on the roles change agents can play to catalyse this diffusion. Descriptions from the early stages of two IS programmes in different regions of the UK follow, before discussing their strengths and weaknesses. In our conclusions, we first emphasize the importance of regional learning as a determinant of both the pace and direction of IS network development.

Then by putting our experiences into the larger 'environmental sustainability' context, we demand attention to be paid to the concern of 'systemic lock-in', and elaborate on the capacity and skill needs of the change agents.

INDUSTRIAL SYMBIOSIS PROGRAMMES – AN INNOVATIVE APPROACH TO REGIONAL ECONOMIES FOR IMPROVED SUSTAINABILITY?

According to Chertow 'industrial symbiosis engages traditionally separate industries in a collective approach to competitive advantage involving physical exchange of materials, energy, water, and/or by products' (Chertow 2000). Although these usually form the core elements of IS networks, such networks go beyond and include cooperation that can be based on the exchange of intangible resources (for example, experience, know how, human resources and so on) and on the shared utilization of infrastructure elements. As these are usually more feasible to develop and operate among organizations physically close to each other, IS networks are better suited for geographically coded regions. Thus, for the purpose of this work IS networks can be defined as a collection of inter-organizational relationships among regional economic activities where collaborative actions bridging local needs with local capacities result in improved resource utilization and associated environmental and economic gains.

According to Dosi innovation is concerned with 'the search for, and the discovery, experimentation, development, imitation, and adoption of new products, new production processes and new organizational set-ups' (Dosi 1988). Moreover, innovation studies place the main interest on the knowledge about, persuasion, or decisions to adopt the ideas and practices rather than the newness of an idea or practice measured by the lapse of time since it is first used or discovered (Rogers 1995). In this understanding, IS programmes can be seen as an object of regional innovation, because:

- IS programmes aim to identify new production processes and products, as well as new organizational set ups that would improve the resource use efficiency at the regional level;
- even if their potential are known, these new approaches are not commonly implemented in most regional economies;
- IS programmes aim to catalyse their integration into the regional activities.

There are additional benefits in studying IS programmes under the innovation and innovation diffusion lenses. One of these is related to the fact that innovation models distinguish the hardware and software dimensions of their objects. They emphasize that although the accomplishment of the tasks in hand is enabled by the hardware in most of the cases it is the soft issues that determine their adoption and effective utilization. This distinction is of crucial importance for IS networks and as we elaborate in the coming sections, has significant implications for change agents aiming to catalyse the development of IS networks. Moreover, the dynamic nature of innovations and the importance of individual and collective learning for their development and diffusion form central themes of innovation and innovation diffusion models. Both dynamism and learning are also central themes for IS networks. Rather than aiming for a strictly defined final form these networks dynamically evolve. This evolution has a lot of room for experimentation and is guided by the learning taking place at different levels. Last but not least, frameworks proposed to study the rate of adoption of innovations and particularly the variables such as perceived attributes of innovation and extent of change agents promotion efforts (Rogers 1995) provide a useful analytical framework for studying IS networks.

According to Rogers, the characteristics of innovations determined by their relative advantage, compatibility, complexity, trialability and observability play an important role in the rate of their adoption. Although all of these are important for the IS programmes to varying degrees, for our purposes here we find relative advantage and compatibility most relevant and limit our focus around these. These are defined as following:

- Relative advantage relates to the degree to which it is perceived better than other alternatives, measured mostly in economic terms but also in social prestige, convenience and satisfaction. The greater the relative advantage of an innovation, the faster its adoption is expected to be.
- Compatibility relates to how consistent the innovation is with the existing values, past experiences, and needs of potential adopters. The incompatible innovations often require a prior adoption of a new value system and therefore take a longer term to be adopted.

DETERMINANT FACTORS OF IS DEVELOPMENTS AND ROLE OF CHANGE AGENTS

A further inquiry into the relevant literature regarding the specific factors that can influence the relative advantage that the IS networks can provide and those that can determine their compatibility with potential adopters' values reveals that these are rooted in technical, political, economic, informational and organizational domains. A number of these factors that can determine the take-up of the concept and their potential influence on the attributes of the emerging network are summarized in Table 4.1.

The importance of change agents is acknowledged by the innovation diffusion literature which defines their role influencing clients' innovation-decisions in a direction deemed desirable by a change agency (Rogers 1995). In the context of IS programmes this change agency is usually referred to as the coordination body and is regarded as having an important role for the development and functioning of IS networks (Ayres 1996; Côté and Smolenaars 1997; Lowe 1997; Baas 1998; Young 1999; Burström and Korhonen 2001). Basically, such coordination bodies help alter the determining factors so that they are more supportive of the development of synergistic linkages. Factors the coordination bodies can particularly influence are related to informational and organizational areas. The main objective of the informational support is to identify possible developments along with their associated benefits for a region and inform relevant parties about these possibilities, whereas the organizational support mostly aims at establishing a fertile institutional environmental for network development.

Informational support mainly involves assessing the needs and capacities of different parties in the region, with the intention of identifying complementarities in their needs and capacities. Here, the main focus is usually placed on assessing the tangible resource in- and out-puts of companies operating in the region, or that can potentially be based there, with the intention of identifying possible exchanges involving material or energy resources. These types of assessments are mostly relevant for operations associated with significant flows of material and energy carrier streams. Acknowledging the fact that synergistic linkages can emerge from areas going beyond tangible in- and out-puts the focus of these efforts can be expanded to assess the processing, logistics, and management related needs and capacities of regional parties thereby uncovering additional opportunities for collaboration. Coordination bodies can also study the main resource flows associated with the focal region and thereby identify areas where major improvements regarding resource consumption or waste reduction is desirable. This is an important exercise and can aid in developing objectives the network should achieve (Chertow 1999), but not always possible due to resource constraints. Disseminating information about technological alternatives and environmentally preferable practices, about markets and their dynamics, and on regulatory issues are also among the tasks coordination bodies can carry out to support IS developments. To assist the functioning of existing synergies and to help new ones to evolve, the informational support must remain a continuous process.

Another area where the coordination role has significant potential to influence the development is related to the establishment of the necessary

Table 4.1 Factors influencing the relative advantage provided by IS networks and their compatibility (adapted from Mirata (2004))

Category	Elements constituting the factors	Potential areas of influence
Technical	• Physical, chemical and geographic attributes of in- and output streams • Processing, utility (energy and water), logistics, and managerial needs and capacities • Availability of relevant, reliable and cost efficient technologies	• Number and diversity of potential symbiotic linkages • Extent of environmental, economic and social gains synergies may provide • Extent of investment and effort required to develop and maintain synergies
Political	• Overarching environmental policies • Nature of laws and regulations • Taxes, fees, fines, levies • Subsidies, credits	• Incentives to develop and adopt environmentally desired technologies and practices and to form symbiotic linkages • Disincentives by rendering synergies illegal (prescriptive) or economically unfeasible (due to 1 high transaction costs)
Economic and Financial	• Costs of virgin inputs, economic value of waste and by-product streams, and the impact of political elements • Cost saving, revenue generation potentials • Amount of necessary investment and cost of maintaining synergies (including transaction and opportunity costs) • Payback time, return on investment (ROI) parameters	• Extent of economic advantage and competitiveness gained • Decisions of private companies • Necessity of alternative source of finance
Informational	• Availability of timely and reliable information from a wide spectrum of areas to the right parties • An information management system systematically monitoring changing dynamics and assessing the desirability and feasibility of options	• Possibilities to identify synergies • Possibilities to operationalize synergies • Risk perception of companies

Table 4.1 (continued)

Category	Elements constituting the factors	Potential areas of influence
Organizational and motivational	• Trust • Openness to each other and to new ideas • Hesitance to disclose information • Risk perception • Intensity of social interaction • Mental proximity • Decision power • Organizational history	• Presence/creation of the necessary institutional framework for collaboration • Development of synergies • Maintenance of synergies

institutional framework. This can be done by identifying the key parties within a region, raising their awareness in related areas, providing communication platforms and thereby facilitating the generation of the necessary common understanding and objectives and collective commitment for their achievement. Such efforts should not be limited to the companies and other parties within the region, and should be expanded engaging regulatory bodies, other policy makers and financiers in the programme, who can help overcome regulatory or financial hurdles, or can facilitate the development of necessary incentives.

The significance of coordination increases when there is limited dependence or communication among regional parties (Boons and Baas 1997), where operations are diverse and traditionally not related (Lambert and Boons 2002), or where there are foreseeable institutional barriers to cooperation (Ayres 1996). It is important to maintain coordination efforts in this nature after the initial set of collaborations become functional for the diversification of interactions and providing further improvement potentials.

As indicated earlier, coordination bodies are in effect representatives of change agencies which have their own values and understandings regarding which direction the development of IS networks should take. The extent of their power to influence this direction varies, and can be quite limited. Nevertheless, it is important to note that their function is not, and should not be limited to effectively catalysing the network's development but should also include influencing the characteristics of the emerging network. They cannot always be in a position to force better environmental performance in individual organizations (Boons and Baas 1997), or in the region as a whole, but it must be among their roles to provide guidance on actions required for longer-term environmental sustainability. This requires having a thorough understanding about necessary transformations as well as about effective means of implementing such transformations. Often this requires thinking beyond existing system elements and the obvious solutions that provide the best economic result in the short run. If that capacity is present, the next challenge becomes the identification of more profound changes that are desirable for longer-term environmental sustainability and facilitating the realization.

INDUSTRIAL SYMBIOSIS PROGRAMMES IN THE UK

In the UK, efforts to consciously catalyse the development of IS networks started in Summer 2000 with the 'Business Council for Sustainable Development – United Kingdom' (BCSD-UK),[2] together with the International Institute for Industrial Environmental Economics (IIIEE) of Sweden's Lund University as their academic partner, taking the role of systematically facilitating an IS network development in the Humber Estuary. Humber Industrial Symbiosis Programme (HISP) sparked interest from other UK regions and similar programmes were initiated in the West Midlands and Mersey Estuary soon after. Along with growing interest from additional regions, a UK wide National IS programme (NISP) has been developed. At the time of writing, there are more than six on-going programmes in different stages of development connected to NISP.

THE APPROACH TO THE DEVELOPMENTS OF IS NETWORKS

In the development of regional programmes a four-stage approach is followed. Following provides an overview of these stages and their objectives.

Awareness Raising and Recruitment

In this phase the key parties whose involvement in the programme is desirable are identified. These parties are then introduced to the programme and its objectives, potential benefits and practical implications using different means of communication. If the necessary resources are available, it is desirable to carry out an initial review in the region to identify potential benefits an IS programme can provide both to the region as a whole and to individual parties.

Interested parties are then brought together in a workshop marking the official launch of the programme where relevant issues are communicated in a collective manner. This event sets the foundation of a communication platform, which has proven to be one of the backbones of the programmes. It also encourages those attending the workshop to participate in the programme and helps to identify other organizations to be recruited. Following this event, formation of a regional steering group is targeted with representatives from identified and interested parties. The regional co-ordinator and the steering group work on gaining commitment from the identified companies and organizations to their participation in the programme.

Data Collection

This phase involves the collection of quantitative and qualitative information from the participating companies in a format that allows synergies to be identified and linkages to be made within the network. Governed by a confidentiality agreement if necessary, data regarding the organization's inputs and outputs, their processes and operational attributes, their needs and capacities in terms of production, utilities and logistics infrastructure, human and information resources is gathered in this phase. These data are fed into a database to optimize its use.

Analysis and Identification of Synergies

In the next phase, the collected data for each company are analysed to identify areas where there are specific needs in terms of the supply of and demand for materials, resources and facilities. Where necessary, the regional co-ordinator assists the analysis by bringing in specialist expertise to identify the synergies and linkages. The most direct linkages are communicated directly and quickly to the interested parties within the terms of the confidential agreement. Other potential synergies are communicated more widely within the network to encourage participants to follow-up the opportunities.

Implementation and Support

The last phase involves facilitation and support to help the network members to realize the identified synergies. This includes identification of barriers (for example, technical, resource and financial) to implementation and the provision of help for overcoming them. At this stage engaging parties other then those participating in the programme and linking into regional or national sources of support and funding may be necessary. The process of overcoming barriers to enable additional synergies to evolve, as well as collecting and analysing data to identify new opportunities continues as part of the programme support. This phase also covers widening the network to include other sectors and/or parts of the region.

RELEVANT ELEMENTS OF THE BUSINESS ENVIRONMENT IN THE UK

Before going into the specifics of various regional and the national programmes, it is useful to touch upon generic elements of business, national policy, and regional governance elements that are relevant to all IS developments in the UK.

In the UK, powerful drivers have combined over the years to greatly increase the demand for clean technologies and processes and the more efficient use of energy, water and raw materials.[3] This need for higher resource efficiency in industry is being driven by a range of factors including:

- Competitive pressures to reduce costs through more efficient conversion of raw materials into products.
- Supply chain pressures to adopt more sustainable practices and to conform to environmental standards such as EMAS and ISO 14001.
- Legislation and regulations to encourage the recycling and reuse of waste materials and to use energy more efficiently. These have been particularly influential where they are backed up by fiscal instruments examples of which are given in the coming section.
- Voluntary measures such as environmental reporting and benchmarking environmental performance.
- Downsizing by companies, especially in traditional sectors such as chemicals, paper and metals, which has resulted in the inefficient use of resources such as land, infrastructure, utilities and services.

Techniques, such as waste minimization, to identify and tackle resource inefficiencies and wastage within individual firms have been in use in the UK for many years. There have also been a wide variety of 'waste minimization clubs' projects in the UK (starting with the Aire and Calder Project) to facilitate companies in a given geographical area and/or in the same sector to implement waste minimization through sharing best practice and learning from each other. Through these schemes some businesses have improved their resource use efficiencies. These programmes, however, have been aimed mainly at management and manufacturing processes within the companies rather than attempting to identify synergies between companies for example, by using one firm's waste material as a feedstock in another's.

Waste exchange schemes in the UK are also quite common but these focus mainly on 'spot' exchanges of waste materials and do not consider the broader aspects of company resources and processes.

RELEVANT GOVERNMENT POLICIES AND ELEMENTS OF LEGISLATIVE FRAMEWORK

There are a myriad of political and regulatory elements that influence the IS networks in one way or another. As thorough accounts of these are clearly outside our scope here, we will only give a background to the ones whose relevance is considerable.

Through a policy guidance document, the UK government defined Resource Productivity (RP) as 'measuring the efficiency of the economy in generating output without using natural resources – including the resource provided by the capacity of the environment to absorb waste and pollution' and identified it as a 'key to change' in achieving sustainable development. This report, among others, recognizes that the important role the Government has to play in creating the right incentives and help overcome barriers to RP including information deficiencies, limited access to finance, and skill shortfalls. Another promising element of this document is that it acknowledges the presence of two different types of barriers: (a) those 'that prevent the take-up of measures already beneficial on both economic and environmental grounds'; and (b) those that 'prevent the take-up of measures that would be environmentally beneficial, and should be economically beneficial, but which for some reason are not' (Cabinet Office, 2001). The points raised in this document constitute important leverage points, also for overcoming the barriers the IS programmes are facing.

The landfill tax, and climate change levy (CCL) are two other important policy elements that are of high relevance for IS programmes. The former applies to almost all kinds of landfilled waste and provides incentives to reduce the amount generated, recover more value from generated streams and find alternative means for their treatment. The CCL, on the other hand, gives incentives to industrial and commercial activities to reduce inputs of selected energy carriers and to switch to environmentally preferable sources. This levy promotes the development of waste-to-energy schemes, combined heat and power (CHP) units and exchange of energy between organizations by allowing exemptions for energy carriers in the form of heat, steam, and 'waste as defined by statute'. Additional support for the co-generation of power and heat is provided by CCL through tax reductions or exemptions for high quality CHP units (HM Customs and Excise 2002). These policy elements incentives for businesses to increase their resource use efficiency and encourage them to create more cyclical resource use patterns. In our work, we have noted distinct, positive influences provided by these two policy elements. Although their implications are more on selected product/material streams rather than general resource use, extended producer's responsibility legislations (applicable to packaging materials, vehicles and electrical and electronic equipment) are also important in terms pressuring affected parties to look for solutions both within and outside of their company walls.

There are, however, also some policy elements that do, or potentially can, counteract the development of synergistic linkages. These include:

- those which are too restrictive such as special waste regulations that make the productive use of certain waste streams very difficult or impossible;
- those which significantly escalate the transaction costs associated with the synergies, such as alternative fuels protocol that requires extensive and costly monitoring, modelling; and
- planning issues and public consultation procedures for example, for cement and lime kilns which intend to burn unconventional fuels.

Clearly, these regulations are there with the intention to protect public health and the natural environment. However, from the viewpoint of companies, they are mostly perceived to be adding to the transaction costs of synergistic linkages, reducing their economic feasibility.

REGIONAL GOVERNANCE BODIES

In England, Regional Development Agencies (RDAs) are assigned the objective of promoting business efficiency and contributing to sustainable development (UK Office of the Deputy Prime Minister 1999). These organizations' prime concern is to foster economic development. However sustainability is increasingly gaining importance on their agendas and they see environmental quality as a prerequisite to attract inward investment (Advantage West Midlands 2001; Yorkshire Forward 2001; North West Development Agency 2002). Consequently, RDAs regard IS programmes as an effective alternative to address the sustainability challenge within their administrative boundaries and provide significant support, including direct finance, for their development. Sustainable development at a regional level is also a key item for the English regional assemblies and government offices. Unlike the RDAs, these organizations do not have direct access to significant funding streams but they can play important roles in the stimulation of regional players to get involved in initiatives such as IS and in the priorities for European Regional Development Funds. The Devolved Administrations in Scotland, Wales and Northern Ireland are all actively involved in sustainable development strategies and are also important potential players in the initiation and funding of IS programmes.

HUMBER REGION INDUSTRIAL SYMBIOSIS PROGRAMME (HISP)

Programme Gestation

The Humber Estuary is located on the East coast of England, dominated by organic and inorganic chemicals, oil refining, food processing, furniture, iron and steel and other metals processing industries that are spread along both the south and north banks of the river.

The idea of initiating an IS programme in the Humber region originated in the year 2000 from a global oil and gas company having major operations in the area. Due to their involvement By-product Synergy (BPS) programme (practically, an IS programme) in Tampico, Gulf of Mexico, managers from this company were informed about IS. This company was also leading a consortium which proposed a large scale CHP in the region and they saw the IS programme, partly, as a means to support this project. Nevertheless, it agreed to be the industrial champion for HISP and help attract other industries to participate. BCSD-UK, by then a less than a year old organization, agreed to support the efforts in the region and was given the coordination responsibility.

Awareness Raising, Recruitment and Data Collection

Initially, persons from the coordinating team from Tampico were brought in to the region to assist raising awareness among regional parties and possibly share some of the coordination responsibilities. In their approach, companies were asked to sign a contract to participate in HISP, which obliged them to channel a certain percentage of their economic gains from possible synergies, to the coordinating body. This, and other elements of their approach, did not gain acceptance by the local companies, leaving BCSD-UK as the sole coordinator, assisted by IIIEE. As a new organization in the region, BCSD-UK first had to concentrate considerable resources on establishing contacts with relevant parties, including local authorities, private companies and business associations and continued the efforts of raising awareness. Concurrently, Enviros Consulting was contracted to perform an initial review in the region. This review focused on two existing proposals in the region: one related to the installation of a 475 to 650 MW CHP plant and the other for a chemical feedstock pipeline bundle to cross under the Humber river, and connect the industries from the north and south banks (Humber Bundle). This study estimated significant economic, environmental and social benefits associated with synergies that can potentially arise with the implementation of these main initiatives, including:

- ~3.3 Mt, 48 kt, and 11 kt annual reductions in CO_2, SO_2 and NO_x emissions associated with CHP generation as compared to conventional methods;
- ~ 750 000 t of hazardous cargo removed from surface transport;
- substantial savings on energy bills for the existing large energy users in the area;
- up to ~ 800 M £/y increase in productive output and up to ~ 2400 new employment possibilities with the development of new businesses taking advantage of feedstocks to be made available by Humber Bundle and access to competitively priced energy from the CHP.[4]

Upon completion of the initial study, HISP was officially launched in a full day event with over 70 participants from a diverse range of regional public and private organizations. The event has not been as effective as desired for generating awareness and commitment for the programme. A limited number of companies decided to get involved in the programme, while some others gained the misleading idea of HISP being something only for large energy users on the south bank and for those who can benefit from the Humber Bundle. Efforts until this stage, however, have helped the CHP consortia to receive their planning permission. Commissioning of the plant, an important step for the development of the IS network, started soon after. In the 18 months that followed the launch, HISP didn't have a strong industrial leader.

Due to the unsatisfactory number of companies joining the programme, following the launch, the coordination team had to combine the efforts of increasing the number and diversity of organizations participating in the programme with those of collecting data from those who agreed to take part. Over the period of six months, more than 150 companies were contacted and one-to-one meetings were held with over 70 of them. In these meetings, managers were thoroughly informed about HISP, and those who expressed interest were provided with a standardized data collection form. This way of interaction with companies had certain advantages as decision makers from various organizations were given more time and attention to acquire a thorough understanding about the attributes of the programme while the coordination team gained access to useful qualitative information about companies and relevant regional dynamics. This approach was weak in demonstrating the collective commitment and created hesitancy among managers partly, because it failed to demonstrate the presence of collective interest and commitment. However, it succeeded in getting over 20 companies from oil and gas, organic and inorganic chemicals, energy and water utilities, food processing, packaging, logistics, ferrous metals, mining, furniture, port facilities and retail sectors interested in taking part in the programme. Additionally, a steering committee was formed by representatives from various regional parties. This management structure is considered important in germinating ownership for the project by its main beneficiaries. HISP however lacked an active industrial champion which encourages others to take part in HISP.

Due to lack of funding the coordination efforts of HISP almost came to a halt for about a year after Autumn 2001. HISP started to regain pace owing to the extra financial support from the region's RDA and funding and momentum gained through the National IS programme and when ex-managers from local businesses were given the project coordination responsibilities. A re-launch event, considerably different in nature than the first one, was organized. The part where managers from local businesses involved in functional resource exchange schemes shared their experiences with other participating parties was particularly effective in engaging more companies from diverse sectors. Within a year after the re-launch the programme had more than 60 registered participants.

Data Analysis

Due to lack of company specific data, two years after its initiation HISP has not yet moved into the formal, detailed and precise data analysis stage. The programme managers report that the companies are very slow in proving required data even though there is a web-based data collection system intending to ease the process. However, the two workshops organized with the participating businesses, one cross-sectoral and one specifically for the food and drinks sectors, aimed at identifying 'low and no cost' synergies have been very productive. Over 160 potential ways of collaboration were identified in these workshops and their follow-up meetings, a great majority of which includes revalorization of waste and by-product streams. Some of these are depicted in Figure 4.1. Other than those involving direct exchange of materials or energy, effluent and waste treatment, analytical services, logistical services, and managerial support to SMEs, were identified as areas where collaborative action can provide benefits in the region.

Implementation and Support

The construction of a 734 Me CHP plant started soon after the initiation of the programme and commenced operation in 2004, fuelled by a mixture of natural gas and waste refinery gas. The plant provides steam to the two adjacent refineries and electricity to other users. Other synergies involving the transformation of wood waste into wood chips and waste edible oils

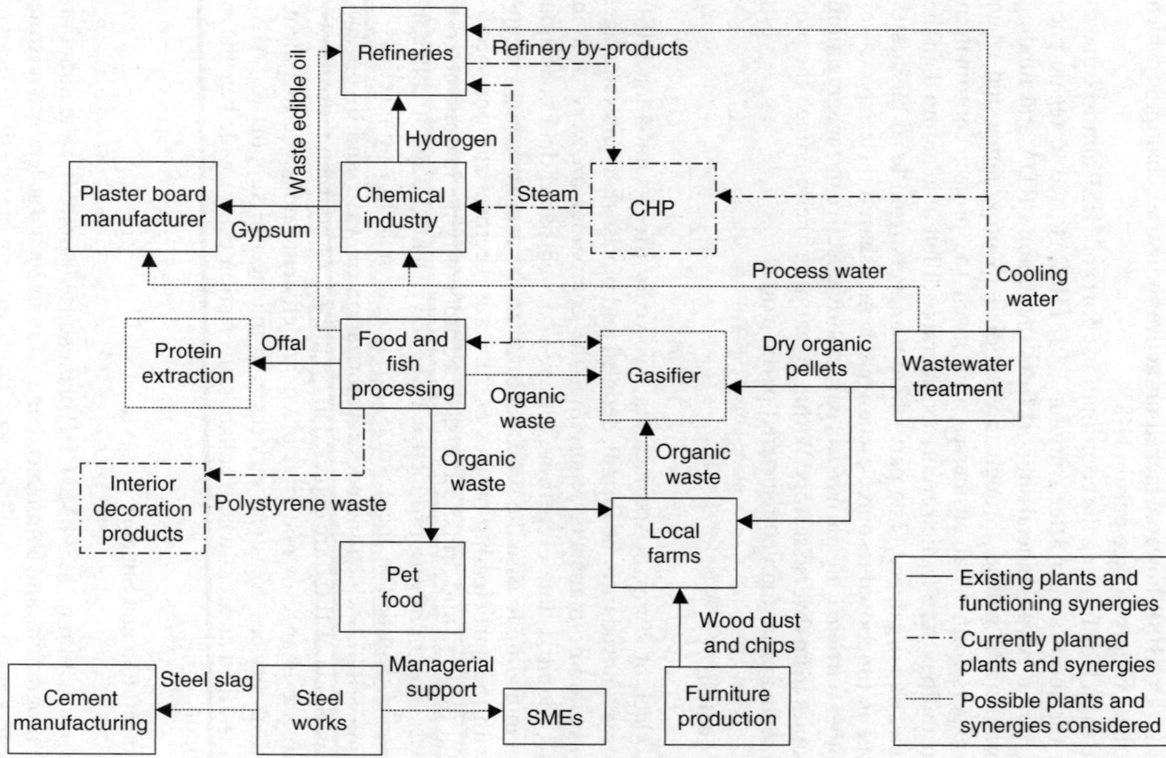

Figure 4.1 Existing, planned and possible synergies identified in the Humber region

into bio-diesel were already operational, turning thousands of tonnes of waste materials into productive inputs. The coordination team is actively working to help overcome the barriers against the development of identified synergies, thereby assisting their realization. This, however, is a slow and resource intensive exercise.

MERSEY BANKS INDUSTRIAL SYMBIOSIS (MBIS) PROGRAMME

Project Gestation

The MBIS project originated early in 2001 following a meeting between BCSD-UK and the North West Chemicals Initiative (NWCI), facilitated by one of the BCSD-UK members, Enviros Consulting, NWCI (subsequently renamed Chemicals Northwest) is an organization, jointly funded by the North West Development Agency (NWDA) and the industry, to improve the competitiveness of the Chemicals Cluster in the region. It manages a number of projects, determined by the industry, that are in line with the NWDA's Regional Economic Strategy.

A review of IS experiences from around the world and the Humber made NWCI quickly recognize the potential to develop an IS programme in the North West. Finding someone with local industry experience to manage the project and particularly to start raising awareness and recruiting participants was seen as a critical early step. This was solved when one of the big NWCI member companies seconded one of its senior managers to fulfil this role. A small project committee was established to guide and support the project manager.

Awareness and Recruitment

It was agreed at an early meeting of the project committee to focus the project initially on an industrial area on the north and south banks of the Mersey that has a concentration of refining, petrochemicals and chemicals companies. There are also a number of companies operating in other process industries in the area including food and paper. This area was selected because:

- There are a large number of chemicals and related industry companies in a compact geographical area hence facilitating the process of establishing the links.
- Recent changes in company ownership, followed by rationalization, have eroded integration.

- There is sufficient local autonomy to enable site managers to make decisions to participate without referring back to their parent organizations.
- The project manager was based in the area and already had good contacts with many of the local firms.

A list of potential participating companies was drawn-up and a profile for the project, covering its objectives and scope, was produced and sent to these firms. This was followed-up by the project manager to determine levels of interest and to arrange initial meetings with some of the key target companies to encourage their involvement. An important objective was to identify a leading figure from the local industry to chair the official project launch workshop and to promote the benefits of participation. The manager of the Shell oil refinery at Stanlow agreed to fulfil this role. The NWCI was able to facilitate this phase of the project quickly due to its high level contacts with its member companies in the area.

Forty-two people representing 34 companies and organizations attended the Project Launch Workshop in October 2001. This proved to be a very useful vehicle for getting the project moving due to:

- The enthusiastic support and involvement of the Chair and other members of the project group who acted as facilitators in the 'break out' session.
- The extent to which many of the delegates had already been 'warmed up' through the initial contacts by the project manager.
- Useful feedback from the 'break out' groups on the experiences of collaboration with other companies, the perceived barriers to the project and the level of interest in participation.
- Financial contributions to the project that were confirmed at the end of the workshop.

A total of 23 companies agreed to take part in the programme. The majority of these are involved in the manufacture of oil, petrochemicals and chemicals products but there are also some suppliers of products and services to the industry for example, energy, water, waste management and analytical services.

Data Collection

The data collection efforts were facilitated through the use of a specially designed questionnaire. The project manager visited each of the participating companies to explain the project in more detail, to go through the questionnaire and to discuss and agree timescales and responsibilities for completion. Although the participating companies were then followed-up to encourage them to complete and return the questionnaires as soon as possible, this part of the process took much longer than anticipated (instead of the projected three to four months, around eight months in practice). This was partially due to lack of time and resources available to the participating companies to gather the required information and to complete the questionnaires, and partially due to the low priority allocated to the above tasks due to competing responsibilities associated with core business activities.

It should also be noted that during the early stages of the project, attempts had been made to obtain public sector funding to support the project management and facilitation but this was not forthcoming. The project was funded through the in-kind contributions associated with the secondment of the project manager, cash from the participating firms and administration support from the NWCI. This arrangement had its strength as it manifested solid commitment from relevant parties. However, the resources that were allocated to the project were more limited than had been anticipated particularly for the data collection phase to support the companies in the completion of the questionnaires.

Data Analysis and Identification of Opportunities

The analysis of the information from the questionnaires was completed towards the end of 2002 and reports were then sent to each of the participating network members. Around 100 opportunities for synergies have been identified including specific opportunities for the individual firms as well as general opportunities for the network as a whole. The project manager provided a brokerage role in putting network members in touch with each other where there are opportunities for links to be made. The opportunities are summarized under the following headings:

- Materials recycling covering opportunities to re-use materials such as waste acids and alkalis, sodium nitrate and potassium fluoride.
- Local sourcing of materials such as carbon monoxide, xylene, hydrogen peroxide and ferric sulphate which are currently purchased from outside the area hence reducing the costs and environmental impacts of transport.
- Energy supply and recovery including the potential for CHP and the use of flammable materials as fuel.
- Water and effluent including the re-use of waste water streams, increased extraction from boreholes, increased use of underutilized facilities for water purification and effluent treatment.

- Sharing of infrastructure, utilities and facilities such as warehouses, storage tanks and silos, pipelines, transport operations and serviced land.
- Provision of services and related facilities including training, analytical laboratories, conference facilities and advice on safety, health and environmental issues.
- Manufacturing facilities based on the use of spare capacity of reactors, filtration and drying plants and packing equipment.

Implementation and On-going Support

- To-date about 40 per cent of the potential synergies are being followed up by the participants and work is still in progress to help the process along.

DISCUSSION

The IS programmes reviewed in the chapter are still in their early stages of development. Possible synergistic connections are only partially identified, and only a small fraction of these are acted upon. Nevertheless, they provide valuable lessons in various areas. One of the important ones is that they help demonstrate how the technical, political, economic, informational and organizational factors influence the relative advantage and compatibility attributes of IS networks and thereby affect its take-up by regional parties as an innovative way of organizing their activities. Next, they provide useful insights regarding the efficiency and effectiveness of different approaches used by different coordination bodies to catalyse the programmes' development. Last, but not least, they give indications about the nature of different synergistic linkages that may develop enabling a speculation on their sustainability profiles.

Before going into specifics of the regions it is useful to touch upon the influence political and regional administration functions have on the diffusion of IS network developments. The UK government is under increasing pressure to meet the targets set forth by the EU landfill directive, and it has been giving increasing importance to resource productivity issues. Therefore, various policy elements, including fiscal instruments, were introduced that place pressure on industries to reduce their waste and emissions. As most of the industries have been trying to address this challenge by realizing efficiency improvements within their company boundaries and many have already taken the actions that are not prohibitively costly, they demonstrate an interest in looking into inter-organizational interfaces for additional efficiency gain opportunities. Likewise, the regional administration bodies which are faced with the challenge of balancing the environmental and social dimensions of the regional development activities with the economic ones appear highly receptive to the idea of catalysing the development of IS networks in their regions and provide considerable support for this purpose. These elements significantly support the diffusion of IS concept in the UK, which is manifested in the national IS programme and its increasing number of regional programmes initiated in a relatively short period of time.

Although the development of both programmes reviewed here followed an approach involving the same stages there were noticeable differences in the nature and pace of their developments. An inquiry into the characteristics of some of the determining factors listed earlier in this chapter helps to provide a partial explanation to such differences.

With HISP the development process has been particularly slow and although there were interested parties from a diverse range of sectors, their sense of commitment to the programme and their intensity of interaction with others remained weak for a long time. HISP gained significant momentum in terms of actively engaging industrial parties, identifying and implementing synergies almost three years after its first official launch. Why has the development been very slow and gradual with HISP? What made it gain momentum? Answering these questions requires considering both the inherent dynamics that were present in the region and different coordination related factors.

Oil and gas refining and chemicals are two of the major sectors in the Humber region and were therefore the main foci of HISP. However, mainly due to the region's fragmented industrial development history, the level of compatibility and integration within and among these sectors regarding their main and by-products and their feedstock needs is however as compared to other UK regions where such industries are concentrated. This fact constrains the technical possibilities for material exchanges. In addition to limiting the possibilities for material connections these technical incompatibilities were also responsible, at least partially, for the weak presence of organizational elements favouring collaboration action. In other words, the chemical and oil and gas companies in the region do not have organizational cultures familiar with direct inter-firm cooperation, causing hesitance against participation in IS programmes. Unfavourable organizational cultures were also present in the food-processing companies, which represent another important sector in the region. These companies have been competing for the same market segments and therefore are strongly against collaborating with others in the early stages of the HISP. Another important attribute observed in the Humber region is related to limited decision-making power of the facility managers. Most facilities, particularly in the chemicals, oil and gas sectors, belong to national and multinational corporations whose headquarters, where decisions pertaining to IS programmes can be taken, are located elsewhere. This lack of local decision power presents additional hurdles for securing the necessary commitment for the programme, and for establishing a cooperative business environment.

With regards to coordination related factors, the fact the BCSD-UK taking the coordination role as a new organization with new personnel in the region and trying to secure commitment of existing companies for a new idea was not very effective. The lack of strong industrial leadership, limiting the highly valuable peer pressure, further hindered the diffusion of the concept. The HISP coordination team wanted to engage parties from diverse sectors in the programme, which was a very important and well-intentioned decision. However, the official launch of HISP was counteractive for this objective as it placed too much focus on chemical and oil and gas sectors, and industries with large energy usage. This left the companies outside this range of industries with the misleading perception that IS was not for them. Changing this misperception required individual company meetings afterwards that took a long time and other resources.

This slow and lengthy development process with HISP, nevertheless, allowed for a lot of highly valuable learning. First, the one-to-one meetings with company representatives allowed the coordinators to clear the type of collaboration that the IS programme intended to develop and thereby helped erode the hesitation. Second, it highlighted the importance of marketing the approach in a way that a diverse group of actors can relate to. Third, it underpinned the importance of having a local industrial champion that could exert some peer pressure and employing locally recognized and trusted coordinators. Last but not least, it made the coordinators realize that it is the open discussion and intensive interaction among the organizations that enables the identification and implementation of synergies in a more effective and less resource intensive way rather than a third party doing the work for them.

MBIS, on the other hand, displayed a more rapid development in terms of getting the relevant actors committed and necessary analyses performed. This is mostly attributable to the different industrial and coordination characteristics. Although dominated by similar industry sectors as those in the Humber region, the industrial development in Mersey Banks has taken place under common ownership and in a much more integrated fashion. Consequently, there are both more possibilities for technical compatibilities and inherited organizational cultures that are more accustomed to inter-organizational collaboration. This made the IS concept more compatible with existing industry cultures. Higher concentration of decision making powers, and strong business leadership to take the programme further are other positive characteristics of this region which resulted in faster developments. The diffusion of the concept was further supported by both the organization supporting coordination and the project manager having a long history and a credible profile in the region. It is recognized, however, that the industrial focus in this programme was confined to a narrow segment of sectors (mainly chemicals and refining) and that there is a need to extend the initiative to other industries in the Mersey Banks areas as well as to other parts of the North West region. Different attributes observed in documented cases which influenced the nature and pace of the developments in these regions are summarized in Table 4.2.

Table 4.2 Observed characteristics of different UK regions that influenced the development of IS programmes

Programme attribute	Humber Region	Mersey Estuary
Industry structure	• Result of a relatively recent, fragmented industrial development. • Low levels of integration within and among sectors. • Technical constraints to integration.	• Long history of common ownership in chemicals sectors. • Higher levels of integration within chemicals and oil and gas sectors.
Position of coordinating body	• New in the region. • Not involved from the very beginning.	• Well established part of a uni-sectoral network. • Initiated the development.
Project championship	• No sustained industrial championship at the early stages.	• Strong industrial championship.
Original institutional framework	• Decision centres outside the region. • Limited familiarity with relevant cooperation/ collaboration. • Diverse group of interested companies.	• Decision centres in the region. • Historical familiarity with materials exchange. • Limited diversity in interested companies.
Awareness raising and commitment	• Centred around specific initiatives. • Fragmented and gradual.	• Generic (without a sector). • Fragmented followed by collective.

ON THE SUSTAINABILITY OF IS NETWORKS

In Roger's words: 'it should not be assumed that the diffusion and adoption of all innovations are desirable' (Rogers 1995). As good as it may sound for contributing to sustainability efforts, IS is only a tool as good as the intentions to use it. Relying on the relevant literature compounded by some of the development taken place so far under the reviewed programmes we would like to attract attention to some of the pertinent sustainability issues of IS networks.

It should be clarified that in the context of IS networks two different dimensions of sustainability need to be considered. The first dimension relates to the sustainability of the network members and the network itself. The main sustainability concern of the network members is rooted in their economic performance and therefore will entail the take up of those ideas that provide sufficient financial returns in a sufficiently short time. Sustainability of the emerging network, as well as to varying degrees the sustainability of individual members, however, is linked to the interdependencies that are created within the IS networks. In other words, it depends on the life span and endurance of the formed inter-organizational linkages. The risk such dependencies present for individual companies' sustainability varies depending on the synergistic linkage's importance for the core business of the involved parties and is usually not as serious as usually thought. One of the reasons for this is that almost all synergistic linkages are governed by legal contracts that provide the involved parties with certain security. In the famous Kalundborg case, for example, every synergy is governed by a contract that requires all the parties to inform all the dependent ones about projected changes in advance, allowing enough time for them to take measures so as not to be adversely effected by the changes. Nevertheless, the sustainability of the network greatly depends on the open and effective communication among the network members about relevant issues.

The other dimension of the sustainability relates to the interplay between the network member's operations and the longer-term environmental sustainability. In the relevant literature reflecting on the nature of actions necessary to achieve long-term sustainability, the need for fundamental changes, more far reaching than incremental improvements in the existing production and consumption systems, is emphasized (McDonough and Braungart 1998; Solem and Brattebo 1999). These require the development and wide-spread application of innovative solutions (Ehrenfeld 1997). The solutions that solely focus on retrofitting the existing unsustainable systems with necessary elements to enable revalorization of waste or by-product streams are likely to fall short for realizing adequate progress in reaching sustainability. Put differently, as certain attributes of existing production and consumption systems are the main reasons underlying the problems we are faced with, solutions that leave those attributes intact cannot be regarded satisfactory for reaching longer term sustainability. Instead what is needed is the development and implementation of innovative substitutes in a wide range of areas to replace the undesired elements. Finding the right balance between more radical, longer-term changes versus some 'quick fix' solutions can become a rather frequent challenge facing IS practitioners. The danger lies in promoting actions based on inadequate considerations, narrow and short-sighted visions, and in persuasion of certain short term gains at the expense of preventing or delaying the identification and implementation of other vital changes. These concerns are very valid for IS applications, because the linkages they promote may:

- reduce the incentives for implementing preventive measures;
- prolong the lifetime of inefficient technologies and increase the viability of unsustainable industries by providing justifications, and even economic benefits, for apparent inefficiencies; and
- weaken the incentives for innovation and lock-in existing practices making necessary change more difficult due to higher interdependencies.

It is a repeated fact that the development of currently operational IS networks (for example, those in Kalundborg, Styra, Jyväskylä) are dominantly motivated by their economic benefits (Ehrenfeld and Gertler 1997; Schwarz and Steininger 1997; Korhonen *et al.* 1999; Korhonen 2001). It is also commonly, and quite rightly, stressed that the evolving networks will, to a greater extent, be shaped by economic considerations (Chertow 1999) and should provide sufficient benefits measured in conventional monetary or competitiveness terms to their members (Côté and Cohen-Rosenthal 1998; Lowe 1997; Cohen-Rosenthal 2000; Esty and Porter 1998). However, it should be borne in mind that taking action for one kind of development as part of an IS network can mean that the possibility to realize other alternative developments will be lost or delayed. It is therefore of utmost importance to assess each and every decision's long term implications not only from an economic, but also from environmental and social points of view. Addressing these concerns requires giving more weight to the assessment of environmental performances of proposed actions and to their conformance with longer term sustainability requirements.

The learning that has been taking place so far with the industrial parties and the coordinators of the IS programmes is in the nature of 'single loop learning'. Ability to address more pertinent sustainability issues, however, requires a 'double loop learning to take place'. In Argyris' words

> single loop learning takes place when a mismatch is detected and corrected without changing the underlying values and status quo that govern the behaviour. Double loop learning occurs when a mismatch is detected and corrected by first changing the underlying values and other features of status quo. Single loop learning remains within accepted routines. Double loop learning requires that new routines be created that are based on a different conception of universe. (Argyris 1997)

So far, the IS programmes have been too preoccupied with symptoms such as waste generation and inefficiencies in existing energy systems and mostly been dealing with those by developing and implementing economically feasible solutions such as diverting the wastes of existing operations from landfills, or increasing the thermal efficiency of fossil fuel based energy systems. For these programmes to make a more profound contribution to the long term sustainability they need to go beyond dealing with the symptoms, start addressing the very roots of problems, and think of fundamentally new ways of organizing regional economic activities. The coordination bodies of IS programmes are of great importance to catalyse such double loop learning. One good starting point in doing this would be helping the development of a long term sustainability vision, identifying the necessary developments that need to take place for reaching that vision, and facilitate the realization of such desired developments. The options such as exercise will identify will most likely face significant barriers for their development. But even their identification can be a valuable feedback for governmental policy development efforts, which, as mentioned earlier, intends to help overcome the economic barriers to the implementation of actions with far reaching environmental benefits. Ability to catalyse such actions, however, necessitates the coordinators to have a thorough understanding regarding the requirements of long term sustainability. Moreover, solid knowledge about promising ways of approaching such a sustainable state and about effective strategies and tools to facilitate the process will be needed. Currently, BCSD-UK, the other regional coordinators, and RDAs do not possess such capacities. This gap, however, can and should be filled by active participation of bodies, such as university departments or research institutions.

ON THE IMPORTANCE OF A NATION WIDE IS PROGRAMME

As indicated earlier, different regional programmes in the UK are now brought together under the umbrella of a national IS programme (NISP). This is an important and valuable development as it enables a central coordination body to receive and provide feedback to various regional programmes. More specifically NISP provides the IS work in the UK with the following strengths:

- Enables the testing and refinement of a consistent approach that can effectively catalyse the development of IS networks.
- Centrally compiles data that is, more or less, uniformly collected from different regions. These data can be used by the central coordinators to carry out analysis that the regional coordinators are unable or unwilling to perform.
- Enables the identification of inter-regional collaboration opportunities.
- Helps to inform policy makers and technology developers about the development bottlenecks experienced by a wider group of actors.
- Disseminates information about good practices and facilitating relevant learning to take place in other regions.

NOTES

1. Mr Mirata has been seconded from the International Institute for Industrial Environmental Economics at Lund University, which has been an academic partner of BCSD-UK between July 2000 and June 2003. He took an active part, as an action researcher, in formulating the methodology to be followed in the development of IS programmes and actively contributed to the efforts of developing regional programmes in the Humber Estuary and West Midlands as well as the nation wide programme within the stated period. Mr Pearce, on the other hand, has been actively involved in various important phases of regional programmes in the Humber and Mersey Estuaries.
2. BCSD-UK is a business association and is the regional affiliate of World Business Council for Sustainable Development. The main mandate of BCSD-UK is to contribute to the sustainability profile of businesses via the development and implementation of practical and profitable projects.
3. Enabling Business in Resources Management. A report for the Innovation and Growth Team for the Environmental Goods and Services Sector: DTI November 2002.
4. These were based on certain assumptions, and their value can fluctuate depending on such assumptions' validity.

REFERENCES

Advantage West Midlands (2001), '*Creating Advantage*', *The West Midlands Economic Strategy*, Birmingham: Advantage West Midlands – The Development Agency.
Argyris, C. (1997), 'Initiating change that perseveres', *American Behavioral Scientist*, **40**(3), 299–310.
Ayres, R.U. (1996), 'Creating industrial ecosystems: a viable management strategy', *International Journal of Technology Management*, **12**(5–6), 608–24.

Baas, L.W. (1998), 'Cleaner production and industrial ecosystems: a Dutch experience', *Journal of Cleaner Production*, **6**, 189–97.
Boons, F.A.A. and L.W. Baas (1997), 'Types of industrial ecology: the problem of coordination', *Journal of Cleaner Production*, **5**(1–2), 79–86.
Burström, F. and J. Korhonen (2001), 'Municipalities and industrial ecology: reconsidering municipal environmental management', *Journal of Sustainable Development*, **9**(1), 36–46.
Chertow, M. (1999), 'The eco-industrial park model reconsidered', *Journal of Industrial Ecology*, **2**(3), 8–10.
Chertow, M.R. (1999), 'Industrial symbiosis: a multi-firm approach to sustainability', presentation to the Eighth International Conference of the Greening of Industry Network.
Chertow, M.R. (2000), 'Industrial symbiosis: literature and taxonomy', *Annual Review of Energy and Environment*, **25**, 313–37.
Cohen-Rosenthal, E. (2000), 'A walk on the human side of industrial ecology', *American Behavioral Scientist*, **44**(2), 245–64.
Côté, R.P. and E. Cohen-Rosenthal (1998), 'Designing eco-industrial parks: a synthesis of some experiences', *Journal of Cleaner Production*, **6**, 181–8.
Côté, R.P. and T. Smolenaars (1997), 'Supporting pillars for industrial ecosystems', *Journal of Cleaner Production*, **5**(1–2), 67–74.
den Hond, F. (2000), 'Industrial ecology: a review', *Regional Environmental Change*, **1**(2), 60–69.
Desrochers, P. (2000), 'Market processes and the closing of industrial loops – a historic reappraisal', *Journal of Industrial Ecology*, **4**(1), 29–43.
Desrochers, P. (2002), 'Industrial ecology and the rediscovery of inter-firm recycling linkages: historical evidence and political implications', *Industrial and Corporate Change*, **11**(5), 1031–57.
Dosi, G. (1988), 'The nature of the innovative process', in G. Dosi, C. Freeman, R. Nelson, G. Silverberg and L. Soete (eds), *Technical Change and Economic Theory*, London: Pinter.
Ehrenfeld, J.R. (1997), 'A framework for industrial ecology', *Journal of Cleaner Production*, **5**(1–2), 87–95.
Ehrenfeld, J. and N. Gertler (1997), 'Industrial ecology in practice: the evolution of interdependence at Kalundborg', *Journal of Industrial Ecology*, **1**(1), 67–79.
Esty, C.D. and M.E. Porter (1998), 'Industrial ecology and competitiveness: strategic implications for the firm', *Journal of Industrial Ecology*, **2**(1), 35–43.
HM Customs and Excise (2002), *A General Guide to Climate Change Levy*, London: HM Customs and Excise.
Korhonen, J. (2001), 'Co-production of heat and power: an anchor tenant of a regional industrial ecosystem', *Journal of Cleaner Production*, **9**, 509–17.
Korhonen, J., M. Wihersaari and I. Savolainen (1999), 'Industrial ecology of a regional energy supply system: the case of Jyväskylä region, Finland', *Greener Management International*, **26**, 57–67.
Lambert, A.J.D. and F.A. Boons (2002), 'Eco-industrial parks: stimulating sustainable development in mixed industrial parks', *Technovation*, **22,** 471–84.
Lowe, E.A. (1997), 'Creating by-product resource exchanges: strategies for eco-industrial parks', *Journal of Cleaner Production*, **5**(1–2), 57–65.
McDonough, W. and M. Braungart (1998), 'The next industrial revolution', *The Atlantic Monthly*, **282**(4), 82–92.
Mirata, M. (2004), 'Experiences from early stages of a national industrial symbio-

sis programme in the UK: determinants and coordination challenges', *Journal of Cleaner Production*, **12**(8–10), 967–83.
North West Development Agency, (2002), 'Regional economic strategy review', consultation document, Warrington.
Rogers, E.M. (1995), *Diffusion of Innovations*, New York: Free Press.
Schwarz, E.J. and K.W. Steininger (1997), 'Implementing nature's lesson: the industrial recycling network enhancing regional development', *Journal of Cleaner Production*, **5**(1–2), 47–56.
Solem, K.E. and H. Brattebo (1999), 'Industrial ecology and decision making', presentation for the First International Symposium on Environmentally Conscious Design and Inverse Manufacturing, Tokyo, Japan.
UK Cabinet Office (2001), 'Resource productivity', London.
UK Office of the Deputy Prime Minister (2002), *Regional Development Agencies: General Information*, also published 1999, London: UK Office of the Deputy Prime Minister.
Yorkshire Forward (2001), *Yorkshire and Humber Region – Regional Economic Strategy*, Leeds: Yorkshire Forward, p. 62.
Young, R. (1999), *By-Product Synergy: A Demonstration Project Tampico, Mexico, Altamira, Gulf of Mexico*, Business Council for Sustainable Development, p. 23.

5. Industrial ecology: a new planning platform for developing countries

Ramesh Ramaswamy and Suren Erkman

WHY DEVELOPING COUNTRIES?

A great deal of manufacturing for the global market is increasingly moving to developing countries, and many countries such as China and India are experiencing rapid growth. Therefore, it is now a crucial time to influence their choice of an industrial development path. While it is an enormous opportunity to improve the living standards in these countries through increased employment and business opportunities, it is a serious load on the local resources (such as water, energy, land, and so on), whose availability to the populations of these countries is very poor.

A development path that is based on resource availability could create industrial growth that uses the resources more efficiently and judiciously, with minimal local and global impacts. A less careful industrial development plan that uses up scarce resources could spell danger to the very survival of over 80 per cent of the planet's population that lives in the developing world.

GROUND REALITIES OF DEVELOPING COUNTRIES

It is important to understand some aspects of life in the poor countries that are very much at variance to what is seen in the developed world. Among the many specific aspects that have to be borne in mind, is the fact that the pattern of resource flows in developing countries and hence the resultant environmental threat could be very different from what we see in the industrialized West.

Typically, the flows of materials through large, organized manufacturing facilities in developing countries could be very small in relation to the overall material flow. Table 5.1, showing the comparative use of water in different countries by different segments of the socio-economy, is very revealing. If any action has to be taken to preserve water in India, for

Table 5.1 Fresh water use in different countries

Country	%age Agriculture	%age Industry	%age Domestic
USA	27	65	8
India	92	3	5
Sri Lanka	94	1	5
Bangladesh	86	2	12

Source: World Development Indicators, World Bank 2002.

example, or stop the deterioration in its quality, the action may have to go far beyond the large, formal manufacturing facilities.

Similarly an understanding of the other resource flows through the system would immediately point to directions for action. Figure 5.1 could be a possible format for aiding such an understanding.

Another point that needs elaboration is the definition of 'industry'. In the developing world, the informal 'industry' of small traders and manufacturers plays a key role and forms a very significant portion of the economic activity. Table 5.2 gives the relative importance of the small-scale sector in different countries.

Typically, such units:

- do not use high technology and cannot afford sophisticated pollution abatement systems;
- are too numerous to be effectively policed by the state; and
- employ large numbers of people (and no democratic government would risk any potentially disturbing action).

In fact, even reliable data about the number and activities of such small scale industries is often lacking. As a consequence, environmental laws are not strictly enforced although the collective consumption of material by small scale industries and the resultant threat to the environment, could be much higher than the large industrial units.

In India, there were an estimated 3.37 million modern SSI units as of the end of March 2001, providing direct employment to around 1.86 million persons. This does not include the multitude of unregistered units (SIDBI 2001).

Figure 5.2 attempts to list a number of possible characteristics of developing countries that may need to be taken into account while preparing an action strategy. Any strategy that does not take into account the local constraints is bound to fail.

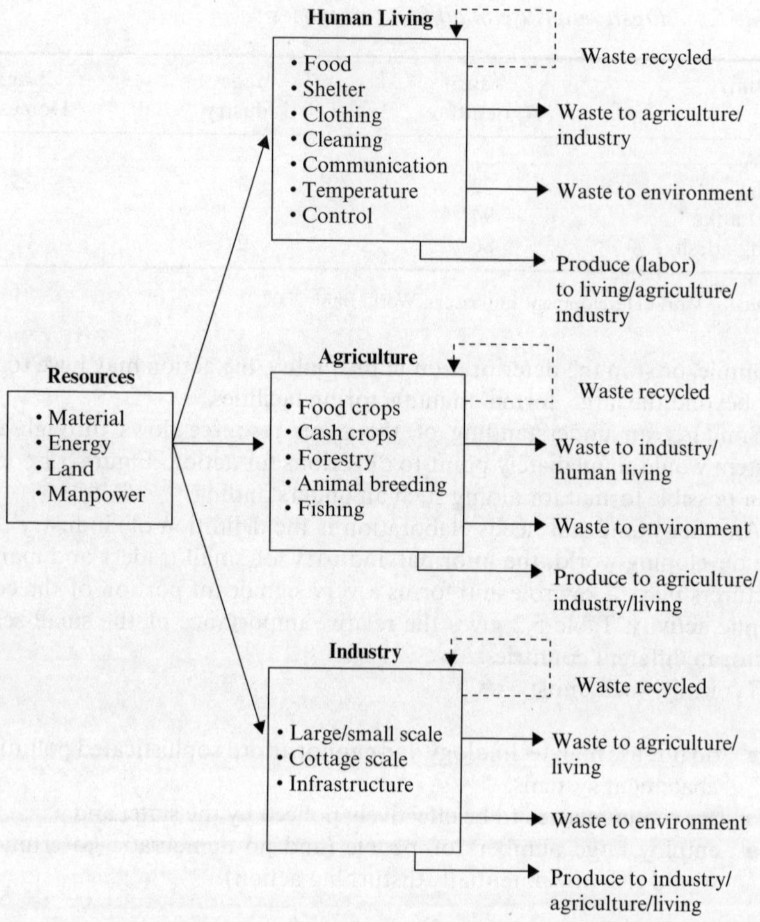

Figure 5.1. Flow of resources through an economic system

A NEW PLATFORM FOR PLANNING – INDUSTRIAL ECOLOGY

Industrial ecology has often been caught in the Kalundborg trap. Although Kalundborg has been a wonderful example of Industrial Ecology at work (Ehrenfeld and Chertow 2002), the excessive accent on the symbiosis there has led to the danger of industrial ecology being narrowly identified as a synonym of Industrial Symbiosis or more narrowly of 'waste exchange'.

The scope of industrial ecology goes well beyond waste exchange. The larger message from Kalundborg is that one of the key elements of policy

Table 5.2 Indicators of SMEs in selected economies in the mid-1990s

Country	As % of total enterprises	% share in Employment	% contribution to GDP	% share in total exports
Japan (1994)	99.0 (excl. primary industry)	78.0	56.0 (of total value-added in mfg. Sector)	14
Korea, Rep. of (1996)	99.1	78.5		42
Taiwan Prov. of China (1994)	97.8	81.1		56
People's Republic of China	98	70 (mfg. only)		40–60
Indonesia	97	42		10.6
Thailand	98	n.a.		10
Malaysia	96	40		15
Philippines	99	45	28% of mfg. Value added	20
Singapore	89	42		16
Vietnam	83	67		20

Source: Mikio Kuwayama (2000), 'E-Commerce as a tool of export promotion for SMEs: comparison between East Asia and Latin America', Kuala Lumpur: Asian and Pacific Development Center.

that could be both economically viable and environmentally sustainable is the optimization of resources flowing through the entire economic system.

This is immediately relevant in developing countries where the availability of resources to the people is very poor. A policy platform that is based on the optimization of resources would also appeal to every citizen in these countries and this would ensure their involvement – so critical to the implementation process.

More importantly, in the industrial ecology perspective, the environmental agenda is not seen as separate and distinct from the economic development agenda. The programs for economic development have to be designed not to damage the environment.

The long-term economic well-being of societies can only be ensured if, at the level of the local community, the state, the country, they acquire a good understanding of the resources available. This understanding alone will enable them to make an objective assessment of their relative strengths and weaknesses. Based on that assessment, they will have to leverage their strengths to develop economically strong and sustainable entities.

Obviously, the Industrial Ecology conceptual framework, which was originally formulated in the USA, does not directly apply to the vastly different context of the developing world. In contrast to industrialized nations, characteristics of developing countries may include:

Infrastructure
- Poor transport network
- Poor telecom links
- Non-availability of reliable data due to the existence of a huge informal sector
- Accelerated obsolescence of infrastructure due to climatic conditions and population pressure

Land-related issues
- High population density, which makes land a very vital resource
- Low per capita availability of arable land
- Low agricultural yields

Labor
- High levels of unemployment
- Low labor cost
- Low labor productivity
- Poor work ethic
- High level of daily wage earners with no job guarantees
- Low level of skills
- Poor working conditions and inadequate social security

Water-related issues
- Low per capita availability of fresh water (either surface or ground water)
- Lack of treated drinking water
- Lack of an adequate sewerage system
- High cost of central water treatment and disposal systems
- Widespread dependence on untreated groundwater

Social
- Low levels of education and consequently, poor awareness of health hazards from pollution or industrial accidents
- Sometimes less concern for social issues (as jobs are often more important than whether a long-term environment problem is caused)
- Low concern for global issues that do not have an immediate bearing on the society
- Lower 'social cost' of law breaking
- Often high levels of corruption among law-keepers
- Very poor social security

Economy
- Restricted availability of raw materials caused by limited financial resources
- Low levels of technology
- Smaller scales of manufacturing
- Existence of millions of informal businesses
- High inflation
- Higher cost of capital
- Restrictions on import
- Volatile foreign exchange rates
- High need to export and earn hard currency
- Low brand equity with many small units doing job work for large domestic or foreign companies
- Perverse subsidies which often encourage wastage of resources

Legal
- Laxity in laws governing environment and worker safety, and hence low costs of disposal of wastes
- Lax enforcement of laws
- Ineffective or slow legal system

Figure 5.2 Characteristics of developing countries

For long it was considered adequate to understand only the flow of money through the economic system, as ordained by conventional financial logic. Development objectives were set in monetary terms; for example, to increase the foreign exchange earnings, often without a rider on resource availability. This often gave rise to the growth of unsustainable economic activities requiring resources that the local community could ill afford. This is clearly not enough.

Industrial ecology offers a new platform for developing strategies that leverage the resources of different societies in various contexts, and ensure long-term well-being and prosperity. This platform facilitates an understanding of the flow of resources through the system (material, energy, land and manpower). Such an understanding could help societies to assess the opportunities available to them, which maximize the productivity of the limited resources available to them, and to more fully assess the threats from their use (or misuse).

This is, of course, of much greater significance in developing countries, where resources are often scarce, and where the nominal value of resources is fixed on the basis of the affordability to the local population rather than on their intrinsic value to the community.

The cases detailed later in this chapter, strengthen the argument against traditional planning systems, where issues of monetary economics blind the planner to many other equally critical elements that should be part of the development process.

A REGIONAL PERSPECTIVE

The example of industrial symbiosis in Kalundborg is simple to understand. A few entities share their resources to increase their net gain from their commercial activity. In effect, a handful of entities have a limited number of transactions with one another (Figure 5.3). It is easy for them to identify and quantify the waste or unused resource, and predict the normal availability of the resource, which is essential to planning. However, the matter becomes extremely complex, when the same exercise is attempted at a regional level where there are hundreds or thousands of entities having multiple transactions with one another (Figure 5.4). The availability, collection and analysis of data necessary to plan such resource exchanges become considerably more difficult. This requires a new perspective and new methods for collecting, presenting, analyzing and using the available data for planning.

SUMMARY CASE STUDIES

A few cases have been summarized below which demonstrate how new perspectives could be generated with the help of a 'system' study and an understanding of the material flows in a region. All the studies were

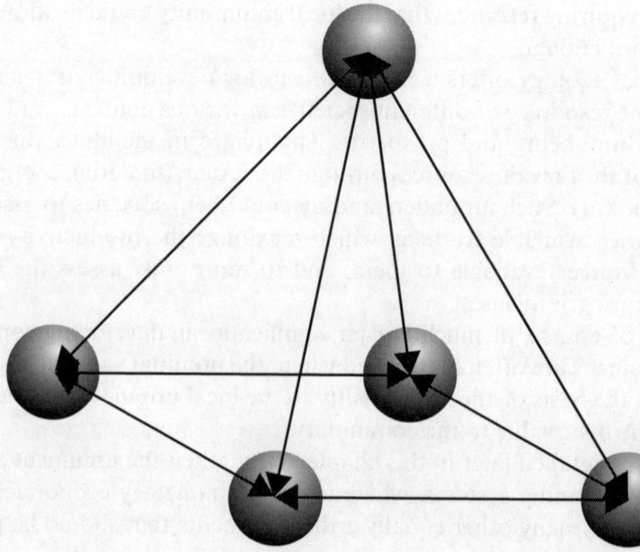

Figure 5.3 Few entities, limited transactions

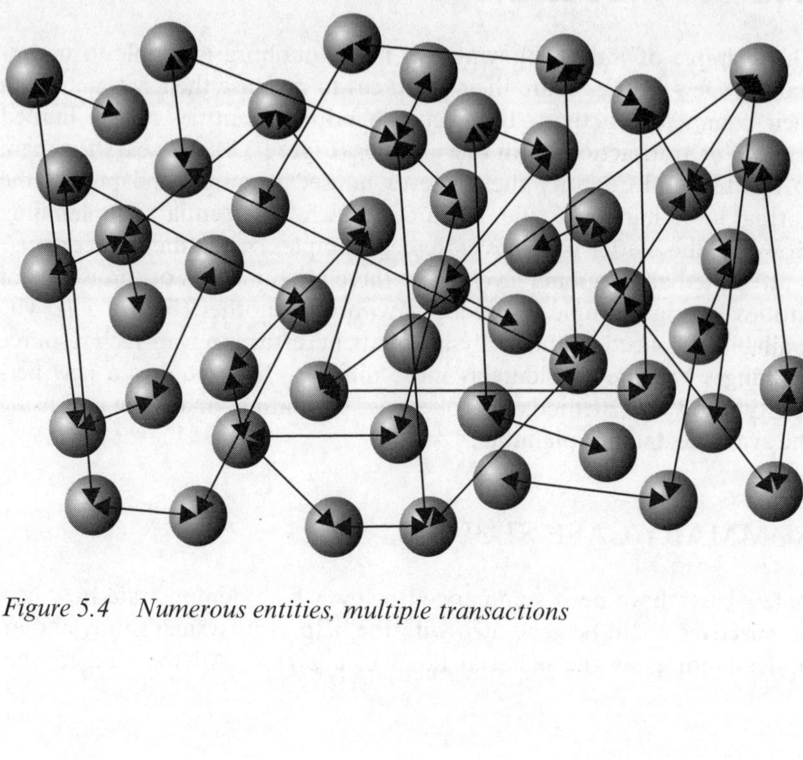

Figure 5.4 Numerous entities, multiple transactions

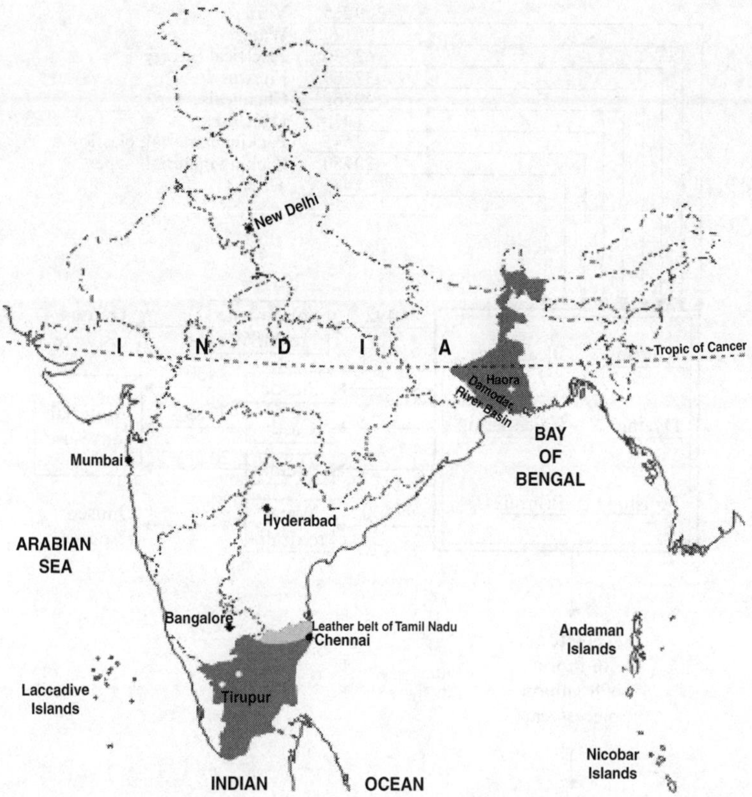

Figure 5.5 *Sites of the case studies*

undertaken during the period 1996–1998. Although the data have not been updated, the core issues remain unchanged over the years. Updating the data would not dramatically alter the key issues in the cases presented here. For the convenience of the reader, a summary of the case studies is given in this chapter. Figure 5.5 indicates the sites of the case studies in various parts of India.

Case Study of the Textile Industry in Tirupur

A Resource Flow Analysis (RFA) was undertaken for the town of Tirupur, in the south of India, which stood as an example of how a Regional Resource Flow Analysis could be effectively used.

The RFA for Tirupur is shown in Figure 5.6. Tirupur is a major center for the production of knitted cotton hosiery. The town is located in the south of

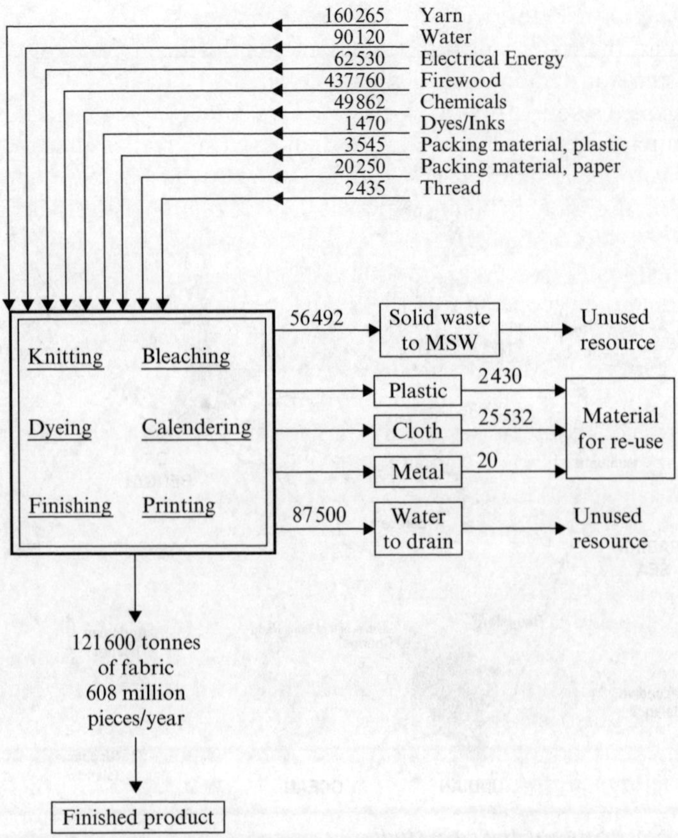

Figure 5.6 Resource Flow Analysis for Tirupur Town (units: water – thousand liters per day, *electrical energy – thousand kWh per year, others – tonnes per year)*

India and has a population of about 300000. The 4000 small units in the town specialize in different aspects of the manufacturing process. The aggregate annual value of production in the town is around US$700 million. Much of the produce is exported, bringing in very valuable foreign exchange.

Water is scarce in the area and the wet processing of textiles has rendered the ground water unusable. A large quantity of salt is used in the dyeing process and the process wastewater (90 million liters per day) is highly saline and is contaminated with a variety of chemicals. As there is hardly any source of fresh water nearby, trucks bring in water from ground water sources (which are yet to be polluted) as far as 50 km away at an enormous cost. A massive US$30 million project is under way to treat the wastewater at Central Effluent Treatment Facilities. After such expensive treatment, the water will still be unusable, as the facility does not include any system for desalination of the wastewater.

A detailed resource flow analysis was carried out for the town. Only when the figures were aggregated did the industrialists realize that they were collectively spending over US$7 million annually on buying water and in addition, the annual maintenance cost of the effluent treatment plant would be an enormous burden. The aggregate figures immediately showed that water could be recycled profitably. On the basis of the study, a private entrepreneur developed a water recycling system, which could be installed in each dyeing unit. The system used the waste heat from the boilers already working in the dyeing units for the recycling process. This is a relatively low cost system, which is gaining popularity in the town.

As a second outcome, the study highlighted the fact that the calorific value of the solid waste (garbage) was high, since it contains large quantities of textile and paper wastes. This could be used effectively to partially replace the 500 000 tonnes of scarce firewood being used in the town (there is grave concern over rapid deforestation in India). Since the use of the firewood is distributed over nearly 1200 points, it was not obvious that such large quantities of firewood were being used. The possibility of setting up a central steam source (needed by some of the industries) is also under serious consideration in order to reduce the consumption of firewood.

The case illustrates the significance of the industrial ecology approach in the context of a developing country. For many years a number of research and development institutions have carried out 'pollution control' studies in Tirupur to minimize water use, minimize use of dyes and to improve the quality of the effluent. There appear to be no studies aimed at evaluating the possibility of profitably recycling the wastewater in the town, which should have been the first priority, from the point of view of industrial ecology. Again, since water pollution was seen as the only issue, no attempt appears to have been made either to minimize the use of scarce firewood or to leverage the high calorific value of the solid waste in the town.

Case Study of the Foundries in Haora

There are nearly 500 cast iron foundries in Haora, a suburb of Kolkata (formerly Calcutta), in Eastern India. The air pollution from the foundries has been a source of concern. The pollution control authorities have been insisting on the foundries installing pollution control systems to mitigate the emissions. The poor health of the engineering industry in the eastern region of India has affected the financial health of the foundries here, which now subsist on manufacturing very low value-added products like manhole covers.

Since pollution from the foundries was a major source of concern for the state authorities and a matter of public debate, a number of agencies had launched studies to develop and set up technologies and equipment for limiting air pollution.

One of the governmental research agencies had developed a process to burn natural gas instead of using coke (the major cause of pollution) as is now the case in the foundries. This process was in an advanced stage of development. It was considered likely that the environment protection authorities might insist on the foundries switching to this new technology to eliminate the pollution problem. Since natural gas is not available in the region, the use of this new technology could substantially increase the cost of production and the foundries would not be competitive.

An RFA of the region showed that the industry could adapt the new technology to use coke oven gases instead of natural gas. As the eastern region is a major coal producing area and as there are many independent coke ovens, coke oven gas is easily available locally and is often wasted. Depending on the economics, either the foundries could be relocated near the coke ovens or the coke oven gases could be transported to the foundries.

The study highlights the relevance of an RFA to an industry planner, as it would point to unused resources (by-products or wastes) in a region. The industry (or a group of industries) could consider how they might leverage the availability of any of these unused resources to their advantage and for their sustained operations. This can be done by establishing new linkages between industries in different sectors (like foundries and coke ovens), which is far from obvious, without an RFA that helps in the detection of such resources in a systematic way.

Case Study of the Leather Industry in Tamil Nadu

Tamil Nadu, a state in the south of India, is the premier center in India for the processing of leather. Water is scarce in Tamil Nadu. India has traditionally been a major center for the export of hides and skins. In the 1970s, the government of India banned the export of raw hides and skins with a view to improving the value added of production, and thereby enhancing the inflow of scarce foreign exchange. Environmental issues were not considered seriously in India in those days.

This boosted the leather processing activity in India in general and in Tamil Nadu in particular. The industry is a major foreign exchange earner and important to the economy of the state and the country. Meanwhile, compliance with strict environment regulations has rendered the processing very expensive in the developed world.

The leather industry (which is made up of thousands of small industries) is a major user of water, as each tonne of hide/skin needs 30 000–50 000 liters of water for processing. This is a large volume, as the average per capita water availability for human settlements in India is estimated at around 30 liters per day. The sludge from water treatment, estimated at 250 kg per 1000 kg of hide processed, continues to be a problem. The sludge is carelessly dumped and the pollutants leach into the groundwater. The industry often buys water in trucks at a high cost. The growth of the industry has resulted in extremely high water pollution in the regions where the tanneries are concentrated.

The leather industry has been under pressure from the pollution control authorities and many have subscribed to a central effluent treatment plant. The water after treatment continues to be unusable, as it is very saline. The sludge from water treatment continues to be a serious problem.

A detailed study in the context of industrial ecology helped in redefining the problem, which till then had been only viewed as a pollution control issue as the effluents did not meet the specifications laid down by the law. Many academic studies have been undertaken to ensure that the effluent quality comes as close as possible to the standards using the best available technology.

However, the problem is much more serious. The tanneries are using a resource, water, which is extremely scarce in the region. The industry is also contaminating the groundwater resources of the local community, which is causing great hardship to the population, as it is depriving them of desperately needed water. It will not be long before the social pressure and the law courts bring the leather industry to a halt.

In the context of industrial ecology, the first priority is to focus on the use of the local resource, water. The local community cannot afford to spare water for the industry. The perspective of industrial ecology opened up totally new options: One of the options could be:

- relocate all the tanneries along the coast;
- set up a power plant close to the tannery cluster;
- use the waste heat from the power plant to desalinate water;
- set up a central effluent treatment system for the waste water from the cluster;
- re-use the wastewater in the power plant; or
- incinerate the solid waste in the power plant.

Figure 5.7 gives a schematic view of a possible sustainable system.

The study points to a new strategy option for sustainability of the leather industry in the region. The study gives an example of how a redefinition of a problem from a perspective of resource-use could drastically alter the

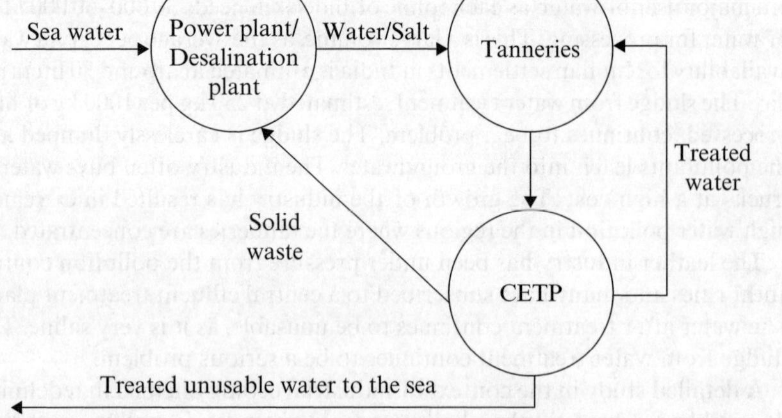

Figure 5.7 Concept of an ideal system built around the leather industry

approach. This could lead to new possible strategy options, more effective (and less costly) than traditional approaches.

The study also points to the need for industries or industry groups to carry out studies on resource availability in the region, while establishing new plants or expanding the present plants. This could be critical to their long-term survival and their peaceful coexistence with the local community.

This is still an idealistic perception. Considerable work needs to be done in ascertaining the technical and economic feasibility of the concept. However, the essence of this case study is that re-definition of a problem from the perspective of industrial ecology can generate new systemic solutions.

It must be mentioned that such relocation (though it may be a plausible and feasible option in India) cannot be achieved in a very short time. It involves the movement of thousands of families, their homes and their work. If such a scheme as suggested were feasible, it would provide a long-term goal to the industry planner. It is possible to develop a long-range plan (say over a decade) and create a suitable road map to achieve such a goal.

Case Study of the Damodar Valley Region

The basin of the River Damodar, in the eastern part of India, covers a vast area. This mineral-rich region (near Kolkata) is the source of much of the coal produced in India. Coal is a major energy source in the country. Many large power utilities and steel plants are located here, in addition to industries associated with coal, such as coal washeries and coke ovens. The region is considered very highly polluted.

An industrial metabolism study was undertaken in the region. The quantities of the flow of two of the major local resources, the waters of the River Damodar and coal, were studied. The results of the study gave a good overview of how the waters of the river and coal are used in the system.

Since agriculture consumes nearly 85 per cent of the waters of the river, it is critical to estimate the impact on the agricultural produce, of the thousands of tonnes of potentially toxic wastes dumped into the river, resulting from the high levels of industrial activity upstream.

All along, to reduce the high levels of air pollution, the policy of the regulatory authorities had been to focus on the 'major' polluters, which in their opinion were the steel and power plants. These plants have access to some of the best available technologies for controlling their pollution.

However, a study of the flow of coal gave surprising results. Huge quantities of coal are consumed in millions of homes and in the informal sector. In this sector, coal is used in very inefficient combustion systems, obviously without any pollution control systems, which makes the whole area extremely polluted. It was obvious that if the air had to be clean, a new fuel policy would have to be evolved. Some new systems of transportation of coal also needed to be designed to minimize the spillages during transportation, a major contributor to the dust levels of the region.

This case highlights the importance of a quantitative study of the resource flows in a region. Even a broad understanding of the flow of the resources serves as a guide to the policy maker and gives a new perspective and a clearer direction for policy making.

HOW CAN THE CONCEPTS OF INDUSTRIAL ECOLOGY BE IMPLEMENTED?

While industrial ecology, in principle, sounds like an attractive option, how can this be implemented? By definition, it calls for a broad system level outlook. This requires the cooperation of different sections of society and the approach has to be multi-disciplinary.

The reader could well ask: What can I do to put into practice the concepts of industrial ecology?

While implementation of these concepts in their full perspective could take some time, it may be useful to start thinking of industrial ecology as an elegant philosophy or a framework. This philosophy can be applied in whatever work group in society that one may belong to – be it management of agriculture, industry, environment or any other.

Listed here are some possible uses of this new planning platform by a few identified user groups. Neither the list of user groups nor the application possibilities outlined are intended to be exhaustive. They are more illustrative.

If different sections in each local community can adopt this philosophy, it will be easier for system level planners to use the concepts at a macro level to plan more sustainable societies.

ENVIRONMENT PLANNERS

Resources Impact Assessments

In most parts of the world, an Environment Impact Assessment (EIA) has become mandatory. While there is no uniform format for an EIA, very often these reports do not reflect truly the impact of an economic activity on the resources of the region. Thus it may be useful to have Resource Impact Assessments. These would give the environment planners a clear idea of the current demand on the available resources. Based on such data, it will be easy to predict the likely future impact on the resources of a region. This could be the basis for licensing new activities. For example if the environment managers receive a request for locating a steel plant, they will be able to decide whether they can meet the water requirement of the steel plant over the next few decades. They could also calculate easily the likely secondary demand on water resources as an outcome of locating the steel plant (for instance, the increase in population). Based on this analysis, the environment planners may choose to either refuse permission or insist that the industry find its own sources of water (set up desalination plants and use desalinated sea water, for example).

Carrying Capacity of Regions

Since a regional Resource Flow Analysis (RFA) would clearly highlight the resources consumed and the wastes generated locally, the carrying capacity of a region can also be determined.

Waste Source Identification

Again, since the data on the use of different materials are easily available, it will be possible to identify sources of any specific wastes or pollutants. For instance, if there is a case of high cyanide content in any lake or stream, it is necessary to identify waste streams that are likely to contain cyanide and start an investigation.

Promoting Systematic Recycling of Waste Resources

The RFA, if prepared, would provide clear data to assist the environment planner to develop possible waste exchange programs.

Setting an Agenda for Action

Since the relative quantities of waste generation can be made available from an RFA, the planner can set priorities, elaborate specific agendas for action, and prepare well-directed action plans. The aim would be to first target the activity most harmful to the region. This can be illustrated with reference to vehicular pollution in cities. (In many developing countries there could be more than 25 different kinds of motorized vehicles on the road.) From a Resource Flow Analysis (RFA) and a Resource Utilization Map (RUM) for different fuels, the contribution of each type of vehicle to the total pollution load in a city can be precisely assessed and analyzed. To improve the air quality in a city, targeting the largest polluter could be made a priority. It will also be possible to make specific quantitative assessments of the likely results from such action, set clear quantitative targets, and evaluate any progress achieved.

Industry Planners

Data from the regional RFA could be effectively used by industrial development agencies in developing countries.

Evaluating the Merits of Different Industrial Activities

Given the limited resources in the region, local agencies could promote those industries that give maximum returns per unit of resource consumed. These parameters will have to be locale-specific and meet with the overall objective of the local government. Some of the parameters used to evaluate the relative merits of different optional industrial activities could be:

- per capita income per kiloliter of water consumed or kWh of energy consumed/per acre of land used;
- employment generation per kiloliter of water consumed or kWh of energy consumed/per acre of land used;
- foreign exchange earned per kiloliter of water consumed or kWh of energy consumed/per acre of land used.

Using Waste Resources

Since the RFA would clearly identify the waste resources in a region, the industry planner could specifically promote industries that use waste resources. New business and employment opportunities could emerge from this resource optimization strategy, in addition to contributing to the environmental sustainability.

COMPANIES AND BUSINESS

Sustainability Studies

Data from an RFA of a region can be of immense use to industries while locating a new commercial activity. A detailed analysis of resource availability is essential for the long-term survival of the business in any area. In addition to assessing the availability and prices of resources as they are today, it is necessary for companies to make an assessment of their availability in the future. Even if the industry can afford to pay for the higher cost of a raw material, caused by rising demand, if it is overusing a scarce resource, it would not be able to exist in harmony with the local community. Such studies would also be in line with business fulfilling its social responsibilities.

New Business Opportunities

Studying the data on the wasted resources in a region from an RFA could be the starting point for setting up new commercial ventures that effectively use these wasted resources.

Substituting Inputs

An understanding of the waste resources could also help to find cheaper or better substitutes for inputs/raw materials by using the wastes available (either in the same form or after processing) in the region.

Product Design and Innovation

Data on the resource availability in a region (including a forecast of its likely future availability) could help companies to develop products that use less of any resource than is or could become critical. If coal is likely to become scarce in a region, or if its use is considered as creating unacceptable pollution issues, the company manufacturing coal-fed boilers has to start redesigning the product to use other fuels. An RFA could serve as an early warning system and allow the company to strengthen its assets and increase its competitiveness.

PUBLIC UTILITIES

Planning and Demand Forecasting

Understanding how resources are used would be essential to planning and forecasting demand. None of the cases mentioned above fully illustrate the concept of RUM for planning utilities, but a typical RUM for a city could be as depicted in Figure 5.8.

Control of Wastage

Data generated from an RUM could also be used for planning effective distribution of resources and for plugging leakages from the system. Such data would immediately focus the utility manager's attention to the areas of

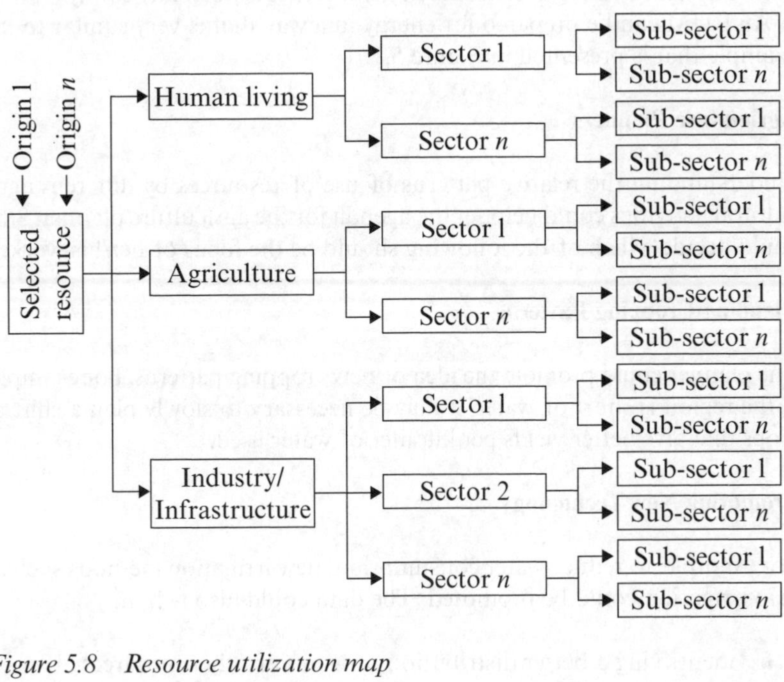

Figure 5.8 Resource utilization map

maximum consumption and would help him in planning meaningful and effective action.

Help Consumers Improve Resource Productivity

From an analysis such as this, the utility manager will also get a clear picture as to which of his consumers need help, advice and support in improving their resource productivity. Similar analyses could be used to understand how any other resource such as energy or fuel is used in a defined area.

Energy Managers

A detailed RUM would be extremely useful to energy managers to know how and in what form energy is being used in a region where they are operating. Not only will this help them in planning and forecasting demand for energy by different sectors among their consumers, but also it would help them to target specific sectors for promoting new or renewable energy sources. For example, if the energy company can estimate the part of the energy that is used by their domestic consumers to heat water, they could promote solar heating systems, in areas where it is normally sunny.

An RUM can be prepared for energy in a way that is very similar to the example that is presented in Figure 5.1.

Agriculture Planners

Understanding the relative patterns of use of resources by different agricultural activities could help set the agenda for the agriculture planner. S/he could decide which of the following should be the focus of her/his work.

Planning Cropping Patterns

The planner could promote the idea of new cropping patterns. For example, if the region is short of water, it may be necessary to slowly plan a shift to crops that give better yields per kiloliter of water used.

Promoting New Technology

For example, to reduce water consumption, new irrigation methods such as drip irrigation could be promoted. The data could also help in:

- identifying a better distribution of water or other resources;

- setting an agenda for improving yields per unit of resource used (water, land, pesticide);
- setting an agenda for promoting new farming practices; and
- providing a better assessment of use of pesticides and fertilizers (per unit of production) and their impact, if any, on land or water sources.

Possible New Parameters

The RFA and RUM methodologies would allow the agricultural planner to develop new parameters (beyond traditional yield or output) like: employment per acre of land, kiloliter of water or unit of energy, per capita income per acre of land, kiloliter of water or per unit of energy, foreign exchange per acre of land, kiloliter of water or unit of energy. Such new parameters would help to develop innovative, integrated complexes, combining agricultural and industrial activities. They could be directly beneficial to farmers and local communities, while improving the health of the rural ecosystem.

Land Use Planners

The data from an RUM for land could help the land planner in:

- understanding the use of land by different sectors;
- planning the allocation of land for different sectors; and
- planning the location and spatial distribution of different activities.

Development Agencies

National and international development agencies and funding institutions would be one of the major users of data from an RFA. They could use the data to:

Prioritize work
By studying data on resource flows, while planning work in a given region, such institutions could focus their efforts in fields that directly impact the critical resources in the area. This will bring maximum benefit to the local community. There will be greater appreciation for their work and will ensure the involvement of the community. Even if the focus of the institution is on rural development, the planners could focus on optimizing the resources of the region and work towards greater productivity of the local resources.

Evaluate options to maximize resource productivity

Data from an RFA of the region could be used as a decision-making tool. Preference would be given to projects that could potentially maximize resource productivity. For example, in a sun-drenched area, if the choice was between a project to introduce solar energy in local industry versus improving the efficiency of the present oil-fired heating system, the choice could be for promoting solar energy solely or in combination with conventional fuels to maximize resource productivity.

Transport and city planners

RFA and RUM would also be useful for transport and city planners. Such studies could effectively help to 'dematerialize' the transportation system. They could create an RUM for the transport infrastructure (the identified resource) and gain an understanding of why people travel (for example, the number of kilometers traveled by people going to school, office, post offices, railway stations, and so on). The total load on the transport infrastructure and the total fuel consumption in a region could be reduced by either bringing services closer to the people or by planning self-sufficient suburbs, thereby eliminating or reducing the need for people to travel.

CONCLUSION

It is crucial to focus attention on developing countries for the following reasons:

(a) Because of the sheer magnitude of their population and economies, developing countries' consumption of resources and generation of waste will soon surpass that of today's already high levels in industrialized countries. Besides, the general pattern of economic activities in the developing world (a large number of small entities and a significant informal sector) makes these impacts more problematic.

(b) The globalization of the economy has resulted in the location of a large part of the production bases of rich countries in developing countries, which are also providing a major part of the raw materials. Thus, industrial ecology has to be implemented at the global scale for maximum effect.

(c) The growth trajectory of developing countries is different from that of industrialized countries. Contrary to the historical development path of industrialized countries, the major developing countries (like India and China) are undergoing a rapid and large population increase along with a strong and quick economic growth in a context of globalization. Their growing population, on average, is becoming more affluent with more purchasing power available for more resource-consuming lifestyles (an increase in the number of cars, white goods, and so on.). Thus, the potential for rapid resource depletion and environmental degradation in developing countries is much more serious than that historically experienced by industrialized countries. Since the beginning of the Industrial Revolution, these countries went through a relatively slow and gradual increase of their population and economic growth, spread over more than a century. This allowed them to adapt progressively to the problems caused by industrialization, in particular with curative means like end-of-pipe systems.

Among the various messages that this chapter intends to deliver, we would like to emphasize two of them.

First, the perspectives of planners must necessarily be different in developing countries. The objectives of a program and the strategies required also need to be different for these countries and have to be tailor made for each country or region keeping in mind the local contexts and constraints.

The second important message is that industrial ecology models take into account the whole system (instead of a narrow perception of problems), and this is better suited for developing countries as it integrates environmental concerns in the process of planning development.

BIBLIOGRAPHY

Ayres, R.U. and U.E. Simonis (eds) (1994), *Industrial Metabolism, Restructuring for Sustainable Development*, Tokyo, New York: United Nations University Press.

Berkhout, F. and J. Hertin (2001), 'Impacts of information and communication technologies on environmental sustainability: speculations and evidence', report to the OECD for the SPRU-Science and Technology Policy Research Unit, University of Sussex, Brighton, p. 25.

Chiu, A. (2002), 'Ecology, systems and networking: walking the talk in Asia', *Journal of Industrial Ecology*, **5**(2), 5–8.

Christensen, J. (1999), 'Industrial symbiosis: a profitable potential for environmental benefits', in P. Pangotra, S. Erkman and H. Singh (eds), *Proceedings of the Workshop on Industry and Environment*, pp. 56–77.

Côté, R.P. (1997), 'The environmental management of industrial estates', UNEP Industry and Environment technical report No. 39, United Nations publication 92-807-1652-2, Paris.

Côté, R.P. and C.E. Rosenthal (1998), 'Designing eco-industrial parks: a synthesis of some experiences', *Journal of Cleaner Production*, **6**(3/4), 181–8.

Ehrenfeld, J. and M. Chertow (2002), 'Industrial symbiosis: the legacy of Kalundborg', in R.U. Ayres and L.W. Ayres (eds), *A Handbook of Industrial*

Ecology, Cheltenham, UK and Northampton, MA, USA: Edward Elgar, pp. 334–48.

Erkman, S. and R. Ramaswamy (2003), *Applied Industrial Ecology – A New Strategy for Planning Sustainable Societies*, Bangalore: AICRA Publishers.

Erkman, S., C. Francis and R. Ramaswamy (2001), *Industrial Ecology: An Agenda for the Long-term Evolution of the Industrial System*, Alliance for a Responsible, Plural and United World.

Francis, C. and S. Erkman (2001), *Environmental Management for Industrial Estates. Information and Training Resources*, prepared for UNEP–DTIE by ICAST, Paris: United Nations Publication.

Frosch, R.A. and N.E. Gallopoulos (1989), 'Strategies for manufacturing', *Scientific American*, **261**(3), 94–102.

Frosch, R.A. and N.E. Gallopoulos (1992), 'Towards an industrial ecology', in A.D. Bradshow et al. (eds), *The Treatment and Handling of Wastes*, London: Chapman and Hall, pp. 269–92.

Herman, R., S.A. Ardekani and J.H. Ausubel (1989), 'Dematerialization', in J.H. Ausubel and H.E. Sladovich (eds), *Technology and Environment*, Washington, DC: National Academy Press, pp. 50– 69.

Hileman, B. (1995), 'Eco-industrial parks offer economic and environmental advantages', *Chemical and Engineering News*, p. 34.

Lowe, E.A. (2001), 'Eco-industrial park handbook for Asian developing countries', report prepared for Asian Development Bank.

Mont, O. (2002), 'Functional thinking – the role of functional sales and product service systems for a function-based society', Swedish Environmental Protection Agency report 5233, July, Stockholm.

Nakicenovic, N. (1997), 'Freeing energy from carbon', in J.H. Ausubel and H.D. Langford (eds), *Technological Trajectories and the Human Environment*, Washington, DC: National Academy Press, pp. 74–88.

Nemerow, N. (1995), *Zero Pollution for Industry. Waste Minimization Through Industrial Complexes*, New York: John Wiley & Sons.

Small Industries Development Bank of India (SIDBI) (2001), Economic Planning and Research Cell report on small scale industries sector, New Delhi, India.

Socolow, R. (ed) (1997), 'Fuels decarbonization and carbon 18 industrial ecology: an introduction, sequestration: report of a workshop', Princeton University Center for Energy and Environmental Studies report no. 302, Princeton, NJ.

Stahel, W.R. (2003), 'The functional society: the service economy', in D. Bourg and S. Erkman (eds), *Perspectives on Industrial Ecology*, Sheffield, UK: Greenleaf Publishing, pp. 264–82.

Tibbs, H. (1993), *Industrial Ecology. An Environmental Agenda for Industry*, Emeryville, CA: Global Business Network.

Weizsäcker, E.V., A.B. Lovins and L.H. Lovins (1997), *Factor Four, Doubling Wealth, Halving Resource Use*, London: Earthscan Publications Ltd.

PART 3

Innovation systems: perspectives on transformation and variety

6. Transformations in food consumption and production systems: the case of the frozen pea

Ken Green and Chris Foster

INTRODUCTION

Technological innovation and the changes in supporting economic and social structures that come with it (collectively known as 'innovation') must be central to the achievement of sustainable production and consumption in all areas of human activity. If current systems of production and consumption are unsustainable in terms of their resource usage, ecological impact and long-term environmental effects, then new systems of provision are needed, entailing new processes, new products, new services and new management practices; if these do not exist, they will have to be invented and launched into social economic use. Conversely, new forms of social relationships that are innovated with environmental improvement as their goal will inevitably use products and processes in new ways. There is thus a strong relationship between innovation in socio-economic arrangements and innovation in the material products and processes in which they are entwined – 'sociotechnical systems of provision' as they might be called. Consequently, understanding the processes that are likely to underpin these developments is crucial for policy intervention to achieve desirable forms of sustainability.

In this chapter, we explore socio-technical systems for the provision of food. As an example, we explore the dynamics of the system for the production and consumption and production of the frozen pea in the UK. We look at various alternatives to the current system – represented through 'advocacy strategies' by proponents of the alternatives – that are argued as being more 'sustainable'. We analyse the frozen pea system to identify the sources of technological control and possible innovation solutions in dealing with the system's 'unsustainabilities'. We might expect that similar analyses could be done on the systems of provision of other types of food (and indeed these are the subject of a research project from which this chapter is derived).

FOOD SYSTEMS AND 'TRANSFORMATIONS'

Introduction

As one of us has argued elsewhere (Green *et al.* 2003), the notion of 'sustainability' in its broadest meaning, as opposed only to the reduction in the environmental impact of individual products or agricultural or industrial processes, requires thinking in 'systemic' terms.[1] Transforming human activities with respect to food implies a focus on the whole system of agricultural, industrial, retailing and household 'sectors' and their interrelationships, with their strongly connecting regional, national and international dimensions. In addition, systemic thinking is concerned with more than the production of food, in agriculture and food processing factories; it also includes distribution and the preparation of final meals whether this be in individual households or in more communal arrangements whether commercial or non-commercial. We can thus define Food Consumption and Production Systems (FCPSs) to include the whole 'chain' of human-organized activities concerned with the production, processing, transport, selling, cooking and eating of food and the disposal of the wastes of such activities.[2] Thinking 'systemically' allows a focus on an important, if neglected, aspect of sustainability, namely the intimately connected relationships of production with consumption.

System Strategies

We would claim from the literature on food and sustainability that it is possible to identify different system 'strategies' for the 'organization' of Food Systems. Strategies for new systems are usually described in opposition to the supposedly dominant institutional forms of food production, distribution and consumption to be found in the OECD countries and said to be the form that is diffusing most rapidly into developing countries. This 'industrialized/modern' FCPS is based on 'Fordist' principles of seeking high labour productivity and economies of scale in all elements of the system, especially in agriculture and food processing. Fordist principles have been increasingly extended to distribution, with the domination of supermarkets in retailing and mass catering in eating-out. Household consumption is based on a wide variety of mass produced commodities with a historically high consumption of animal products. Agriculture and food processing is the subject of continuous innovation, based on scientific understandings. There is a constant search for innovation in products and agricultural/factory processes. This 'industrialization/modern' form FCPS is much caricatured by critics, not just for the quality of the food it provides (with rising concerns about food safety and hygiene) but also for its insensitivity to environmental and animal welfare concerns.

What alternatives to 'industrialization/modern' systems are there? For developed countries, there are only two contenders. The first can be labelled the 'organic' strategy.[3] Advocates of organic systems focus on food production that engages with natural systems and cycles in agriculture and processing. They approve of the proposed dismantling of 'industrialized' systems that are prevalent in rich countries and their replacement with methods of agriculture, food processing and distribution that emphasize social sustainability. Much cultural significance is given to 'natural' products and production methods as a means of ensuring health – of humans, of farm animals and of the eco-system in general. Agricultural pollution (especially damage to soils and water courses) can be minimized by the avoidance of 'chemical' inputs into agriculture (synthetic fertilizers and pesticides) with the use of closed nutrient cycles (with much waste recycling). The use of GM seeds would be completely ruled out. Some advocates of this strategy go further than a concern with agricultural methods; they see it as part of a socially and ecologically responsible approach to the production and distribution of food, with a strong bias to bioregionalism and against large-scale world food trading, though some organic food grown in large farms for international export can be countenanced.[4]

The second strategy we have called 'new industrial': 'new' because it is advocated as a restructuring of the 'industrialized/modern' strategy to take account of a number of scientific and technological developments of the last 20 years.[5] It takes seriously criticisms of the environmentally-destructive nature of post-1945 methods of high-productivity agriculture. This leads to the introduction of new methods of crop management, often using Information Technology, and diversification of agriculture into new materials. The strategy could readily incorporate the technical and certification features of the 'organic' strategy, though not the other, more social and bioregionalist aspects of the organic movement. Second, it allows the use of genomic knowledge to develop new seed varieties both through genetic engineering and traditional breeding methods enhanced by a better understanding of a crop's molecular biology. This is seen as a huge jump from mere 'Monsanto-type' genetic modification, which used genome knowledge linked only to changes in the use of agrichemicals. This knowledge also presents the opportunity to improve crop protection technologies, through a better understanding of crop pathogenicity. There are a number of benefits to be gained from better understanding of the full genetic makeup of crop plants and food animals, as part of extending the benefits of the Green Revolution beyond the basic crops of maize, soya and rice. Third, it takes on board the notion of foods as a way of delivering health care, through the development of functional foods and 'nutraceuticals'.

The strategy is still based on high outputs in agriculture and processing within internationally-organized production and trade. It continues the strong twentieth century emphasis of the industrial/modern system on high output and low labour agriculture and innovation in agriculture and food processing based upon science. Farms would still be large with high productivity (and low labour inputs) but with new developments in soil and pest management that allow more eco-sensitive approaches to biodiversity. Greater attention would be paid to hygiene and quality, especially in relation to animal products with the development of new non-soil methods of food production (for example, fungal protein). The strategy thus continues the focus on producing large quantities of food for rapidly expanding urban populations. It seeks to respond to the undoubted environmental degradation that twentieth century agriculture has caused by the application of new technologies, but by the application of further modern technologies – especially in biotechnology – that are considered risky by many environmentalists.

In the next part of the chapter, we explore a system of provision of one particular food, seeking to identify the potential for the alternative strategies that are advocated as preferable to Fordist systems. The production of frozen peas can been seen as *the* example of a Fordist production and consumption system. Sustainable alternatives to it need to be considered to give peas a chance.

GIVE PEAS A CHANCE

Introduction

According to Robert White, ex-President of the US National Academy of Engineering, industrial ecology is 'the study of the flows of material and energy in industrial and consumer activities, of the effects of these flows on the environment and of the *influences* of economic, political, regulatory and social factors on the flow, use and transformation of resources' (1994, emphasis added). The direction of flow between the 'physical'/'material' world and the 'social/economic/political' world is, in this definition, in which the social 'influences' the physical. But – as work in innovation studies continues to show – it is possible to see the physical-social relation in a different way, with the process of innovation being 'embedded' in structures of social relations (including those that inform consumption patterns and practices), inter-industrial relations, technological relations, and capital/investment relations. A key idea is how we can re-think the link between the flow of materials, a flow which industrial ecology is especially skilled at analysing, with the social, economic, and organizational structures which cause physical flows to be and become 'clumped' (concentrated/dispersed) in particular ways. We can also proceed to identify empirically the location(s) of actual innovative change within those structures. We can further identify potential sites for innovation, together with, importantly, constraints to change and reasons resisting change.

To elaborate, we can conceive four structural domains which together provide organizational logic to the system. They are: the structuring of materials flow; the structuring and organization of economic activity together with the pecuniary redistributions which arise from the processing of those materials; the social structures and structuring of relations (including power relations) which demarcate classes of agent and, finally, the production of structures and meanings of knowledge including how that knowledge (and its associated symbolic significance, the ways meanings are produced and interpreted) is generated and applied.

Peas: Industrial Ecology and Innovation

Figure 6.1 presents a 'system map' for the frozen pea in the UK. The frozen pea is especially important, symbolically if not quantitatively or nutritionally, for the UK diet. It is *the* green vegetable, the first one to be available in a frozen form in the 1950s and the first to have its consumption, in a 'fresh' form, detached from its seasonality. It symbolizes other things as well – as something that might be considered 'unsustainable', both in growing it and in freezing and distributing it.[6] As such it has become the subject of examination by its major processor in the UK – Unilever/BirdsEye – through its work, in partnership with the Forum for the Future of the Sustainable Pea.[7] The focus of the FftF/Unilever initiative is in making the agricultural methods of pea production more sustainable by, for example, reducing the quantities of chemical inputs suggesting that there are or should be other, and more 'organic', methods of agriculture. However, there are other aspects of the frozen pea's ecological impact which also need to be considered. We need to consider all the resource inputs and ecological impacts before assuming that pea-growing – the agricultural part of the system – is the (only) problem. And we need to identify the sources of technological control and possible innovative solutions in dealing with any of these unsustainabilities; in particular we need to take account of the apparently only fixed point in the whole pea system map: the continued place of frozen peas, conveniently purchased year-round at a low price, in UK meals.

A food system is thought of here as a sequence of activities, starting with the production of plant seed, that link together to bring food to consumers'

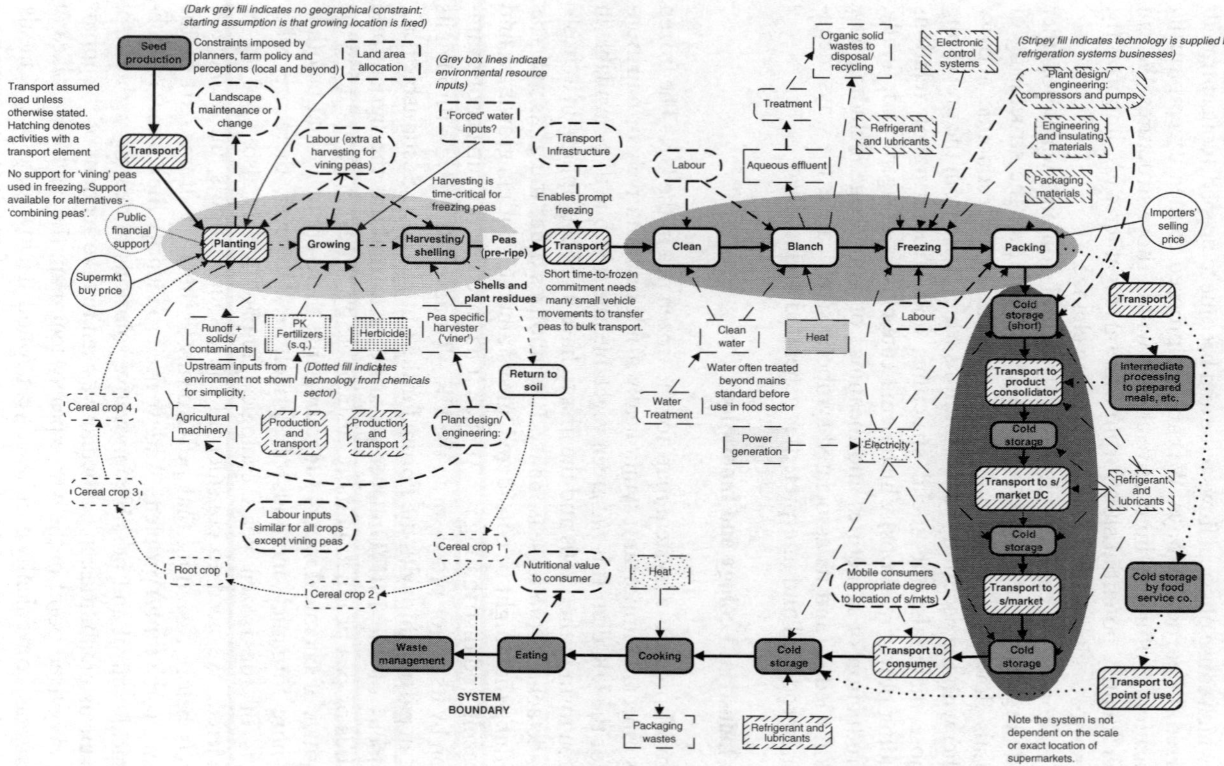

Figure 6.1 Frozen peas in the UK – a system map

mouths. If we want to analyse the implications of the existence of a certain food system for society, the environment and technology we must start with three questions:

- What characteristics of society, technology and the environment enable the system to exist as it does?
- What are the consequences of its existence?
- What tensions within the system exist between pressures for change and pressure for stasis, and how are these resolved as outcomes/processes of adjustment and co-evolution?

The overall system map included as Figure 6.1 shows a string of basic activities.[8] However, we have not just drawn a flow diagram of the elements of the pea agricultural, processing and distribution system (something that we would expect from a straightforward IE-type study). We have added those elements that indicate how the system is controlled by a number of 'core' organizations, with inputs from and outputs to its socio-economic environment, the 'technosphere' and the natural environment. By technosphere we mean the set of human activities which transforms naturally-occurring resources into the forms used in the system under study, and turns wastes from that system back into substances that are released into nature. The catalogue of inputs and outputs is not exhaustive: there would not be space in a graphic representation of this sort for such a listing. We have tried to focus on 'critical' inputs and outputs, namely those without which the system could not exist in this form.

The Pea Consumption and Production System: the Materials Flow

The basic materials flow within the full system diagram is extracted in Figure 6.2. The system is centred on growers in the UK. The UK is both the largest grower and consumer of immature, or vining, peas (as distinct from dried peas) in Europe. Some 35 000–40 000 hectares are dedicated to their cultivation in this country, with this area tending to fall with time. Because of this selected focus, the geographical locations of some of the activities in this sequence are defined or constrained. Such activities are shown in Figure 6.2 as boxes with no shading. Many activities in the sequence entail transport or motor-powered vehicles: these are denoted by hatched boxes. Boxes with grey shading then denote activities which are static but are not geographically-constrained by virtue of our focus on UK grown peas. The core activities are shown in boxes with bold frames and linked by solid arrows in Figure 6.2 and subsequent figures.

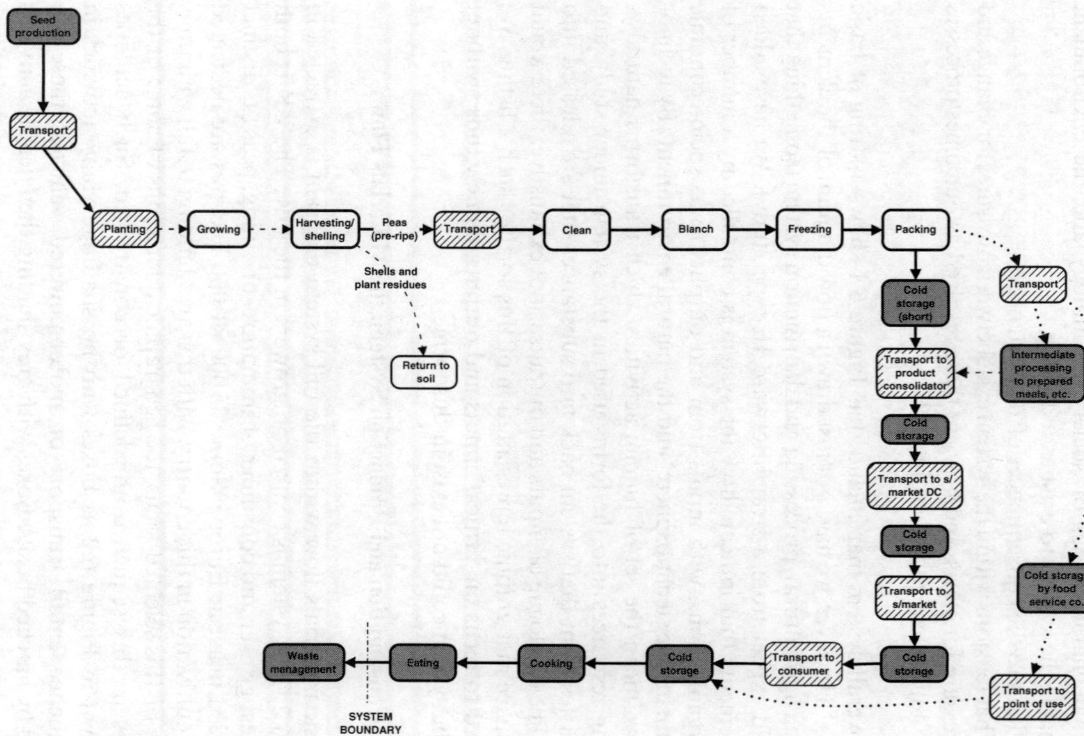

Figure 6.2 Frozen peas in the UK – basic activities

On the farm

There are a number of seed suppliers from whom growers can source seeds for peas. By definition the chain of activities from planting to harvesting is geographically fixed, but we draw attention to the fact that planting and harvesting are dependent on the use of motorized equipment. A key aspect of the freezing of immature peas is the time that elapses between picking and freezing. This is portrayed as being a critical factor in determining the taste of the finished product. Indeed, the idea that no pea is packed more than 150 minutes after it has been picked features in the marketing of some brands. (As an old advertising jingle put it: BirdsEye peas are 'Fresh as the moment when the pod went pop'). While other processors have no specific commitment, all apparently aim for similar levels of performance. This has two implications for the system:

- harvesting involves many small vehicles to transfer peas quickly from field to bulk road haulage container; and
- the location of processing plants is geographically constrained to being reasonably close to the farms. We have not done sufficient research to establish a specific radius: however, since the 150 minutes must include time to fill a 40-foot trailer, and time to offload, wash, blanch and freeze the peas as well as actual travelling time, it seems unlikely that this would be greater than 100 km.

The harvesting equipment (known as a *viner*) also separates the peas from their pods and the remainder of the plant. These residues are later returned to the soil.

Into the freezer

On arrival at the processing plant, peas are cleaned and checked, then blanched (partly cooked by immersion in very hot water, before being frozen and packed. Fluidized bed freezers are used to allow efficient heat transfer from cold air to pea.

Through the distribution chain

The activities that follow processing are common to most food ingredients. A proportion will be shipped on to other food businesses that produce prepared foods such as ready meals, soup, and so on. A further proportion goes to 'food service' businesses – operators of canteens, restaurant chains, commercial caterers, and so on. The remainder is delivered to shops for sale to individual consumers.

It is generally held that supermarkets account for 80 per cent of all food sales in the UK, so it is assumed that most peas pass through their logistics chain. These start with delivery to a product consolidator (a logistics firm), who feeds goods from a number of suppliers into a distribution centre from which they are sent out to the stores themselves. The last few activities in the sequence, those undertaken by individual consumers, will be familiar to all of us. With the exception of those lost in processing, all the peas that leave the farm pass through these activities, whether they reach the consumer via the supermarket directly, in a prepared food product or via a food-service business.

The Pea Consumption and Production System: Core Organizations

Any analysis of the implications of implementing one or another definition of sustainability must consider potential changes in the balance of power between organizations at different points in the 'value chain'. One of the contradictions associated with the promotion to business of, earlier, environmental and, more recently, sustainable good practice has been that it offers competitive advantage to all – 'win-win'. In the case of sustainability, different definitions have different implications for different actors: for example, stressing organic production would appear to favour organic producers and all those involved in moving products to consumers, while stressing local production would appear to provide opportunities for UK farmers and pose a number of threats to the existing food distribution system centred on chains of supermarkets with centralized purchasing.

Figure 6.3 shows the sections of the chain of basic activities in the pea system that are under the control of three groups. Farmers, or more accurately, 'growers' groups' – formal co-operatives bringing together up to 50 farms and controlling cultivation of up to 4000 hectares – control the planting, growing and harvesting activities (light shading in Figure 6.3). These growers' groups own the equipment needed for these activities and, for the most part, have in-house agronomy expertise. One large, well-known processor eats into this sphere of control by having its own agronomists work alongside producers contracted to supply its peas. There are reckoned to be some 10–15 of these grower groups in the UK now, and the tendency is for them to concentrate further in pursuit of economies of scale.

Moving downstream, the current level of concentration appears to be greater still. There are reported to be only three large pea-freezing operations in the UK, as well as a handful of smaller independents. Their sphere of control is shown by the medium shading in Figure 6.3. One of the large freezers produces branded peas under its own label, leaving the rest to cover other brands and all supermarket own-brands. (A single cannery also takes in some pea production.)

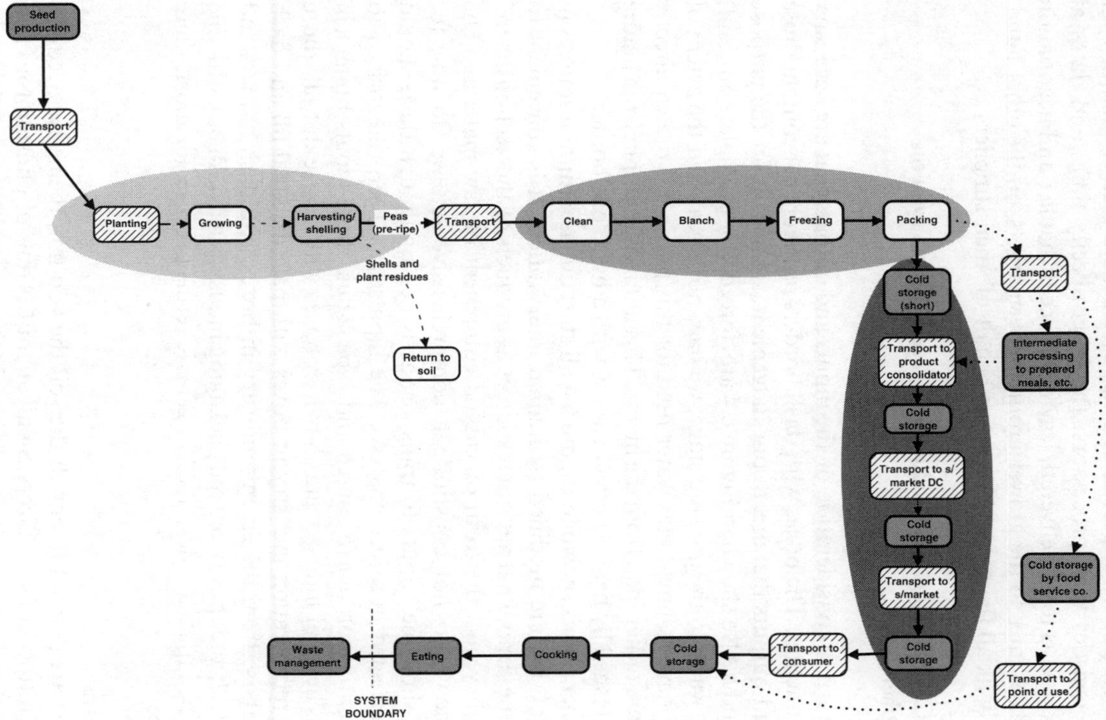

Figure 6.3 Frozen peas in the UK – core organizations

Despite this high level of concentration, power seems to remain with the supermarkets, which control those activities contained within the dark shaded ellipse in Figure 6.3: the economic forces that account for this are discussed on p. 147. Supermarkets appear to have greater control over inbound logistics than processors: the latter specify times and dates at which product is to be delivered, leaving choice of haulier and negotiation over haulage rates to the grower group. Supermarkets, on the other hand, commonly fix all these parameters 'on behalf of' their suppliers.

The Pea Consumption and Production System: Inputs from the 'Technosphere'

We now turn to consideration of the inputs and outputs that are necessary for it to function. The boxes with dash borders in Figure 6.4 contain those inputs and outputs that are, in our judgement, significant for the purpose of this study. Also shown in Figure 6.4 are 'forced' (for example, non-rain!) inputs of water to the growing stage: we have not researched the extent of these but have assumed that water used for this purpose is drawn directly from nature rather than from the mains. This unmodified input from nature is distinguished by being shown in a box with a light tone border.

The other inputs shown in Figure 6.4 all start out as natural resources in some form, but are modified by human intervention. It is convenient to think of these modified natural resources as products of the 'technosphere' whether they take the form of capital equipment or raw materials. The inputs shown do not constitute a comprehensive set: we do not, for example, show fuel inputs to transport activities – although these should not be neglected in future analysis. The inputs have been categorized to some extent according to source and type. Thus, those inputs bought in from the chemical industry are shown in boxes with coarse dot fill; those from the refrigeration industry in boxes with pale diagonal fill and dash borders; those from the energy industry in boxes with dash borders and light grey fill, and those from the packaging industry in a dash box with fine dot fill. The inputs are described in generic terms because, for most, there is a choice.

On the farm

Since peas are planted to enrich the soil they do not, themselves, require inputs of nitrogenous fertilizers. Small quantities only of phosphorous and potassium fertilizers may be used to maintain mineral balances. Selection and application rates of crop protection chemicals (herbicides, fungicides, and so on) is case-specific and is often determined by drawing on suppliers' expertise.

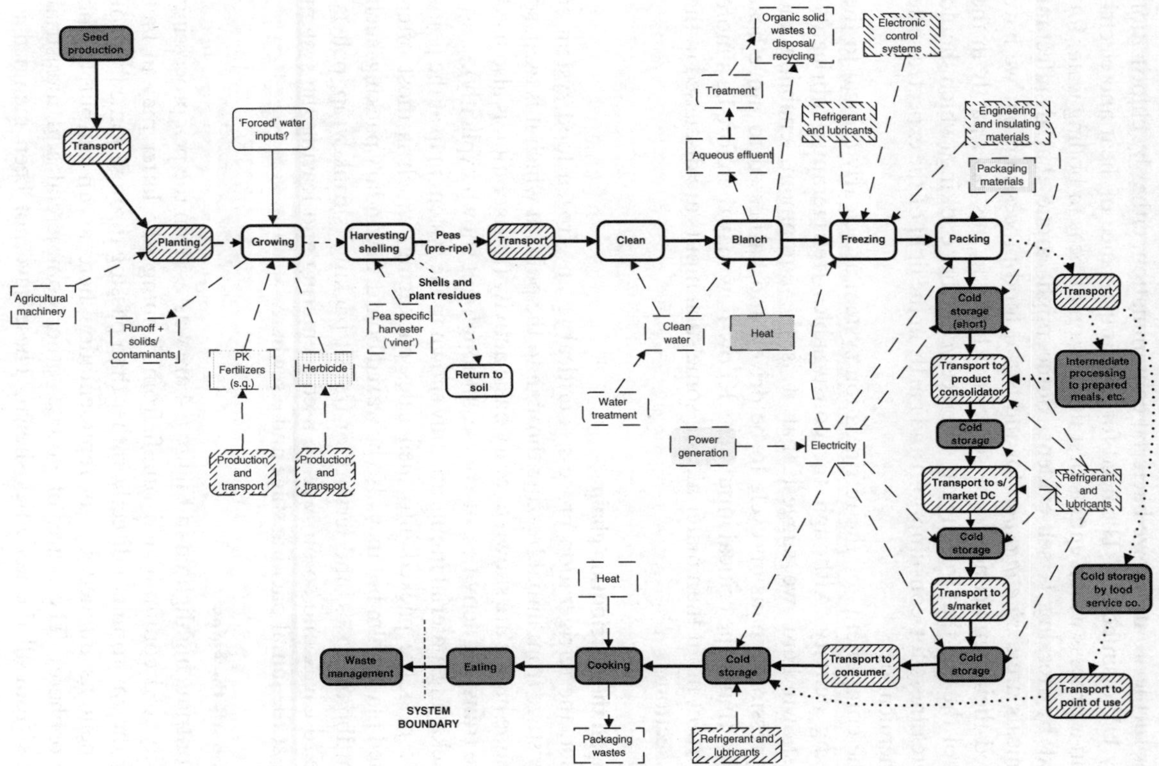

Figure 6.4 Technosphere inputs and outputs

Into the freezer

Clean water is used in large quantities in industrial-scale food processing, both for cooking and for cleaning. It is common practice to treat mains water further to minimize bacterial contamination, either by chlorination or by UV disinfection. The need for, and importance of, heat and electricity in a process which entails first immersion of peas in boiling water followed by freezing is self-evident. Some might argue that disinfectant chemicals are a *sine qua non* for industrial food processing, but we have judged their significance to be somewhat lower in this case, partly in the light of the fact that frozen peas will receive further cooking (which should ensure fitness-for-consumption) and partly in the light of the expected scale of chemical use.

The components of the refrigeration system are clearly critical to the freezing activity. Although the consumables (refrigerants; lubricants) are shown here, we suggest that it is the equipment, enabling the compression-expansion cycle to be driven and harnessed to move heat energy, that is the critical input here. Know-how may therefore be a more important input than material, and some such inputs are discussed in the next section.

Through the distribution chain

In fact, the refrigeration process is critical to any frozen food system at every stage from initial freezing through to the point at which it is used, so the same inputs are shown to every basic activity (refrigeration is also used in the transport activities, of course, although not shown explicitly).

Packaging material inputs are only shown in the system map at the point where peas are packed into their sales packaging, which is most often printed plastic film but may also be waxed board. Secondary packing, such as cardboard cases, and tertiary or transit packing (shrink-wrap, pallets, wheeled cages, and so on) will be used – entering and leaving the system both at the initial packing stage and elsewhere.

Falling off the sides

The outputs highlighted in Figure 6.4 are wastes from the pea processing activity and contaminated runoff from farming (the latter may in fact enter the environment directly, rather than passing through some form of treatment as implied by its representation here as an output to the technosphere). There are, of course, other commercial and industrial wastes from all the activities shown. These have not been included in the system map – partly for want of space, and also because they are judged to be of less significance to a study of food systems' particular characteristics.

Socio-economic Inputs and Structures

While physical resource flows are needed for the operation of the frozen pea system, a range of economic and societal inputs are also necessary. It is possible to identify a variety of these. Some, such as labour, can be seen as a potentially substitutable input, exhibiting differential mobility and different degrees of fixed/flexible supply, depending on the labour class involved, the levels of skill (and therefore training) involved, and the terms and conditions for hiring labour. The supply of labour does not typically nowadays have significant implications for natural resource consumption. Others are decisions, such as the decision to allocate land to agriculture or the decision to build transport infrastructure. Decisions like the latter obviously lead to natural resource consumption, so that the provision of road transport infrastructure could be represented as an input of built road from the Technosphere to the pea system. However, it is widely acknowledged by workers in the field of life cycle assessment (LCA) that the inclusion of capital goods in product systems makes no significant difference to the results of LCAs, because the impacts associated with the production of these goods is spread so thinly over their lifetime use. On the other hand without decisions to build the transport infrastructure in something like its current form the pea system as shown here could not function. In particular, we suggest that in this case the road network is essential for fulfilment, by a small number of processing centres, of the short time-to-frozen commitments that appear to be common in the industry. Further exploration of this aspect of the system may well be worthwhile as the project moves forward.

Figure 6.5 shows the chain of basic activities in the system with inputs from and outputs to society shown in ovals with dash borders and economic 'inputs' (forces might be a better term here) in circles. Labour inputs at the farming and food-processing stages are shown, because the existence of jobs in rural areas is a significant factor to some parties in the sustainability debate. It seems, however, that labour inputs to pea cultivation are not very different from labour inputs to the cultivation of other crops, although the need for very rapid collection at harvest time requires some additional labour for a short period on any individual farm (grower groups stagger planting across the land they operate so that harvesting continues for a period of weeks).

Land allocation is a direct input to the system from nature (denoted in the system map by a box with a light tone border) but has been included as a socio-economic input because the *decision* to allocate the land is seen as a significant factor, as much as the occupancy of land by pea cultivation.

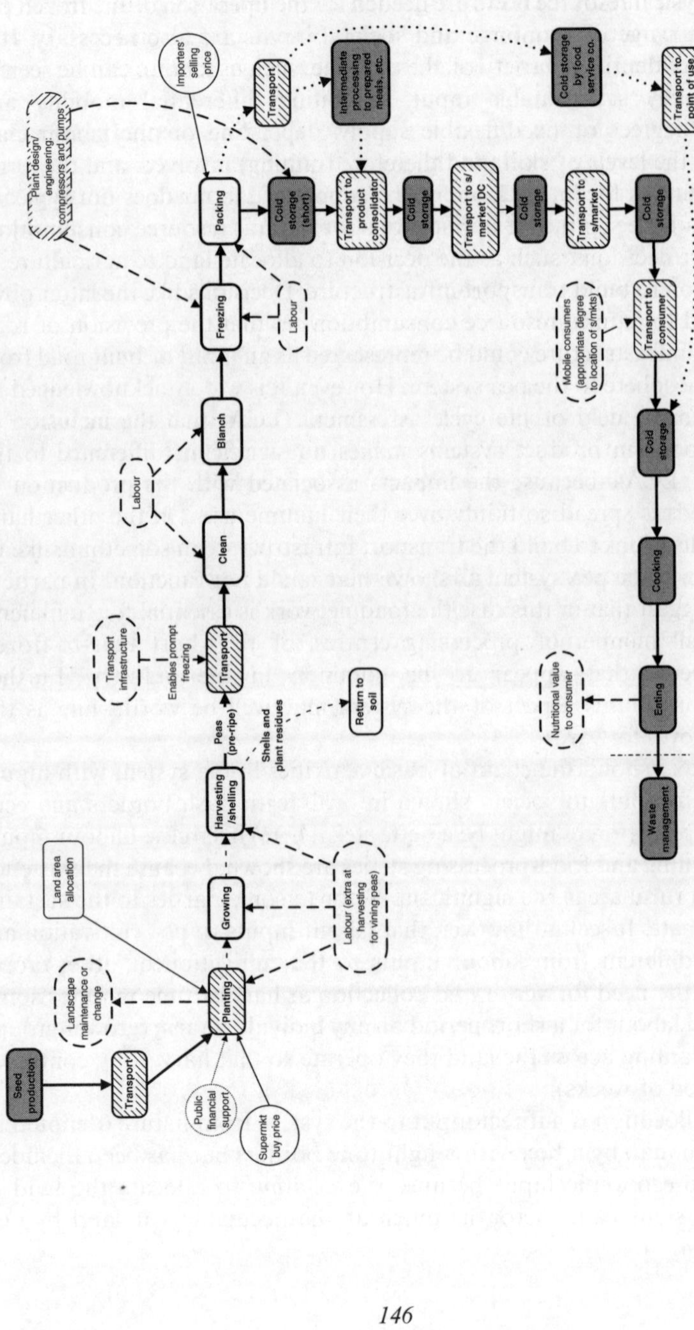

Figure 6.5 First order socio-economic inputs and outputs

The importance of compressor and pump technology to the refrigeration process has already been mentioned, and is shown here. Consumer mobility is also an important factor, although the penetration of the market for vegetables by the frozen form pre-dates the move of supermarkets to out-of-town and edge-of-town locations, so the car-bound consumer is not judged to be critical.

Only three economic factors are shown, all inputs (support payments, supermarket buying price and the price of imported frozen peas). Clearly the balance between what supermarkets, as buyers, are willing to pay growers for their product and what growers could receive for alternative crops would be expected to be an important factor influencing crop selection. The support payments available to growers for combining peas and field beans amount to some £260 per hectare currently. The selling price of combining peas is in the region of £80 per tonne, with crop yields of the order of 5 tonne/hectare, so that a hectare of this alternative crop may yield some £650 income, of which 40 per cent is support payment. This alternative might reasonably be expected to set some lower limit on the price to which supermarket buyers can drive frozen pea growers down. The price at which imported peas are available imposes an upper limit on the price that growers and processors can obtain from supermarkets, although it has been reported that the supermarkets' desire to be seen to be supporting UK farming may allow growers in this country a slight premium for peas destined for direct sale to consumers. It should be noted that intermediate processors and food-service businesses, with lower public profiles, have no such sensitivities.

A further crucial economic input (though it is not shown in the Figures) is the availability and access to finance. Modern market economies only exist according to the precondition that there exists a flow of investment capital and credit facility to 'lubricate' the productive system, enabling production to take place in the absence of, but in the expectation that, consumption will follow in the future. A working and workable integrated financial system is often taken for granted and rendered invisible in resource-flow models. History shows however that when financial systems enter crisis this can have catastrophic and often amplifying contagious effects across the system.

Economic outputs have not been included in the system map. While they can readily be identified (payments to workers, business profits, taxes), investigating their relative significance (say in terms of which organizations get which proportions of the selling price of a pea, and how much is profit in each case) would require more detailed research.

CONCLUSION

The map of the UK frozen pea system in this chapter has been presented not just in terms of materials flows but also of the particular institutional, technological and economic factors that influence and indeed structure it. As such, it presents the opportunity for further research into the implications for the system and the actors within it of working towards different definitions of sustainability. In drawing up a description and graphic representation that cover all elements of the frozen pea system from seed to consumption but are at the same time reasonably concise, some judgement and selectivity has been essential. This selection process has endeavoured to focus on factors (which we believe can usefully be classified as inputs or outputs) that either enable or constrain the system as it operates now. It has tried to pick out technological knowledge, societal characteristics, resource flows and economic conditions which, if changed significantly or taken away, would cause peas – if they were grown at all – to be handled very differently.

The continued survival and reproduction of the UK frozen pea system we have described depends on a number of conditions that are both social and technological. It is clear that, at the level of system actors, if there is to be one actor with a central structuring role and qualitative asymmetric power it is Unilever. This is especially important when we look at the sources of knowledge in the system. Unilever's expenditure on R&D and its ability to mobilize knowledge of agriculture, the freezing process and the logistics of pea distribution make it the key location for any innovation within the system (or the breaker of other innovations that might adversely change the system). Unilever is thus the key agent in producing and interpreting knowledge about 'sustainability', in the sense that it is Unilever that is the agent that considers what is worth investigating and acting on to bring about more 'sustainable' pea production. Unilever's interest in sustainability is connected with the maintenance of its power in the pea system. So far, this interest in sustainability has been confined to an investigation of agricultural practices of pea-growing. This can be seen either as the 'first step' in an examination of the sustainability of the pea system as a whole, or as an attempt to define sustainability as just being about agriculture.

However, as we have sought to show, there are a number of features of the pea system that deserve investigation if we are to think more systemically about sustainability. These could be called the 'bottlenecks'/'pinch-points' that would have to be subjected to change for any sustainable reconstruction of the chain; they are:

1. The influence on the system of the notion that peas have to be moved from 'field to frozen' in a relatively short period of time.

2. The central position of the pea in the everyday eating habits of the UK populace.
3. The centrality of the refrigeration process, at numerous sites as well as in transit.

However, if we can identify one element of the system that structures the rest of it, it is the transport infrastructure for the necessary prompt freezing of the pea. This in turn is set by the instituted consumption practice that puts the frozen pea as a cheap, year-round convenient component of green vegetables in the average UK diet. Sustainable reconstruction of the chain might depend on basic changes in some of the current system conditions. These include:

- the possibility of higher prices (necessary if all peas were to be 'organic');
- a shift back to seasonality for the vegetable (a contrary trend at the moment for virtually all fruits and vegetables); and
- the assumption that delivery of peas requires long food chains can be altered.

All of these would certainly require some change in the place of the pea in UK diets. Organic advocates would expect that some of these changes would be necessary throughout UK agriculture and food consumption practices. However, there are other more 'neo-industrial' strategies that can also be imagined. In this strategy, you could envisage new varieties of peas that travel better, overcoming the transport/prompt freezing bottleneck. This might come from better knowledge of the pea genome and the ability to use that knowledge to create or engineer new varieties. This would then reduce the need for peas to be grown very near to freezing plants thus opening the possibility of changing the economies of scale of the industry, opening up the possibility of local agriculture and local freezing. Such ideas are purely speculative at the moment.

Further research might also be conducted into the underpinning structures and meanings that inform consumption as *practice* (Warde 1996) and which then have an iterative or complicit effect on production. We have already identified that pea consumption has a geographic structure, peas being a 'staple' of the UK diet. We can also conjecture social class, age, and 'occasion' dimensions of the structuring of pea-eating practices. We have identified the pea as a 'stand-by' freezer food, therefore integrated and dependent for its existence and meaning on a whole range and combination of household domestic appliances, notably freezers and cookers. We can also conjecture that peas are eaten primarily as a complement to other, equally taken for granted foods (chips, fish, chicken, burgers) as staples of the UK diet. Perhaps they are more likely to be eaten as a mid-week rather than weekend meal, as a children's rather than an adult meal, and for everyday occasions rather than special candle-lit dinners. All these ways of appropriating peas into the mundane everyday lives of ordinary people have profound impacts on the way peas have come to be used, understood, bought and stored (and thus produced, and most importantly, transported). Furthermore, producers do not passively accept these structures of consumption, rather, through their marketing 'segmentation' and communication strategies they proactively seek to reinforce stratified consumption patterns. Alternatively, producers may use product differentiation and product variety generation strategies to push appeal into new segments and ratchet up total consumption. Thus, although peas have arguably not been subjected to the same variety generation processes as the 'humble tomato' (Harvey *et al.* 2002), we are nevertheless familiar with the distinction, exaggerated by producers, between the ordinary 'garden pea' on the one hand the special 'petit pois' on the other.

Giving peas a chance to be the product of a sustainable system clearly requires consideration of more than agricultural practices – it focuses attention on the intimate connection between consumption and production and the way this connection 'crystallizes' into particular technological practices across the system. Innovating to transform these practices is essential for sustainability.

ACKNOWLEDGEMENTS

A version of this chapter has previously been published in *Technology Forecasting and Social Change*. It is an output of a project on 'Technological Transformations in Food Consumption and Production Systems', funded by the UK Economic and Social Research Council's Sustainable Technologies Programme. The project is a joint one between UMIST and the University of Cardiff. The chapter was originally presented, in a longer version, to the IHDP Open Science Conference, Montreal, October 2003. Thanks to Andrew Flynn, BRASS, University of Cardiff and Sally Randles, CRIC/Institute of Innovation Research for comments on earlier drafts.

NOTES

1. See Lifset and Graedel (2002) for a justification for this.
2. For details of the elements of the Food System, see Tansey and Worsley (1995), Millstone and Lang (2003), present current information on global food production, trade and consumption in atlas form.

3. The arguments here are based on Browne *et al.* (2000), Soil Association (2001), Wright and McCrae (2000).
4. At the moment, organic food is internationally-traded and sold through supermarkets, whose sales of such food is rising rapidly in the richer countries. This is unacceptable to many supporters of organic agriculture – notably those in the 'organic movement' – whose broader agenda is bioregionalist.
5. This account is based on Ford (2000), Conway (1997), Manning (2000), Heasman and Melletin (2001).
6. Its symbolism might be illustrated by the dust-jacket of the recent book by Felipe Fernandez-Armesto (2001). Fernando-Armesto is an Argentinian academic, who works in the US and Europe, and is especially critical of modern food consumption practices. His publisher has chosen to depict an opened pea-pod on the front of the book, despite the fact that the pea is only briefly mentioned in the book as part of Fernando-Armesto's denunciation of frozen foods in general.
7. Forum for the future/Unilever, no date.
8. The picture of the frozen pea system presented here has been developed by reference to a variety of published material supplemented by interviews with growers' representatives, processors and a small number of other food industry sources. For discussion of Life Cycle Assessment, see 'The Eco-indicator 99: A damage oriented method for Life Cycle Impact Assessment, Methodology Report', PRé Consultants B.V. 2000.

BIBLIOGRAPHY

Browne, A.W., P.J.C. Harris, A.L. Hofny-Collins, N. Pasiecznik and R.R. Wallace (2000), 'Organic production and ethical trade: definition, practice and links', *Food Policy*, **25**, 69–89.

Coe, S.D. and M.D. Coe (1996), *The True History of Chocolate*, London: Thames and Hudson.

Conway, G. (1997), *The Doubly Green Revolution: Food for all in the 21st Century*, London: Penguin.

Diamond, J. (1997), *Guns, Germs and Steel: The Fates of Human Societies*, New York: Norton.

Dicum, G. and N. Luttinger (1999), *The Coffee Book: Anatomy of an Industry From Crop To Last Drop*, New York: New Press.

Evans, L.T. (1998), *Feeding the Ten Billion: Plants and Population Growth*, Cambridge: Cambridge University Press.

Fernandez-Armesto, F. (2001), *Food: A History*, London: Macmillan.

Ford, B.J. (2000), *The Future of Food*, London: Thames and Hudson.

Forum for the Future/Unilever (n.d.), *In Pursuit of the Sustainable Pea*, London: Forum for the Future.

Goodman, D. and M.J. Watts (eds) (1997), *Globalizing Food: Agrarian Questions and Global Restructuring*, London and New York: Routledge.

Green, K., M. Harvey and A. McMeekin (2003), 'Transformations in food consumption and production systems', *Journal of Environmental Planning and Policy*, **5**(2), 145–63

Hall, P.A. and D. Soskice (eds) (2001), *Varieties of Capitalism: The Institutional Foundations of Comparative Advantage*, Oxford: Oxford University Press.

Harvey, M., S. Quilley and H. Beynon (2002), *Exploring the Tomato: Transformations of Nature, Society and Economy*, Cheltenham, UK and Northampton, MA, USA: Edward Elgar.

Haverkort, K. and W. Hiemstra (eds) (1999), *Food for Thought: Ancient Visions and New Experiments of Rural People*, Leusden, Netherlands: ETC/COMPAS.
Heasman, M. and J. Melletin (2001), *The Functional Foods Revolution: Healthy People, Healthy Profits?*, London: Earthscan.
IHDP-IT (1999), *Science Plan*, Bonn: IHDP.
Lifset, R. and T.E. Graedel (2002), 'Industrial ecology: goals and definitions', in R.U. Ayres and L.W. Ayres (eds), *A Handbook of Industrial Ecology*, Cheltenham, UK and Northampton, MA, USA: Edward Elgar.
Manning, R. (2000), *Foods' Frontier: The Next Green Revolution*, New York: NorthPoint Press.
Marsden, T.K., A. Flynn and M. Harrison (2000), *Consuming Interests: The Social Provision of Foods*, London: UCC Press.
Millstone, E. and T. Lang (2003), *The Atlas of Food: Who Eats What, Where and Why*, London: Earthscan.
Mintz, S.W. (1986), *Sweetness and Power: The Place of Sugar in Modern History*, London: Penguin.
Parrott, N. and T. Marsden (2001), *Organic and Agroecological Farming in the Developing World*, Cardiff: Department of City and Regional Planning, Cardiff University.
Polanyi, K. (1957), *The Great Transformation: The Political and Economic Origins of Our Time*, Boston: Beacon Press.
Pretty, J. (1995), *Regenerating Agriculture: Policies and Practice for Sustainability and Self-reliance*, London: Earthscan.
Pretty, J. (2002), *Agri-Culture: Reconnecting People, Land and Nature*, London: Earthscan.
Ritzer, G. (1998), *The McDonaldization Thesis*, London: Sage.
Schlosser, E. (2001), *Fast Food Nation: What the All-American Meal is Doing to the World*, London: Penguin.
Soil Association (2001), *Principal Aims of Organic Agriculture and Processing*, accessed at www.soilassociation.org.
Tansey, G. and T. Worsley (1995), *The Food System: A Guide*, London: Earthscan.
Warde, A. (1996), *Consumption, Food and Taste*, London: Sage.
White, R. (1994), 'Preface', in B.R. Allenby and D.J. Richards (eds), *The Greening of Industrial Ecosystems*, Washington, DC: National Academy Press.
Wright, S. and D. McCrea (2000), *Handbook of Organic Food Processing and Production*, Oxford: Blackwell.
Zuckermann, L. (1999), *The Potato*, London: Macmillan.

7. Sustainable technologies and the construction industry: an international assessment of regulation, governance and firm networks

Paul Dewick and Marcela Miozzo

INTRODUCTION

This chapter examines the factors facilitating and hindering the adoption of sustainable technologies in the domestic sector of the construction industry in Europe, particularly emphasizing the role of government policy and inter-firm relations. The chapter provides an international comparative analysis of the role of government as client, regulator, market-broker and promoter of sustainable technologies. The cases of two technologies suitable for reducing domestic sector energy consumption are discussed: thermal insulation, for reducing the energy consumption of space heating, and active solar heating, for reducing the energy consumption of water heating. The chapter also examines the role of long-term inter-firm relations, which have been advocated as an integral part of encouraging the introduction and diffusion of technologies in the fragmented and often adversarial construction industry.

Although increased household energy efficiency (combined with improvements in the energy efficiency of electrical appliances) has contributed to reduced energy consumption per dwelling in most EU countries since the mid-1980s, final energy consumption from the domestic sector has increased. This can be explained by an increased number of households, a higher average size of household (in m^2), a reduction in the average number of persons per household and falling domestic electricity prices.[1] Across Europe, the domestic sector accounts for over a quarter of final energy consumption (EEA 2001a), the vast majority of which is required for space and water heating: 84 per cent of EU household energy consumption stems from space and water heating (EEA 2001b). To reduce total energy consumption across the domestic sector, attention has been focused on the role of technology in improving the energy efficiency of space and water heating. Technical change is seen as the cost-effective solution to increasing energy efficiency whilst maintaining economic growth, securing the future competitiveness of the industry, adding to the strength of the economy's productive structure and maintaining the employment and skill level. However, technological change is only one facet of an integrated approach to increase the energy efficiency of buildings. Because of the nature of the construction industry (fragmented, conservative, mature and with low profit margins), the characteristics of the final product (immobility, uniqueness, complexity and costliness) (Nam and Tatum 1988; Gann 1994) and the operating environment (highly regulated, high liability and litigation risk) (Pries and Jansen 1995; Blackley and Shepard 1995) technological change needs to be accompanied by an active role of government policy and changes in inter-firm relations.

Drawing on our research in Dewick and Miozzo (2002a and 2002b), this chapter looks at the role of government in facilitating the adoption of technologies capable of reducing significantly domestic energy use from space and water heating. For space heating, the external temperature and the level of thermal insulation primarily govern the heating requirement of buildings. This chapter questions the energy efficiency of so-called 'natural' or sustainable thermal insulation materials over the life-time of a building and highlights the key role of regulation in stimulating innovation and increasing the energy efficiency of domestic space heating.

More high-tech solutions are being implemented to reduce the energy efficiency of water heating. Active solar heating (ASH) systems, like thermal insulation technologies are suitable for widespread use across new and existing buildings in the housing stock and have the potential to significantly reduce the energy requirement of water heating and contribute to sustainable building. The chapter explores the international differences in the diffusion of ASH systems and considers the wider role of government in promoting their adoption through legislation, fiscal and financial incentives and through disseminating information both to the different actors in the construction industry and to the general public.

Perhaps more fundamentally, notwithstanding the extent of regulation and the role of other institutional factors, the fragmented structure and project-based nature of the construction industry means that the effective adoption of innovation (and particularly of environmental innovation) requires the participation and collaboration of all parties in the industry. Drawing on evidence from Dewick and Miozzo (2004) based on interviews undertaken with clients, contractors and designers working on sustainable housing projects (including specifications for water and space heating technologies), the chapter describes the organizational factors hindering the adoption of sustainable technologies.

The chapter is structured as follows. The following section briefly reviews the previous literature on factors facilitating and hindering innovation (and sustainable technologies in particular) and examines the initiatives fostered by government and industry across Europe. The third section explores the relationship between regulation and sustainable innovation and considers the case of thermal insulation as a solution for increased space heating efficiency. The fourth section explores the wider role of government and other institutional and structural factors affecting the diffusion of ASH systems, an energy efficient water heating technology. The fifth section presents evidence from interviews undertaken with the actors in the building chain regarding how the relations between actors in the building chain can determine the success or otherwise of projects using sustainable technologies. The final section draws lessons for policy makers looking to increase the diffusion of sustainable technologies in the construction industry and reduce energy consumption in the domestic sector.

SUSTAINABLE TECHNOLOGIES, THE ROLE OF GOVERNMENT AND INTER-FIRM RELATIONS IN THE CONSTRUCTION INDUSTRY

The barriers to innovation in the construction industry reflect the risk and cost factors attached to adopting a new technology. For sustainable technologies, these barriers seem to be exacerbated as they are perceived to be more risky and more costly. The risk of adopting any new technology in the construction industry stems from the use of an untested product or process about which little is known. Also, safety considerations for those who build, use or occupy the building add to the equation. Market imperfections where environmental and social costs are not considered in the cost of technologies mean that sustainable technologies are at a further disadvantage. Capital cost is the first consideration, both in the private sector building trade where profit maximization is the owner's objective and in the public sector where maximizing value with limited public resources is the objective. This gives rise to the well documented trade-off between ecology and the economy, with social benefits on one side and private costs on the other (see Porter and van der Linde 1995; Wubben 1999). Moreover, these costs are borne not by industry but by the ultimate owner of the building (see Malin 2000; Bordass 2000).

The construction industry is heavily regulated: technical regulations, governing products and processes; planning and environmental regulations, governing the finished product; and labour market regulations, governing the welfare of workers during the construction process (Gann 1999). Although there is no empirical analysis that offers convincing evidence to support the assertion that environmental regulation stimulates innovation (see Jaffe *et al.* 1995; Welford and Starkey 1996), the building industry offers good examples of increased resource productivity and lower finished product total cost in the presence of stricter environmental regulation. For example, in Sweden, the Netherlands and Germany, where there is considerably stricter environmental regulation, total building costs are below those in the UK, despite higher material costs and labour costs. In these countries, construction processes have been improved to out-weigh the component costs of building.

The government can be very influential in facilitating sustainable development targets through its role as largest single client of the building industry, by using fiscal and regulatory measures to stimulate innovation and by acting as a broker in markets for environmental technologies.[2] With sustainable technologies in particular, the government also has an important role as the market leader, prototyping innovative solutions through demonstration projects and as chief educator and disseminator of information (both to the industry and to the general public).[3]

Also, the role of the firm and inter-firm collaborations are important because the construction industry can be seen as archetypal network system where a coalition of organizations (including contractors, the government, clients, designers, sub-contractors, suppliers) come together on a temporary basis to undertake each project (Winch 1998; Gann and Salter 2000). Each of the parties may have their own distinct roles and responsibilities for encouraging innovation but it is the relationships and interactions with each other that determine the success of innovative projects (Dewick and Miozzo 2004). This interdependency requirement for the effective diffusion has been hindered by a 'vicious circle of blame' whereby each actor in the industry blames each other for not building environmentally friendly buildings (Cadman 1999).[4] Facilitating best value and encouraging long-term inter-firm relations were recommendations of two important government reports (Latham 1994; Egan 1998) published during the 1990s whose objectives are also consistent with innovation and sustainability.[5] Not only have these recommendations been implemented by projects funded by large clients and employing large national construction firms, policy guidance has diffused down to the local level to encourage long-term inter-firm relations and sustainability in smaller projects funded by smaller clients and involving local and regional construction firms (for example, Scottish Homes 2000a, 2000b, 2000c).

SUSTAINABLE INNOVATION AND REGULATION: THE CASE OF THERMAL INSULATION

Here we draw on Dewick and Miozzo (2002a) to describe the case of innovation in thermal insulation. One method of reducing energy consumption suitable for widespread application in new and retro-fit building is through better thermal insulation. The performance of insulation materials depends primarily upon their ability to trap still air and although cavities and surface resistances are important, the thermal resistance of construction materials is the most significant factor in energy efficiency. Thermal insulation is a hidden innovation (in as much as it has no aesthetic properties) and therefore is likely to be adopted only if there is a performance (for example, economic) advantage of doing so. The performance of a material is a reflection of its thermal conductivity (or K-value), where high performance is related to low thermal conductivity. Therefore, for a given thickness of material, a thermal insulation technology would offer a credible alternative to the incumbent technology if it had a better (or no worse) thermal properties at a better (or no worse) cost.[6] This suggests that substitution is determined largely by capital costs. However, sustainable building and regeneration requires one to look beyond capital costs to consider the material's life cycle cost or social and environmental cost (for example, the so-called triple bottom line).[7]

The current incumbent technologies are glass wool and rock wool and plastic foams such as phenolic foam.[8] The widespread use of these materials can be largely explained by their low K-values, efficiency and relatively low cost, combined with the construction industry's preference for tried and tested materials, the performance of which has been monitored and proven over many years. However, if one considers the 'triple bottom line' costs, these materials fit uncomfortably alongside the concept of sustainability as they are responsible for a significant environmental impact during their production and there are question marks against their safe use. Table 7.1 shows a comparison between conventional materials (plastic foams and mineral wool slabs) and natural materials (for example, cellulose fibres). The trade off between performance (thermal conductivity) and environmental impact is clearly shown.

The use of natural insulation materials alleviates many of the environmental problems caused by the production and use of more conventional insulation materials. However, in achieving sustainability targets through increasing the energy efficiency of the domestic sector, there are two contrasting issues that must be addressed. The first issue concerns the thermal insulation material's performance in limiting heat loss and reducing the compensating energy requirement for space heating. The second issue relates to the energy intensity and other environmental and social impacts

Table 7.1 Performance and environmental impact of insulation materials

Product	Performance (thermal conductivity)	Environmental impact description
Plastic foam: Phenolic foam	0.022	• Extraction of crude oil • The raw materials, oil and natural gas, are non-renewable resources and their use, associated with emissions of oils, phenols, heavy metals and scrubber effluents, account for over half of all toxic emissions into the environment • Sulphur oxides and nitrogen oxides are produced contributing to acid rain and causing photochemical oxidants • Phenols cause hazardous vapours during in-situ foaming
Mineral wool slab: Glass mineral wool	0.031–0.037	• Mining is required to extract the raw materials • Production process is energy intensive, creating emissions of fluorides, chlorides and particulates and releasing solvents and volatile organic compounds such as phenol and formaldehyde • Sulphur oxides and nitrogen oxides are produced contributing to acid rain and causing photochemical oxidants • Non-biodegradable properties: atmospheric pollutant • Contains small traces of carcinogenic properties (glass fibre)
Natural material: Cellulose fibres (organic fibres)	0.034	• Made from processed waste paper and treated with borax (sodium tetraborate) to guard against fire and insects • Production causes zero pollution and has a relatively low embodied energy • Only negative environmental impact stems from the energy used in the production of the materials

Table 7.1 (*continued*)

Product	Performance (thermal conductivity)	Environmental impact description
		• No toxic by-products during manufacture and no health problems during installation • Fully biodegradable (i.e. contain no toxic or synthetic chemicals)

Notes:
1. Thermal conductivity measured in W/mK at 10°C.
2. Also, refer to Note 6 for a note regarding a material's suitability for purpose.
3. See Dewick and Miozzo (2002a) Table 1 for a more detailed table comparing other conventional and natural thermal insulations.

Sources of environmental impact: Curwell and Mach 1986; Curwell *et al.* 1990; Doran 1992; Harland 1993; Woolley *et al.* 1997; Thermal Insulation Manufacturers and Suppliers Association 2000; Construction Resources 2000.

of the insulation material's production and use. Many conventional materials have high embodied energy and have properties that affect health and prevent the materials biodegrading or being reused. Some natural insulation materials offer a credible alternative with significantly fewer negative externalities. The range of natural insulation products available today demonstrates that there is no lack of innovation in thermal insulation materials. Table 7.1 shows that insulation made of cellulose fibres has comparable thermal conductivity properties (at similar thickness) to more conventional insulation materials such as glass wool slab.

However, there is evidence that the energy savings in terms of a natural insulation material's embodied energy do not offset energy savings over a conventional material's lifetime performance. The over-riding benefit of thermal insulation is that the energy consumed and pollution emitted during its production is vastly outweighed by the energy savings and pollution reductions attained through its use.[9] The best natural insulation materials do not match the necessary cost effectiveness and performance in terms of energy conservation of conventional materials over a 50-year lifetime. Heath (1999) estimated that despite the higher embodied energy and higher capital cost of conventional materials such as plastic foams, their far superior insulation performance results in positive net energy, environmental and financial benefits when compared with their fibrous alternatives.[10] A comparison of two materials is shown in Figure 7.1.

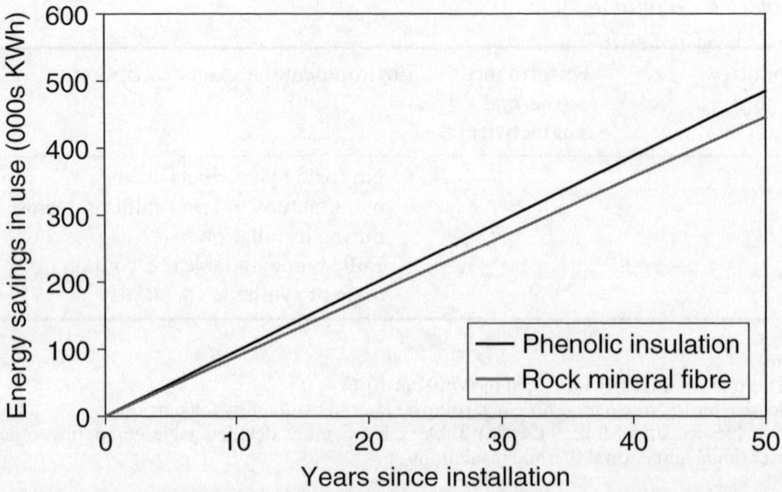

Notes: The energy savings are compared in the use of plastic (in this case phenolic foam) and fibrous (in this case, rock mineral fibre) insulation materials. The figure shows that phenolic foam delivers net energy savings of 485 000 kWh over a 50 year period compared with 446 000 kWh for rock mineral fibre: a saving of 39 000 kWh, equivalent to 11 tonnes of carbon dioxide over the 50 year lifetime. When one considers the embodied energy involved with producing phenolic foam and rock mineral fibre – 6100 kWh and 2200 kWh respectively – the energy savings of using phenolic foam over 50 years become clear. In addition, it is worth noting that the energy savings associated with using phenolic foam over rock mineral fibre equated to a financial saving of £1073.

Source: Data from Heath (1999).

Figure 7.1 Energy savings in the use of plastic and fibrous thermal insulation materials over a building's lifetime (compared with zero insulation)

Since the natural insulation materials have similar thermal conductivity properties at a particular level of thickness to the materials tested (for example, rock mineral fibre and expanded polystyrene), one can assume the results are commensurate with natural insulation's performance. In addition, conventional materials have proved their reliability against the key factors affecting their field performance, for example, the settlement of loose-fills, ageing of gas-filled foams, effect of air on glass-fibre and the effect of moisture on the thermal performance of all insulations.

Natural thermal insulation materials, despite their low embodied energy do not increase the energy efficiency of buildings. To achieve more energy efficient buildings in terms of space heating one must increase the minimum insulation levels. However, because of the private costs and social benefits the building industry has no real incentive to build above the minimum standards. A regulatory tightening, announced with a sufficient time lag to implementation is needed to stimulate innovation. And we are not simply referring to new products: innovation is needed in the design and construction stage if regulations change since one cannot simply increase the thickness of thermal insulation because of space considerations. These conclusions are supported by what has happened to building regulations across Europe during the 1990s. In Denmark in 1995 for example, new building codes were introduced to cut heat demand in buildings by 25 per cent (Kerr and Allen 2001). In Germany, also in 1995, the federal government reviewed thermal insulation requirements to 'limit carbon dioxide emissions by the more efficient use of energy' (IBC 1998). In 2001, the UK also introduced more stringent building regulations governing heat loss that aim to reduce building emissions between 25 per cent and 30 per cent (DETR 2000; Building 2000).

SUSTAINABLE INNOVATION AND INSTITUTIONAL FACTORS: THE CASE OF ACTIVE SOLAR HEATING

In this section we draw on Dewick and Miozzo (2002b) to illustrate the factors facilitating the adoption of a high-tech technology, active solar heating (ASH) systems, suitable for reducing domestic energy consumption for water heating. Although active solar heating technologies can be seen as part of a bundle of technologies suitable for solar buildings (with passive solar design and active photovoltaic technologies), on their own they offer significant environmental savings (in terms of lower energy consumption) and economic savings (in terms of lower energy bills) if one considers life-cycle costs/payback times.

There are essentially three types of technology common to ASH systems in Europe. Glazed solar collectors are the most common types of ASH system, accounting for 82 per cent of the total surface area of solar collectors in Europe in 2000.[11] 'Simplified collectors' or 'solar carpets' are used predominantly to heat water in outdoor swimming pools and account for 16 per cent of all solar collector applications. Vacuum solar collectors, capable of carrying water of a higher temperature than other types of solar collector, remain relatively expensive and do not have a great market presence outside Germany. The market penetration of these three technologies across Europe is shown in Table 7.2.

Costs and payback periods differ across the countries because of the different average size of collectors, ratios of collector to storage tank, average solar irradiation levels, usage profiles and electricity prices. ETSU

Table 7.2 Solar thermal collectors across Europe

Country	Total cumulative surface area by collector type (thousands m^2) in 1999				Per capita surface area (approx. m^2 per 1000 inhabitants)
	Glazed	Non-glazed	Vacuum	Total	
Germany	2130	400	220	2750	32
France	321	332	5	656	11
Denmark	291	2	0.5	294.5	57
UK	132	75	–	207*	4
Netherlands	116	90	4	210	14
Sweden	135	15	1	151	17
EU	7764	1549	–	9314*	26

Note: *Surface area totals in the UK and EU do not include those accounted for by vacuum solar collectors.

Source: Dewick and Miozzo (2002b).

(1999b) studied the performance and cost measures of ASH systems in Denmark, the Netherlands and Sweden and found that purchase and installation cost varied between 500 euros per m^2 of solar collector (in the Netherlands and Sweden) and 1000 euros per m^2 (in Denmark), with a payback period between 5.5 and 16 years.

ASH systems have been developed over the last 30 years following the oil crises in the 1970s and their market acceleration has been facilitated by national and international initiatives. For example, there are global initiatives such as the Solar Heating and Cooling (SHC) Programme, established by the International Energy Agency (IEA) (an autonomous body within the Organization for Economic Co-operation and Development), within which countries collaborate to develop solar technologies to heat, cool, light and power buildings (Bosselaar 2001).[12] In Europe, the European Commission set a target of 100 million m^2 of solar collectors to be installed by the end of 2010 (RES 1997). The Soltherm Europe Initiative, an international collaborative project encompassing existing initiatives across ten European countries, was subsequently established to install 15 million m^2 of solar collectors by 2004 by developing a framework of large demand satisfied by a sales and installation infrastructure (Van der Leun 2001).

However, as Table 7.2 shows, there are wide variations in the size of domestic markets for ASH across Europe, reflecting institutional factors such as national government initiatives to stimulate the market (for example, advertising campaigns and other dissemination strategies), private sector acceptance of solar energy and wider supportive public opinion. As Table 7.2 shows, Germany accounts for 27 per cent of collectors installed across the EU excluding vacuum solar collectors (though Denmark has the largest surface area of solar collectors in per capita terms). The wide diffusion of ASH systems in Germany has been facilitated by a combination of government (financial incentives) and industry (acceptance/promotion of solar from the traditional heating industry), assisted by an active national solar promotion programme (for example, 'Solar na klar') that has gained support from private individuals and local authorities (Van der Leun 2001; Systemes Solaires 2000). 'Solar na klar' was initiated by Baum, a group of green entrepreneurs representing small and medium sized solar firms, and had federal state backing from Gerhard Schroder, the former German Chancellor and Jurgen Tritten, the Environment Minister. The campaign raised public awareness through advertisements and PR work – 65 000 people requested information in 2000 – and is funded by the private sector (for example, the solar industry and other private companies) and public sector (for example, federal and state funding) (van der Leun 2001).[13]

Whereas in Denmark and Sweden there has been added emphasis on installing ASH systems in multi-house units/apartment blocks using integrated collective solar systems, in Germany, ASH systems have been primarily adopted by single family residences (and predominantly in retrofit properties). Sales of systems in Germany (and more recently in the Netherlands) have been enhanced by the incorporation of ASH systems into the traditional heating industry: over one half of solar systems sold in Germany in 2000 were sold in combination with a new boiler (Van der Leun 2001).

When the market is directed towards individual households (and retrofit as opposed to new build), payback calculations are very important. With an emphasis on providing a cheaper, more widely available technology the government has an important procurement role to play, creating markets and encouraging innovation by acting as a broker in a 'technology procurement' strategy. Across Europe there has been no shortage of innovative projects using solar thermal technologies in the public housing sectors where it has been common for social housing to prototype ASH technologies and act as demonstration schemes, largely funded by the EU, national governments, local authorities and housing agencies.[14]

What Table 7.2 doesn't show is that Denmark, Germany, the Netherlands and Sweden increased their total installed solar collector surface area year on year from the mid-1990s, whilst the cumulative surface area of collectors in the UK and France has fallen consistently over the same period (Systemes Solaires 1999, 2000; EurObserv'ER 2000). Dewick and Miozzo

(2002b) argue that the wide variations in the adoption of ASH systems across Europe can best be addressed by adopting similar initiatives to those successfully implemented by governments in countries such as Germany, the Netherlands and Scandinavia. For example, government may introduce grants and fiscal incentives, channelling funds towards R&D and facilitating economies of learning and experience, beginning with demonstration projects and continued through information dissemination. For instance, Denmark and Germany promote fixed price schemes for renewable technologies and together with the Netherlands and Sweden offer direct capital grant support and tax incentives for renewable energy projects. Denmark and Sweden also offer net metering to encourage small-scale renewable energy production (Thorp 2000). Some European countries offer low interest loans for solar water heating and others, such as Norway, offer lower rate mortgages to buildings that will improve the quality of the built environment (for example, energy efficient buildings, healthy housing, and so on) (Thorp 2000; Gilbert 2000).

SUSTAINABLE INNOVATION AND INTER-FIRM RELATIONS

This section explores the contradictions between policies aimed at facilitating the adoption of sustainable technologies and processes and the barriers inherent in inter-firm relationships in the construction industry. Unlike in many other industries, innovations in construction are not implemented within construction firms themselves but on the projects on which firms are involved (Winch 1998). The interactions and interdependencies between organizations (including contractors, government, clients, designers, subcontractors, suppliers and tenants) have an important role in shaping the process of production and innovation. The successful adoption of innovations depends, in part, on the efficient and co-operative functioning of the whole network. This means that the management of innovation in construction is complicated by inter-firm co-ordination and demands negotiations along the building chain. Previous research has shown that explicit consideration of implementation activities in construction firms can significantly improve both the innovations and the degree to which they can be used effectively within the construction industry (Slaughter 1993, 2000).

During the 1990s there have been a number of initiatives in the construction industry, particularly in the UK, aimed at promoting closer inter-organizational relations, with the aim of facilitating the successful implementation of innovation and especially sustainable technologies. However, there remain important tensions and contradictions between the interests of the different parties involved in the construction process that may militate against the achievement of these objectives. Concentrating on the role of the client as the one responsible for specifying and, more importantly given capital costs considerations, funding the technologies, this section examines the tensions between the client, the contractor and the design team in facilitating the adoption of sustainable technologies. The evidence presented is based on work in Dewick and Miozzo (2004), based on extensive interviews with clients, contractors and designers (architects and engineers) working on social housing projects using sustainable technologies in Scotland. Public sector housing projects provide a good forum within which to demonstrate new technologies: innovative projects are generally more expensive, for the reasons outlined above, and social housing projects can channel higher funding into a higher building specification. But if the technologies can be prototyped, their use in subsequent projects should unlock future economies, stimulating the market for sustainable construction services (for example, contractors, designers, consultants and suppliers).

Traditionally contractors are selected by competitive tendering on a lowest cost basis and the relationship with the client is characterized by a lack of communication, trust and co-operation (Miozzo and Ivory 2000). Following initiatives promoted at the national (for example, the Latham and Egan reports) and local level (Scottish Homes 2000c), the introduction of alternative procurement forms has made the relationship between clients (that is housing associations) and contractors more important. Although in 2000 the majority of contracts were still procured through a traditional tendering route, many housing associations had prototyped innovative procurement forms, including traditional 'off-the-shelf turnkey', 'design and build' and 'negotiated design and build' (mentored partnering). The adoption of alternative procurement methods by housing associations had been gradual, characterized by early collaborations with 'trusted' contractors. Evidence presented in Dewick and Miozzo (2004) suggests that even with good client-contractor relations, experiences of the housing associations were mixed. Benefits in terms of cost certainty (guaranteed price) were tempered by a loss of control.

In terms of innovation and alternative procurement forms, one important benefit acknowledged by the housing associations and the contractors alike was the earlier involvement of the contractor in the building process. Adding the contractor's construction expertise to the design and specification detail, the programming and the site management and control was considered important in improving the buildability of the scheme. Without the contractor's presence the design team may overlook or give less priority to issues that may have a significant impact on cost, such as the contractor's space requirements.

However, despite the advantages unlocked by alternative procurement forms, housing associations acknowledged enduring tensions with contractors, reflecting their contrasting non-profit and profit motives (and, more generally, low industry profits margins), which, in their opinion, adversely affected innovation. It was the opinion of those interviewed that the contractor was prepared to negotiate on areas of specification (for instance, in a design development project) only if there was an economic incentive. However, as the above analysis has discussed, projects using innovative technologies are rarely more profitable especially if the project is a demonstration scheme or if the technologies are being prototyped.[15] Long-term relationships were recommended to help overcome conservative tendencies, increase trust between the parties and encourage the adoption of new technologies, particularly sustainable technologies.

Although most other actors in the building chain support the development of long-term relationships, not all parties have welcomed the adoption of alternative procurement strategies, particularly in terms of innovation and sustainability. Architects and consulting engineers interviewed almost exclusively considered that innovation was facilitated by the traditional contract form and stifled by 'design and build'. Architects argued that the adoption of technologies can only be generated by the more traditional forms of contract where, with a full remit he can explore innovation, discuss ideas with the client and consider the cost implications before going to tender. On a 'design and build' basis, the architects interviewed argued that innovation is hindered because the contractor is more in control of the end costs and will try and make it easier for themselves by omitting some of the harder (that is more innovative) aspects. Also, architects interviewed expressed the view that 'design and build' procurement inhibits the implementation of sustainable innovation because contractors are happy to build to minimum regulatory standards. Therefore, under a 'design and build' arrangement the contractor contrasts the building regulation requirements with the architect's innovative specification (which is above the building regulations) before diluting the specification to meet the regulatory requirements. In this sense, so-called 'buildability' crowds out innovation.[16]

Architects and engineers argue that their lead role in the construction process is one of the more attractive characteristics of traditional procurement, particularly in terms of implementing sustainable innovations. This is particularly important for sustainable technologies, which, because of the specialized knowledge they require, demand capabilities more likely to be found in an architectural or engineering practice. Their greater involvement, earlier in the process, can be very influential and although it is possible outside traditional procurement practices, it requires integration of the construction team where contractors and sub-contractors work closely with the architect and engineer. Housing associations endorsed this view believing that innovative projects were better procured under a more traditional form.

So whilst there is a lack of agreement regarding the preferred procurement form, there does seem to be agreement among the different parties that long-term relationships are important for the introduction of sustainable technologies since they foster trust, stability and economies of learning and experience. Indeed, policy makers have embraced the idea of facilitating innovation through the development of closer long-term inter-firm relations. Communities Scotland (previously Scottish Homes), the national housing agency in Scotland, have endorsed concepts of formal and informal partnering arrangements including 'project partnering' (in one-off projects) or 'strategic partnering' (in multi-phased projects) (Scottish Homes 2000c) through which closer ties are established between housing associations and contractors, architects, engineers and other parties in the construction process. These policy initiatives lend themselves to sustainable projects since technologies often stem from upstream suppliers with whom housing associations do not tend to get involved.

The results from our study show that many of the different interests of the parties in construction could be reconciled if there was more specific funding channelled towards integrating innovative products (for example, through changing procurement criteria to encourage product differentiation and wider technology adoption) and processes (for example, through promoting modernized production methods) and to establishing procedures to assess these innovations. The different organizations interviewed argued that it would help if public funding bodies had clear and different consideration of costs and time in projects using sustainable technologies from those using incumbent technologies. This re-enforced the view expressed earlier in the chapter that sustainable technologies often need support to make them economically viable. If fiscal incentives are not available then consideration must be given to the environmental and social aspects of adoption. At the firm level, other measures may help reduce the tensions surrounding the implementation of sustainable technologies. First, it may be useful to include the project coalition at an early stage in the construction process. Second, replicating demonstration projects may diffuse learning and experience more easily, especially if the same construction team is used. These practices may reduce some of the tensions between the different aims of the various parties in the construction process and may help to overcome some of the barriers to the achievement of policy aims of sustainability.

CONCLUSION AND POLICY IMPLICATIONS

This chapter began by highlighting the increase in the final energy consumption of the domestic sector and identified the main components of household energy use as space and water heating. An important factor in reversing the trend of increasing energy consumption is the diffusion of energy efficient technologies. However, because of the peculiar nature of the construction industry, its final product and its operating environment, the diffusion of sustainable technologies is complex and relies on a number of institutional and cultural factors that either facilitate or inhibit the process. The roles of government and of inter-firm relations play an important part in promoting the adoption of sustainable technologies.

First, the findings of the research on thermal insulation technologies suggest that without an economic incentive or a regulatory pressure to do so, sustainable technologies are unlikely to be adopted on a large scale. Moreover, questions can be raised as to whether those technologies marketed as sustainable (or 'natural') reduce emissions over the lifetime of the product. Regulation is needed to increase the minimum standards and increase the energy efficiency of the domestic sector by reducing the amount of energy required to heat buildings. It is important to existing properties as well as new build in the new regulations. In the UK, new build housing only increases the existing housing stock by 1 per cent per year and it is estimated that houses built before 1990 need to decrease heat loss by over 50 per cent to meet current regulations (Harper 2000). Therefore, to make a significant impression on domestic sector energy demand, the standards need to be applied to retro-fit properties as well.

Second, for more high-tech innovations, such as active solar heating (ASH) systems for water heating, the higher capital cost and additional risk considerations contribute to its relatively slow adoption. National governments across the EU have used fiscal incentives, facilitated inter-industry collaboration, supported generic R&D and demonstration projects and marketed the technology to the wider public to encourage its widespread diffusion. The role of information and government backing cannot be overstated. Governments seeking to facilitate the further diffusion of the ASH systems need to recognize the difficulties in the implementation process and the need for learning about sustainable technologies across the construction industry and across the wider public. Therefore, another important area for government action is in increasing the capacity of construction firms to identify appropriate sustainable technologies and evaluate their potential costs and benefits.

Third, the experience of inter-organizational networks in the Scottish social housing sector endorses the findings that both regulation and other institutional factors play an important role in facilitating the diffusion of sustainable technologies. Without either, there is little incentive for the contrasting aims of those operating within a firm network to align their interests. The existing regulations and the economic costs of implementation hinder the use of additional thermal insulation. This has meant that alternative insulation technologies or higher specification insulated houses have only been built within demonstration projects where additional funding has been made available, either from national governments or the EU. Similarly, the use of ASH systems has been restricted to demonstration projects. One of the most serious bottlenecks in this process identified through the research was that the technologies used in the demonstration projects have not fed into a general specification for new build and retro-fit housing; thus, economies of learning and experience are being lost.

More generally, long-term inter-firm relations facilitate the adoption of sustainable technologies but there are different views among the construction parties as to which is the most appropriate alternative forms of procurement. Despite the strong policy guidance and support from the funding body in promoting sustainable technologies and closer inter-firm relations, it was not sufficient condition to realize either. In fact, it was the very characteristics of the network form of the construction industry that appeared to conspire against innovation.

In summary therefore, regardless of the wider policy initiatives to promote sustainable technologies and processes, government and industry need to understand the bottlenecks inherent in the network alliance of firms that characterize the construction industry. The different aims of the parties involved in the network may not be easily reconciled and traditional approaches to construction may reinforce these differences hindering the efforts to introduce innovation, and particularly sustainable technologies, in construction networks and projects.

NOTES

1. Although EU domestic sector energy consumption per square metre fell by 8 per cent between 1985 and 1997, final energy consumption increased by 4 per cent between 1985 and 1998 (EEA 2001a; EEA 2001b). One can attribute this anomaly to an increased number of households (up 19 per cent between 1980 and 1995), higher average size of dwellings (up by 5 per cent between 1985 and 1997), less persons per household on average (down 12 per cent between 1980 and 1995) and falling electricity prices (down by 1 per cent per year between 1985 and 1996) (EEA 2001a, 2001b; Enerdata 1999).
2. In Sweden, for example, the government subsidizes municipalities (many of which have their own energy companies) to implement measures that reduce the environmental impact, use energy more efficiently and promote the use of renewables and recycling (Kerr and Allen 2001). In Denmark, high electricity prices (maintained through the levy of additional energy taxes, including taxes to fund an Energy Savings Trust) have

encouraged substitution of electricity for alternative energy sources to heat space and water, for example district heating (UNFCCC 1999). In Germany and Denmark, the government has legislated to guarantee price levels for electricity sourced from renewable energy (EEA 2001a). More generally, national policy has been geared towards improving the energy efficiency of buildings and (at a European level) the electrical efficiency of appliances (Kerr and Allen 2001).

3. Education campaigns to further awareness both in industry and amongst the public have been used effectively across Europe (for example, the Energy Efficiency Best Practice Programme for firms and the 'Are you doing your bit?' public information campaign in the UK) in addition to improved fuel efficiency policies (for example, low sulphur fuel for use in high efficiency boilers in Germany) and energy efficiency Eco-labels (for example, the increased market share of the most energy efficient products bearing the 'A' EU label in Denmark) (Dewick and Miozzo 2002b; Kerr and Allen 2001; EEA 2001a). Through the provision of social housing, the government can forge a sustainable path, driving down the cost of adopting energy efficient technologies by bulk buying technologies (ETSU 1991a).
4. Contractors argue they could provide environmentally efficient buildings but complain that the developers do not specify them. Developers argue they would like to specify more environmentally efficient buildings but investors will not pay for them. Investors argue that they will not pay for these because there is no demand from client occupiers to justify them (Cadman 1999).
5. For example, one finding of the reports was that traditional contractual arrangements between collaborative parties within construction projects inhibited the identification and implementation of new products and processes through mutual distrust, lack of communication and time and cost constraints. This finding could also apply to addressing the requirements of a more sustainable construction industry. For example, appointing an integrated design team at an early stage in the construction of 'green buildings' (Sorrell 2001) and using alternative procurement strategies to address sustainable development issues such as higher environmental standards, eco-design principles and lifecycle implications (Pollington 1999) may help to forge a path towards sustainable building and regeneration.
6. An insulation material's weight, strength to weight ratio, convective heat loss, settling and loss of insulating capacity, thermal and vapour resistivity, water absorption properties and resistance to moisture transmission and fire credentials are also important (Caleb 1997; EREC 1995). Only when one considers all these properties is one able to say whether a material could be deemed a suitable substitute. For example, thermal insulation made of cellulose or cork are not appropriate for brick cavity wall insulation but are widely used in breathing timber frame construction or roof cavities.
7. To calculate the life cycle cost of a material, capital costs must be considered alongside maintenance costs, the materials' availability, installation costs and forecast life span. One can also calculate the cost of the material in terms of its triple bottom line, where environmental and social costs must be considered alongside the economic cost. This should include the externalities generated in both the production and use of the materials and consider the liability and risk issues involved with the safety of those who build, use or occupy the building.
8. The European market for insulation materials in 1995 was dominated by mineral fibre (for example, glass mineral wood and rock mineral wool) and plastic foams (for example, extruded polystyrene foam), which accounted for, for example 90 per cent of the market share in the UK and 85 per cent of the market share in Germany (Caleb Management Services 1997).
9. At the Kyoto conference 1997, a paper submitted to the conference by the international insulation associations (including EURIMA (European Insulation Manufacturers Association) and NAIMA (North American Insulation Manufacturers Association)), estimated that thermal insulation has a ratio of energy saving to energy investment of 12 to 1 per year.
10. Heath (1999) compared the performance of phenolic insulation, rigid urethane insulation, XPS, EPS and Rock mineral wool at constant U-values and constant thickness over

a 50-year life span. The results confirmed that energy savings over the life span of the building dwarfed embodied energy savings, that environmental and financial motives were consistent and that plastic foams performed far better than fibrous insulants. For example, at a constant thickness of 77 mm, over 50 years, the various insulants had net energy benefits ranging from 446 000–485 000 kWh, compared with embodied energies of the insulants that ranged between 2.2–6.1 thousand kWh. In terms of their environmental impact, the difference in the materials performance over 50 years relates to benefits between 127–139 tonnes of CO_2 emissions, the difference in their embodied energies relate to 0.6–1.7 tonnes of CO_2 emissions. In financial terms, the savings produced over the 50 years by using the insulants ranged between £9321–£10 394. In all the tests phenolic foam was the best insulant and rock mineral fibre the worst. Over 50 years, phenolic foam's net energy benefit was 39 000 kWh better than rock mineral fibre, saving more than 11 tonnes of CO_2 equivalent emissions with a financial saving of £1073.

11. Glazed flat plate water collectors work by pumping water between a transparent cover and a black plate with high thermal conductivity properties. Although there are many potential uses, the vast majority of glazed solar collectors are installed for individual water heaters: in 1990, 85 per cent of all installed glazed solar collectors installed were intended for individual water heaters. Other less common applications are for combined sanitary and hot water systems, accounting for 5 per cent of all glazed solar collectors, and thermal solar plants (where hot water is stored during the hot periods and used across local districts in cold periods), accounting for 1 per cent.
12. Fourteen European countries and the European Commission are involved in the SHC programme agreement alongside Australia, Canada, Japan, Mexico, New Zealand and the US. The research is task-based with individual countries funding and conducting their own work within particular tasks.
13. Similar inter-industry arrangements in the Netherlands have been boosted by national ASH promotion schemes such as BelDeZon (Call the Sun) and Ruimte voor Zonnewarmte (Space for Solar) (Van der Leun 2001).
14. For example, projects have been funded in the UK, Germany, the Netherlands, Denmark and Sweden under the European Commission's THERMIE and Ecorenewal programmes, Solar Housing through Innovation for the Natural Environment (SHINE) and Solar Urban New Housing (SUNH) projects.
15. One project from the case study, for example, which included a passive solar design, the use of sustainable resources (for example, borate treated timber, natural water based paints) and other energy efficiency measures (for example, higher insulation, condensing boiler, low energy lighting) increased capital costs by 25 per cent.
16. Those firms interviewed advocated the rationale proposed in the third section of the chapter that regulation is needed to increase the energy efficiency of buildings.

REFERENCES

Blackley, D.M. and E.M. Shepard (1996), 'The diffusion of innovation in home building', *Journal of Housing Economics*, **5**, 303–22.

Bordass, B. (2000), 'Cost and value: fact and fiction', *Building Research and Information*, **28**(5/6), 338–52.

Bosselaar, L. (2001), 'Solar heating – a major source of renewable energy', *Renewable Energy World*, July-August, 219–29.

Building (2000), 'War on waste', *Building*, September, pp. 10–11.

Cadman, D. (1999), 'Environmental audits of the construction industry: what they show and how they can be broadened and acted upon', proceedings of the Construction Confederation Conference: Constructing a Sustainable Environment, Birmingham.

Caleb Management Services (1997), *Thermal Insulation and its Role in Carbon Dioxide Emission Reduction*, Caleb Management Services, Bristol.

Construction Resources (2000), 'Construction resources product data: natural insulation', accessed at www.econconstruct.com/htmlpdf/insulation.htm.

Curwell, S.R. and C.G. Mach (1986), *Hazardous Building Materials*, London: E&FN Spon.

Curwell, S.R., C.G. Mach and D. Venables (1990), *Buildings and Health – The Rosehaugh Guide to the Design, Construction, Use and Management of Buildings*, London: RIBA Publications.

Dewick, P. and M. Miozzo (2002a), 'Sustainable technology and the innovation-regulation paradox', *Futures*, **34**, 823–40.

Dewick, P. and M. Miozzo (2002b), 'Factors enabling and inhibiting sustainable technologies in construction: the case of active solar heating systems', *International Journal of Innovation Management*, **6**(3), 257–76.

Dewick, P. and M. Miozzo (2004), 'Networks and innovation: sustainable technologies in the Scottish social housing sector', *R&D Management*, 34(3), 323–34.

Department of Employment, Transport and the Regions (DETR) (2000), *UK Climate Change Programme*, London: DETR.

Doran, D.K. (1992), *Construction Materials Reference Book*, Oxford: Butterworth-Heinemann.

Egan, J. (1998), *Rethinking Construction: Report of the Construction Task Force*, London: HMSO.

Enerdata (1999), 'Energy efficiency in the European Union 1990–1998', accessed at http://193.54.191.189/SiteOdyssee/rapfra.pdf.

EREC (1995), 'Consumer energy information: loose fill insulation, energy efficiency and renewable energy network', *US Department of Energy*.

ETSU (1999a), *Large Scale Solar Purchasing*, ETSU: ETSU S/P3/00266/REP.

ETSU (1999b), *Active Solar Heating System Performance and Data Review*, ETSU: ETSU S/P3/00270/REP.

EurObserv'ER (2000), 'EurObserv'ER barometer', *Renewable Energy Journal*, **10**, 32–41.

European Environment Agency (EEA) (2001a), 'Europe's Environment – the Dobris Assessment', Chapter 26, in 'Households', accessed at http://themes.eea.eu.int/sectors_and_ activities/household/reports.

European Environment Agency (EEA) (2001b), 'Europe's environment – indicators', accessed at http://themes.eea.eu.int/sectors_and_activities/household/indicators

Gann, D. (1994), 'Innovation in the construction sector', in M. Dodgson and R. Rothwell (eds), *The Handbook of Innovation*, Aldershot, UK and Brookfield, US: Edward Elgar.

Gann, D. (1999), 'Can regulations promote construction innovation?', CRISP Commission, accessed at www.ncrisp.steel-sci.org/publications/show_single_publication.asp?id=11096&PG-1&MODE=9.

Gann, D. and A. Salter (2000), 'Innovation in project-based, service-enhanced firms: the construction of complex products and systems', *Research Policy*, 29, 955–72.

Gilbert, J.D. (2000), 'Scotland the brave: innovation in housing', accessed at www.johngilbert.co.uk.

Harland, E. (1993), *Eco-Renovation: The Ecological Home Improvement Guide*, Darlington: Green Books.

Harper, D. (2000), 'Emission impossible?', *Housing Association Building and Maintenance*, **8**(5), 32–3.
Heath, P. (1999), 'Energy in the balance', *Local Authority Building and Maintenance*, **15**(2), 13–14.
Institute of Building Control (1998), *Review of European Building Regulations and Technical Provisions: Germany*, Epsom, UK: Institute of Building Control.
Jaffe, A.B., S.R. Peterson, P.R. Portney and R.N. Stavins (1995), 'Environmental regulations and the competitiveness of US manufacturing: what does the evidence tell us?', *Journal of Economic Literature*, **33**(1), 132–63.
Kerr, A. and S. Allen (2001), 'Climate change: North Atlantic comparison, the Scottish Executive Central Research Unit', accessed at http://www.scotland.gov.uk/library3/ccna-00.asp.
Latham, M. (1994), *Constructing the Team: Joint Review of Procurement and Contractual Arrangements in the United Kingdom Construction Industry*, London: HMSO.
Malin, N. (2000), 'The cost of green materials', *Building Research and Information*, 28(5/6), 408–12.
Miozzo, M. and C. Ivory (2000), 'Restructuring in the British construction industry: implications of recent changes in project management and technology', *Technology Analysis and Strategic Management*, **12**(4), 513–31.
Nam, C.H. and C.B. Tatum (1988), 'Major characteristics of constructed producers and resulting limitations of construction technology', *Construction Management and Economics*, **6**, 133–48.
Pollington, C. (1999), 'Legal and procurement for sustainable development', *Building Research and Information*, **27**(6), 410–12.
Porter, M. and C. van der Linde (1995), 'Green and competitive: ending the stalemate', *Harvard Business Review*, September-October, 120–34.
Pries, F. and F. Janszen (1995), 'Innovation in the construction industry: the dominant role of the environment', *Construction Management and Economics*, **13**(1), 43–51.
RES (1997), 'The European Commission white paper on renewable energies', COM (97) 599, 26/11/97, accessed at www.agores.org/POLICY/COM_STRATEGY/WHITE_ PAPER/.
Scottish Homes (2000a), 'Sustainable development policy', accessed at www.scot-homes.gov.co.uk.
Scottish Homes (2000b), *Sustainable Housing Design Guide: A Handbook for Practitioners*, Edinburgh: The Stationary Office.
Scottish Homes (2000c), 'Procurement and partnering: policy advice note', accessed at www.scot-homes.gov.co.uk.
Slaughter, E.S. (1993), 'Builders as sources of construction innovation', *Journal of Construction Engineering and Management*, **119**(3), 532–49.
Slaughter, E.S. (2000), 'Implementation of construction innovations', *Building Research and Information*, **28**(1), 2–17.
Sorrel, S. (2001), 'Making the link: climate policy and the reform of the UK construction industry', SPRU working paper series paper no. 67, July.
Systemes Solaires (1999), 'Fin d'eclipse', *Systemes Solaires*, **133**, 1–10.
Systemes Solaires (2000), 'Thermal solar barometer', *Systemes Solaires*, 138, 85–91.
Thermal Insulation Manufacturers and Suppliers Association (TIMSA) (2000), *Insulation Industry Handbook 1999/2000*, TIMSA.
Thorp, J.P. (2000), 'Tax modification: sustainable energy and society', in

Proceedings of Conference C75 of the Solar Energy Society: Renewable Energy for Housing, Oxford: The Solar Energy Society, pp. 39–45.

United Nations Framework Convention on Climate Change (UNFCC) (1999), 'Report on the in-depth review of the second national communication of Denmark', UNFCC.

Van der Leun, K. (2001), 'Soltherm Europe Initiative: joining forces to expand solar markets, fast', *Renewable Energy World*, September–October, 127–37.

Weitzman, M.L. (1997), 'Sustainability and technical progress', *Scandinavian Journal of Economics*, **99**(1), 1–13.

Welford, R. and R. Starkey (1996), *The Earthscan Reader in Business and the Environment*, London: Earthscan.

Winch, G. (1998), 'Zephyrs of creative destruction: understanding the management of innovation in construction', *Building Research and Information*, 26(4), 268–79.

Woolley, T., S. Kimmins, P. Harrison and R. Harrison (1997), *Green Building Handbook*, London: E&FN Spon.

Wubben, E. (1999), 'What's in it for us? Or: the impact of environmental legislation on competitiveness', *Business Strategy and the Environment*, **8**(2), 95–107.

8. Waste incineration for energy: the experience of China

Yuhong Cen, Xiaodong Li and Sally Randles

INTRODUCTION

China is in the process of rapid modernization and industrialization. As a result the country now faces huge, and exponentially rising increases in solid waste (SW) stocks and flows, including municipal solid waste (MSW)[1] and industrial solid waste. Indeed, China now faces very similar problems to those that have been experienced by developed countries for many years. The increases in solid waste can be directly attributed to the triple processes of urbanization, industrialization and rising living standards. The problem is exacerbated because waste disposal facilities across the regions of China are inadequate and technically outdated. Our chapter describes the nature and history of, specifically, the MSW problem in China. It discusses a number of technological platforms developed to deal with MSW, of which one – incineration – is discussed in some detail. Of particular relevance here is the dual process of waste incineration and the simultaneous production of energy from incineration the coupling of which is central to the development of waste-incineration-for-energy (WIE) technologies.[2] The case illustrates the dynamic process of technological development involving a range of actors and agencies including private interests (corporations), government authorities primarily at the regional level, science and technology institutions, and waste producers (primarily domestic and construction). We provide a detailed description of how one Chinese WIE technology developed and was gradually received (and in some respects resisted) by Chinese authorities in different regions. Theoretically the chapter follows a 'systems of innovation' approach with particular attention paid to how in this case new systems emerged in a context of competitive co-existence with other systems, both existing and new. Alternative technologies and groups of actors addressing the MSW problem emerged, and indeed we discuss how new technologies were taken up and diffused in this developing country case context. Finally the case is briefly revisited from an industrial ecology (IE) perspective to offer some critical reflections on the limits of an IE approach when considering processes of dynamic, endogenous, self-organizing techno-socio change and its acceptance and/or resistance.

THE EMERGENCE OF A MUNICIPAL SOLID WASTE (MSW) PROBLEM IN CHINA[3]

Before China opened its doors to international trade and introduced a form of open liberal market economy, there existed a natural circulation economy in terms of the production and consumption of MSW. Family incomes were low almost universally following the Mao-era philosophy of 'egalitarianism' and the related, politically-driven dismantling of China's class structures. Furthermore, there was forced out-migration from the cities to the land both to reverse previous tendencies for urban growth and in line with a policy of privileging productive agricultural activity. At the time also the economy was underdeveloped, and there was a shortage of commodity (let alone consumption) goods. The Mao policy at the time was to use revenues to build up the foundation of China's industrial base through state-owned production units. Therefore domestic 'waste' items such as food packaging, paper, card, plastics and glass were much lower in volume.

In addition the guiding philosophy of the time which emphasized collective responsibility and prudence (and was captured in the slogan 'plan living and build up China through thrift') governed consumption behaviours. Associated with this principle, there was a used-goods reclamation system established by the state. In addition to re-use within the family, small-scale revenues could be generated for the household from the sale of used bottles, paper, and so on to supplement family incomes. This was a routine practice. Furthermore in rural areas organic waste was generally re-used in family husbandry. Also, levels of construction materials 'waste' from urban building programmes was much lower. Effectively, the production of MSW occurred at a much lower scale and there was little need to consider MSW as a waste 'problem'.

Since China opened her doors philosophically to market liberalism and in trade terms to internationalism, with the assimilation of advanced technology and product systems from the West, the country has followed a dominant Western pattern which has entrained a ratcheting upwards of natural resource use, mass production, mass consumption and disposal.

Meanwhile, the development of the Chinese economy accelerated urbanization. New cities and towns planned by regional governments literally appeared on an annual basis to accommodate rural migration linked to urban economic development and an influx of foreign direct investment.

By the end of 1998, there were 668 central cities in China with a population of 360 million, being 30 per cent of the total population of the country. An MSW problem now became evident, linked to the process of urbanization. The problem can be seen in general terms as the result of the effects of two parallel processes: the growth of the volume of garbage, and the linking of this growth to processes of urban economic development. In terms of the former, there has been an increase in MSW whether measured as total volume or per capita levels. Between 1984 and 2000, the amount of MSW produced per annum increased dramatically from 50 million tonnes to 150 million tonnes with an 8–10 per cent annual average rate of increase. The per capita MSW level in 2002 reached 1.58 kg/day, which exceeds that of an average medium developed country, usually around 1.0 kg/day (Cen *et al.* 2003). So, economic development activities in China not only result in the rapid growth of the Chinese economy and improved living standards, but also in a growth of garbage.

In terms of the latter, the capacity of facilities and infrastructure providers (note that waste disposal and management are the responsibility of local governments) to deal with MSW in China has become increasingly outdated and inadequate in many cities. The Public Environment and Sanitary Department of each local authority organizes and undertakes street cleaning, waste collection, transportation and transfer, waste processing, waste treatment and disposal and all other related functions. A first problem in the management of MSW from the 1980s onwards concerned poor standards of littering and dumping clean-up associated with weak regulation and under-developed management systems. A further difficulty lay in a mismatch between local authorities capacity, skills and facility infrastructures to deal with rising MSW levels, and rapidly rising MSW levels themselves. Collection of refuse was patchy and transport and storage capacities were low and poorly equipped to deal with the rising problem. As a result garbage piled up at the sides of rivers, highways and railway lines, inducing a strong negative reaction from the general public and media.

As a second stage 'solution' to the MSW problem, the practice of transferring garbage from cities to urban fringes became more prevalent. Because of the shortage of waste treatment and disposal facilities, large quantities of MSW have been hauled to the urban fringe to be directly stacked or land-filled in open-air pits in river valleys or swamps which cause pollution in those areas. This response to the mounting 'problem' has created a ring of garbage encircling cities and creating political and social tensions between people and authorities and between urban territories and urban fringe territories: people living and representing the urban and rural areas respectively. The outcome, captured in the phrase 'cities cleaned, but countryside polluted' became, a point of resistance and tension when inhabitants of the urban fringes, primarily low income peasants, started to protest against the identification of landfill sites within, and the import of garbage to, their areas. Furthermore, the problem is reaching crisis levels as data from satellite remote sensing shows, indicating that almost two-thirds of Chinese cities are now surrounded by garbage.

Indeed, for some cities, it became increasingly difficult to locate landfill sites near the cities; a third response is therefore to transport garbage to more remote countryside locations which results in a dramatic increase in the cost of transportation and social conflict once again, this time with countryside inhabitants further away. Altogether, this results in mounting social and economic pressures on governments of the cities and regions.

From the middle of the 1980s, there was a further policy response which took the form of dealing with MSW through the planned expansion of waste treatment plants. A number of MSW disposal facilities were constructed in rapid succession. According to the report of the Chinese Engineering Academy, from 1986 to 1995 the number of plants constructed for MSW disposal increased from 23 to more than 900 with the disposal rate raised from 0.9 per cent to 43.7 per cent of the total volume of MSW. Nevertheless, by the end of 1998, still more than half of untreated garbage was stacked across China, and the stock of wastes that were not properly treated on time exceeded 6000 million tonnes, occupying 55 400 hectares. Controls and standards remain both low and poorly enforced. In 1998, only 2.3 per cent of the total garbage being treated met the National standard GB16889-1997, the Standard for pollution control on landfill sites of domestic waste.

The building and utilization of low-cost, low-tech waste disposal operations has serious consequences. First, large tracts of farmland are appropriated for the disposal of part-treated MSW and the soil in those areas becomes polluted. Using untreated or poorly treated garbage as a kind of fertilizer is harmful. The waste outputs are not well broken down and the residues contain pieces of glass, metal, broken tiles and bricks and so on. The soil is in fact destroyed as a result of mixing it with part-treated garbage because the waste products change the soil clump structure and alters other physical or chemical characteristics of the soil. The result is reduced capacity of water retention and fertilizer retention. Second, serious air pollution is caused. Effluvium/odours, rats and flies appear in dumping areas. More than 100 escaping gases could be potentially generated from open-air dumping. Some of them may, according to some sources, cause foetal malformation or cancer.

Third, the leachate from landfill or dumping contaminates the underground water. In the summer of 1983, dysentery broke out in the areas near to the landfill sites of Guiyang City. It was found that coliform bacteria in the underground water was 2600 times higher than the recommended environmental quality standard for surface water, the National Standard GB 3838-83, 1983. Fourth, accidental explosions occur regularly in landfill sites. With the increase of organic content in MSW and the transition from dispersed pile-up in the open air to gathered stacking with simple covering, methane, an explosive gas under certain conditions, is produced. A series of three explosions in the landfill site of Changpin County of Beijing occurred in 1995 for example. Finally, there is an official estimate that the discharge of methane from MSW every year exceeds a million tonnes in China (Zhang *et al.* 2002). With the growing production of MSW heavy in organic content, the discharge rate of methane may skyrocket in coming decades, inducing serious 'greenhouse' effects.

THE EVOLUTION OF SOLUTIONS TO THE MSW PROBLEM IN CHINA

Figure 8.1 illustrates the municipal solid waste treatment systems and typical waste disposal options used in China up to 2003. In generic terms, they comprise three parts: the collection and transportation system, the disposal system and the reuse or recycling system. Public environment and sanitation departments are in charge of collection, transfer, storage and compacting MSW. A variety of technological platforms has been used for disposal of waste. There are channels for reuse or recycling of waste though these are not well fostered or developed. Notably, new waste will be generated during the reuse, recycle and disposal process. The dashed lines refer to the second generation of waste produced in the process. It will go back to the loop or be incinerated or landfilled. Nevertheless, new products, secondary materials and recovered energy are structurally coupled with the different disposal methods. Though this development is still at a stage of infancy compared with equivalents in some developed countries, this evolution is the outcome of government encouragement, financial incentives, and policy as well as involving the active participation of different actors at a micro-level.

This perspective, which chimes with the philosophy and management practice of industrial ecology, is evident in new governmental policies to encourage comprehensive utilization of resources. The slogan 'comprehensive utilization of resources, transfer waste into resources' captures this new philosophy and was first put forward by the government in the early 1980s. In 1985 a policy was instituted in line with this thinking which included an

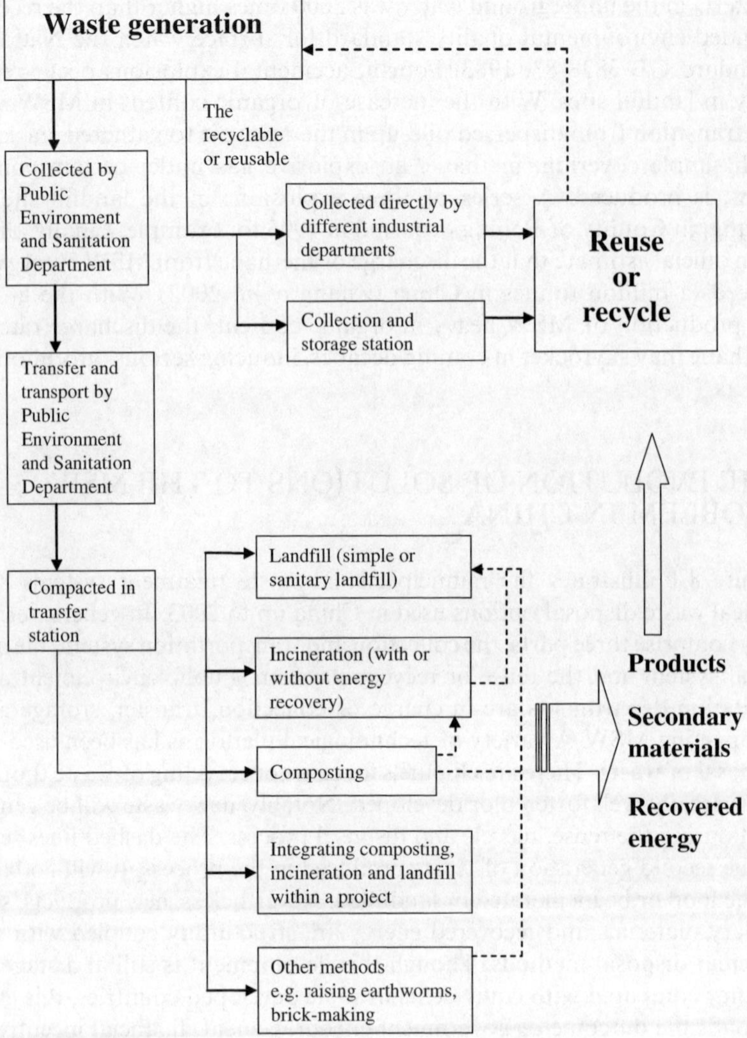

Figure 8.1 The flow chart of waste treatment systems in China up to 2004

'encouragement list'. Issued by the Chinese government, the list initially sought the 'comprehensive exploitation of associated ores in the ore extraction' process. The standard also encouraged the re-use of 'waste' by-products from various production processes, such as a variety of waste slag, liquid, emission, and recovery water, waste heat or pressure from production practices. Used goods arising directly from social consumption processes were emphasized and added later. Essentially, a set of priority resources to be protected have been regulated in favour. The document says 'The main purpose for comprehensive utilization of resources at the time is to reduce dissipation and waste, augment social wealth, gain economic benefits and protect environment'. The policy document further outlines a package of conditional preferential instruments[4] including pre-emptive offers, preferential pricing, and listing of activities in governmental R&D programmes among others. The document is issued to firms to encourage the take-up of the incentives and practices which are seen to underpin the achievement of the policy objective to comprehensively utilize the resources which are on the list.

Up to 2004, landfill was a major waste disposal method constituting about 70 per cent of the total, followed by composting which accounts for less 20 per cent and finally incineration with energy recovery only developed in recent years. The latter disposes of approximately 10 000 tonnes/day of MSW in China, around 3 per cent of the total MSW. The rest is disposed of by a range of other diversified methods, such as earthworm-raising and brick-making, and composting used by peasants in households. Each of these supplementary methods is tailored to very small and specific markets.

Such systems evolved historically and incrementally. The purpose of the following section is to unfold this development trajectory of the three typical waste disposal methods: landfill, compost and incineration.

THE MAIN WASTE DISPOSAL TECHNIQUES IN CHINA BEFORE THE END OF THE 1990s

Simple landfill and composting were the main disposal routes for MSW in China before the end of the 1980s. Even currently, landfill as a major waste disposal method constitutes about 70 per cent of the total, followed by high-temperature compost, which accounts for less than 20 per cent. Landfill sites accepting MSW in China are ranked into three grades, namely 'simple', 'controlled' and 'sanitary' landfill. These vary according to the strength of environmental protection measures regulating a landfill site and whether the site meets the standard for pollution control on landfill sites for domestic use, for example, the National Standard GB16889-1997 and the Professional Standard CJJ 17-2001, 2001, which is the technical code for sanitary landfill of municipal domestic refuse. For example measures in place for a modern sanitary landfill site include lining the site, pre-compacting of waste, intermediate and daily material cover, cut-off drains of rainwater, treatment of leachate, landfill gas collection and pumping extraction and so on. By contrast 'simple' landfill sites have virtually no environmental protection measures. Pollution is inevitable around these kinds of landfill. Controlled landfill provides some protection measures but not sufficient to meet the above mentioned National Standard. Indeed, an investigation[5] into 288 landfill sites in 30 provinces or municipalities found that there were 157 simple landfill sites, accounting for 54.5 per cent of the total 288 sites, and 115 (39.9 per cent) were controlled landfill sites. Only 16 sites met the Chinese sanitary landfill standard. The main sanitary landfill sites were all built after 1987 yet only three of these met the international landfill standard (Cen *et al.* 2003).

The development of composting technologies in China started in the 1950s, based on traditional composting methods used by farmers for orchards or plant nurseries with no specialized composting equipment used in the composting process. With increasing recognition of, and social pressure to deal with, the MSW problem in China from the 1970s, composting gradually received recognition by the Chinese central government. Research and development on composting technologies and associated specialized equipment were listed in the sixth, seventh and eighth five-year plans of key technology R&D Programmes. In the Eighth National Five-year Plan (1990–1995) research on composting became an important theme with the involvement of many research institutes and universities in major cities. Treatment capacity of composting developed from 3713 tonnes across 26 composting sites in 1991 to 5853 tonnes across 32 composting sites in 1997. From 1997 to 2000, 43 more composting sites with a total treatment capacity of 12 110 tonnes/day were planned (Cen *et al.* 2002, p. 22).

Yet, although the sites and total treatment capacity of composting increased, the proportion of solid waste disposed of by composting has decreased compared with earlier years. The rapidly increasing rate of MSW production is one reason, but the defects and problems of composting itself are another. There are several problems, first, technological. The proportion of MSW that is suitable for composting varies from region to region. Indeed the very composition of the MSW varies considerably across and within the regions. Until now MSW in China has generally been collected without source-segregation. This causes difficulties in selecting and combining machinery to form a suitable specialized separation system. A main problem is how to remove plastic films and heavy metals. Because composting technologies are still at a very early stage of commercialization,[6] the design of the systems is not yet standardized due to the absence of specialized manufacturers for the component machines in the composting system. Thus the composting systems used in many projects are *ad hoc* and not reliable. Second, the difficulty of separation consequently results in

MSW that is totally unsuitable for composting methods. Composting products may contain glass, metal or other large additions. Sometimes even, harmful poisons may be mixed in.

In the medium and small cities in Northern China, coal is used as a household fuel. However the ash from coal affects the quality of the compost. The quality of the compost can be improved by using additives. But, the use of additives as well as processes involving many machines for separation increases the cost of composting. Third, the composting product cannot compete with chemical fertilizers because of its higher cost and its slow action on soils compared with organic fertilizers. Fourth, in China, generally, the residue after coming through the composting process is about 20–30 per cent. The residual materials still require further disposal (for example to be landfilled or incinerated). Fifth, composting processes also produce waste and pollutions which need to be carefully controlled.

Composting in China is now recognized as a specific option to deal with MSW with a high organic content, or to use composting in combination with other waste treatment options. Composting is encouraged by official policy[7] as the disposal option to be adopted for MSW with more than 40 per cent compostable material. Indeed currently, several waste disposal plants which integrate composting, incineration and landfill techniques within one plant system have been set up in different regions of China. However, with rapid industrialization more and more poisonous chemicals and macromolecules are being introduced into MSW which increasingly hampers the wide development of the composting option.[8]

THE USE OF WASTE-INCINERATION-FOR-ENERGY (WIE) TECHNOLOGIES SINCE 1988: A HIGH-COST, HIGH-QUALITY OPTION

The development of WIE technologies in China started at the end of the 1980s. The 1990s witnessed rapid development and advancement of these technologies. At present, more than 30 actors, including research institutes, universities and manufacturers conduct R&D in the field. Furthermore this expansion coincides with the changing composition of MSW in Chinese cities. From 1984 to 1996, the annual average proportions of paper, fibre and plastic in Chinese MSW increased by 20 per cent, 20.5 per cent and 48 per cent respectively. Increasing living standards appear to go hand in hand with changing consumption patterns which mirror those of developed countries. Every year about 70 per cent of the packaging materials produced in China are discarded after use.[9] Indeed the volume of discarded packaging makes up 35–40 per cent of the volume of household garbage. Interestingly, as a result of the increasing proportion of waste paper and plastics in MSW in China, the calorific value[10] of MSW increased. The average calorific value of MSW in China is about 3300 KJ/Kg (for example it is about 2670–5060 KJ/Kg in Shenzhen and 2510–4600 KJ/Kg in Shanghai (Chinese Academy of Engineering 2002). It is also estimated that on average, the calorific value in each city will increase by 120 KJ/Kg every year in forthcoming years (Kong and Jiang 2003). The trend of increasing organic, combustible composition in Chinese MSW suggests that this key pre-condition for utilization of available WIE technologies – for example the increased calorific value of the 'fuel' source, is going to rise, which will very likely result in the emergence of a large market for WIE plants and equipment.

From the 1980s, environmental protection became a national policy priority in China. Furthermore a major theme national policy makers developed encouraged multiple and comprehensive utilization of resources. At the time, the Ministry of Construction in China brought forward a technology policy proposing that 'sanitary landfill' and 'thermophilic compost' be considered as two practical technologies and encouraged the development of incineration technologies in certain regions according to their relevance to prevalent conditions. As a result, the first modern WIE plant in China was built in Shenzhen in 1988: a wealthy metropolitan area with a highly developed economy in the southern coastal part of China. The project decision makers selected Martin stoke technology supplied by Mitsubishi Heavy Industries, Ltd, Japan as their preferred equipment.

The government of the Shenzhen metropolitan area is responsible for investing and paying the operations costs as well as initial capital investment of the waste disposal project. The initial contact between the decision making agents and the suppliers can be sourced to 1985 and it took 11 years to finish the whole project. In Phase I of the project, two 150 tonnes/day stoke incinerators and a 500 KW power generator were put into operation in 1992. In Phase II of the project, another 150 tonnes/days stoke incinerator and a 3000 KW power generator plant were commissioned. It is worth noting however that nearly 85 per cent of the equipment of the whole system installed in Phase II was manufactured in China. An effort to facilitate the further uptake of Chinese technology has been made by the authorities by reducing investment costs generally and in particular for this project. The influence of the first WIE project, the Shenzhen project, is far-reaching though the disposal capacity is small compared with many later projects built elsewhere in China in the twenty-first century. A new and potential market emerged, spurred by this project. New and adapted WIE technologies are now being developed or updated by different actors learning from the economic and technological problems illustrated in the Shenzhen project.

THE USE OF WASTE INCINERATION TECHNOLOGIES BEFORE 1999: THE LOW-COST, LOW-QUALITY OPTION

One of the major problems experienced in the Shenzhen project has been its prohibitively high total cost. Even for Shenzhen, one of the wealthiest cities in China, the high operation and maintenance costs of the Mitsubishi system in addition to the high fixed costs for the core equipment became a great burden for the city. It is the only WIE plant in China which uses this particular technology imported from Japan. For most Chinese cities, the Mitsubishi system is way beyond their economic affordability even if the initial capital investment could be reduced by manufacturing the rest of the system equipment in China. One of the major causes of the high operation and maintenance costs is that because the design did not include an MSW separation system situated before the waste enters the incinerator system, blocks and hard substances such as pieces of glass affect the functioning of the conveyor sub-system. The whole incinerator system, consequently, has to be stopped periodically for cleaning up and maintenance.

The Shenzhen Mitsubishi incineration plant was a first for China. Prior to 1999, no other large scale MSW power plants had been installed or operationalized in Chinese cities (see Table 8.1), though some projects were negotiated during the period. The high operational cost has been a prohibitive factor, along with the absence of available viable technology. Importantly however, it was not until this historical juncture that the regulatory trigger operated, in the form of central policy prioritizing the need for metropolitan authorities to deal urgently with the growing MSW problem. However during this period, a competing, technologically simpler and lower-scale solution, in the form of small, hand-fed incineration plants supplied by Chinese boiler manufacturers had been introduced in some places. These small plants had no energy recovery capability, were low-tech and poorly designed, and had poor emissions monitoring and control capabilities resulting in pollution problems. Inevitably their operation produced a poor impression and image of incineration among local governments and people living in the vicinity of the plants.

THE USE OF WASTE-INCINERATION-FOR ENERGY (WIE) TECHNOLOGIES AFTER 1999: THE OPTION SUPPORTED BY PUBLIC, HIGHER EDUCATION RESEARCH INSTITUTES

The institutes which comprise the Chinese Science Community[11] and which study thermal power engineering, have paid attention to the waste disposal problem in China since the beginning of the 1990s. The Institute for Thermal Power Engineering of Zhejiang University (ITPE) is one of the first movers in this area. Two others are the Thermal-Physical Engineering Institute of the Chinese Academy of Science (TPEI of CAS) and Tsinghua University. All of these public research institutes focus on the development of fluidized bed technology.[12] The development and subsequent commercialization of domestic WIE technologies has developed at an astonishing rate since the operation of the first WIE plant in Yuhang in 1999. The technology from ITPE has secured a significant market share, nearly 40 per cent of the total market of large scale MSW power plants (Table 8.1) in 2003. As early as the beginning of 2002, many scholars in the academic circle have held the view that Chinese fluidized bed technology is at the point of scaling-up in terms of R&D spend on this particular technology to address immediate and future pollution concerns and commercial opportunities. By the end of 2003 the technology had demonstrated its diffusion potential with seven WIE plants built and supplying electricity (Table 8.1) and another five at the design or construction stage. During the same time, some big cities have chosen imported technologies from different overseas suppliers, which they assumed to be of superior design, leaving the domestic manufacturing proportion much reduced.

Between 1988 and 2004, more than 60 MSW incineration power plants using WIE technologies had either been constructed or were listed in local development plans in China.

Up to now, WIE treatment has been viewed as a primary solution for MSW disposal in many Chinese cities. However in 2000, a discernible shift in policy positioning could be identified away from two practical technical solutions (composting and landfill) and towards a range of possible solutions suited to local conditions across the cities and regions of China. So, the Ministry of Construction issued a revised technology policy. It states that sanitary landfill, incineration, composting and energy recovery from waste technologies and equipment all have their corresponding applicable situations. Any one of the options or any combination of the options should be selected according to the principle of meeting local conditions and requirements. The principles of feasibility of technology, reliability of equipment, reasonableness in terms of scale and capacity, and comprehensiveness in terms of treatment and utilization were put forward.

Table 8.1 MSW incineration power plants using WIE technologies in China up to 2003

No.	Year	Site	Technology and its supplier	Power capacity (MW)	MSW capacity (t/d)
1	1988 1992*	Shenzhen	Martin stoke, supported by Mitsubishi of Japan	4	3 × 150
2	1993# 1998 2000*	Zhuhai, Guangdong Province	Ladder mechanical grate, supported by US Detroit Co.	12	3 × 200
3	**1999**	**Yuhang, Hangzhou, Zhejiang Province**	**Circulating fluidized bed, supported by ITPE of Zhejiang University**	**6**	**1 × 150**
4	1999	Pudong, Shanghai	Inclined to-and-from ladder mechanical grate, supported by France Alstorm Co.	17	3 × 365
5	**1999 2002***	**Qiaosi, Hangzhou, Zhejiang Province**	**Circulating fluidized bed, supported by ITPE of Zhejiang University**	**12**	**2 × 200 + 1 × 300**
6	1999# 2001	NingBo, Zhejiang Province	Hydraulic ladder mechanical grate, Germany Novel Co.	12	3 × 350
7	1999	Longgang, ShenZhen	Rotary kiln pyrolysis and afterburning, supported by Richway Co. of Canada	6	1 × 300
8	11, 2000	Dongzhuang, WenZhou,	HWM double tiers to-and-fro	15 + 30	160 + 225

Table 8.1 (*continued*)

No.	Year	Site	Technology and its supplier	Power capacity (MW)	MSW capacity (t/d)
		Zhejiang Province	furnace grate, supplied by Wenzhou Wei Ming Environment Protection Engineering Co.		
9	**2001**	**Heze, Shangdong Province**	**Circulating fluidized bed, supported by ITPE of Zhejiang University**	**24**	**3 × 350**
10	2001	Shaoxing, Zhejiang Province	Circulating fluidized bed, supplied by ChongLiang Co., supported by Chinese Academy of Science (CAS)	12	1 × 400
11	2002	Haerbin	Circulating fluidized bed, imported from EBARA Co., Japan	6	1 × 200
12	**2002**	**Zhengzhou, Henan Province**	**Circulating fluidized bed, supported by ITPE of Zhejiang University**	**24**	**3 × 350**
13	**2002**	**Wuhu, Anhui Province**	**Circulating fluidized bed, supported by ITPE of Zhejiang University**	**12**	**2 × 200**
14	**2003**	**Yiwu, Zhejiang Province**	**Circulating fluidized bed, supported by**	**12**	**2 × 200**

Table 8.1 (*continued*)

No.	Year	Site	Technology and its supplier	Power capacity (MW)	MSW capacity (t/d)
			ITPE of Zhejiang University		
15	2003	Jia Xing, Zhejiang Province	Circulating fluidized bed, supported by ITPE of Zhejiang University (supplied by Jing Jiang Group Co.), and CAS (supplied by Zhong Ke Co.)	12	2 × 250 from CAS + 1 × 250 from ITPE later
16	2003	JiangQiao, Shanghai	Ladder mechanical grate, supported by Germany Steinmuller Co. and Spain	24	3 × 500
17	2003	Lingjiang, WenZhou, Zhejiang Province	HWM double tiers to-and-fro furnace grate, supplied by Wenzhou Wei Ming Environment Protection Engineering Co.	12	600 (3 × 225)

Notes:
1. Plants in bold use Chinese indigenous technologies.
2. Years marked # refer to the year when the project started to be negotiated; Years marked * refer to the years when the plants were put into operation, which are not in the same year that the plants are constructed.

To provide some examples of this 'horses for courses' policy approach, sanitary landfill can be seen as the primary choice for a city if there is abundant land and other natural conditions suitable for landfill, whilst incineration can be used in a city where economic standards of living attach to high use/wastage of organic combustible materials such as packaging which provide high calorific value of MSW and where there is no site for landfill.

Furthermore, development of biological treatment technologies and their integration and utilization alongside other options are encouraged. Fortunately, dumping and stacking without control are now prohibited.

WIE IN CHINA – ACHIEVING THE TWIN OBJECTIVES OF WASTE REMEDIATION AND FUEL SUPPLY

Calculated according to average calorific values of current Chinese MSW, every 5 tonnes of Chinese MSW provides the same calorific value as 1 tonne of coal. If one third of MSW could be incinerated for energy recovery, then based on an assumption of 150 million tonnes of MSW produced every year in China, this means that 10 million tonnes of coal can be substituted by MSW so conserving coal reserves.

Moreover, MSW generated in each city being directly transformed into fuel for energy supply saves resources directly and indirectly, and is one of the methods of exploitation of the highest resource value. Especially, it saves the transportation fee[13] for the coal that has been substituted, and reduces the pollution that comes from the process of transportation and using the coal for electricity generation.[14] Currently, incineration of 1 tonne of MSW in China can generate around 200 to 300 KW of electricity. Thus, the costs of MSW disposal can be partially compensated for by revenues earned from the sale of electricity, which is an important advantage for this MWS disposal method.

But it is also relevant to note that currently WIE plants are supported by government through subsidized pricing, tax holidays and other preferential policies. Without the subsidy, WIE plants cannot compete with other types of power plants. Kong and Jiang (2003) have made a comparison of the cost and unit investment between different sources of power generation[15] in China (Table 8.2). Kong and Jiang's data show that the investment required to support a WIE power plant is 5–10 times higher than that of a thermal power plant using coal as fuel. But MSW is a fuel free of charge[16] whereas there is a cost for coal. Kong and Jiang compared the fixed cost of different power plant projects based on 20-year plant life.

They conclude that WIE power plants cannot compete with a normal coal-fuelled thermal power plant if there is no governmental subsidy or compensation for waste disposal, even though the fixed capital cost of a sample fluidized bed WIE power plant using ITPE's technology is approximate to that of a normal coal power plant considering the costs of coal[17] in 20 years. If the variable cost of a WIE power plant is considered, it will be much higher than a normal coal power plant. For example, lime powder and

Table 8.2 Comparison of unit investment between different sources of power generation (Dollar/KW)

Type	Thermal Power (Coal)	Nuclear Power	Hydro Power	Wind Power	WIE Power
Unit Investment	$650–$760	$1445	$840–$1200	$840	$3600–$7230

Note: Adopted from (Kong and Jiang 2003) with 1 dollar = 8.3 RMB.

other chemical substances have to be added in the combustion system during the operation for the treatment of toxic and corrosive gases produced in the combustion process. All WIE technologies use a certain proportion of auxiliary fuel in addition to MSW due to the low calorific value of Chinese MSW. If the calorific value of the original MSW is not sufficient, additional procedures such as separation and pile-up after collection are carried out. This increases the operation fee of a WIE power plant.

According to the ITPE, the waste disposal fee for using their WIE technology is around six dollars per tonne. Simple landfill is, of course, much cheaper than this. Under current policy, the electricity price for each WIE power plant is different from project to project and is negotiated between the developer and the local government authority.[18] Importantly, budgetary pressures on local authorities hinder their capacity to grant subsidies and concessions, hence exerting a dampening effect on the development of the industry. In July 2002, the Chinese government issued a policy to enact a waste disposal charge directly on households. Later that year, two other policies to nurture the waste disposal industry and to restructure and to privatize the state-owned waste disposal plants were enacted. However, it is difficult to fully implement and monitor such policies in most Chinese cities. So, to summarize, though the witnessed reform of waste disposal and incineration activities in China is not smooth, the transformation of the Chinese waste management system is occurring.

THE SEGMENTATION OF CHINESE WIE MARKETS AND THE IMPLICATIONS FOR INNOVATION

At present, the WIE technologies utilized in China can be grouped into three general types: grate incineration, fluidized bed incineration, and rotary kiln pyrolysis and afterburning incineration. Each of these has a variety of suppliers from domestic as well as from foreign countries. This is because WIE technologies provide complete systems and yet technologies of the same family come from different suppliers and exhibit wide and fundamental variety in terms of the engineering concepts and design principles encapsulated in the sub-systems and/or components within the WIE system. In addition to this technological diversity, we witness variety in spatial terms. For example, grate incinerators are mainly adopted in big cities, while circulating fluidized bed incinerators are more widely adopted in mid-sized cities with a proportion of near 50 per cent: 50 per cent, grate: fluidized bed technologies at the end of 2003. We now unpack the detail and explanation of this variety.

Six points contribute to this phenomenon. First, grate technologies imported from western countries are regarded as mature technologies, while circulating fluidized bed technologies were mainly developed indigenously during the 1990s. The level and depth of knowledge about circulating fluidized bed WIE technologies at the time was still developing. Yet, because the new knowledge was developed by new entrants to the waste management industry, the existing specialists in the industry and relevant government department did not fully understand it and resisted the new knowledge, while specialists of fluidized bed WIE technology were not accepted to have a voice in the waste management circle. As a consequence, the new technology is not accepted as a practical technology listed in the technological policy guide which indirectly influences the selection and approval of WIE projects, with the statement in that policy that 'caution and prudence should be paid in choosing systems other than those using grate technologies'. Guided by this policy, local governments or state companies tended to select the more mature and presumed less risky grate technology.

Second, the variety reflects the different features and composition of MSW between big cities and medium or small size cities. As the composition of MSW is mainly influenced by size of city, geographic circumstance, lifestyle and the living standards of residents (and therefore the type of fuel that people use), the organic and inorganic fractions of MSW in big Chinese cities are about 31–36 per cent and 60 per cent respectively; while they are 20 per cent and 65 per cent respectively in medium or small size cities (Cen *et al.* 2002). A comparison between different regions of China also shows that there are more organic components in the MSW from cities in the southern part of China than that from the cities in the northern part of China due to different lifestyles and household fuel used (*ibid.*). More combustible components in the MSW of big cities such as Shenzhen, Zhuhai and Shanghai mean they were the first sites where the grate technologies could be used.

Third, local governments of big cities are more wealthy and have more financial resources to invest in WIE facilities since as discussed above, it is municipal governments who both provided the main investment source and took responsibility for the operation of MSW projects, as part of the management system formed during the 1980s. Fourth, big cities require a bigger MSW disposal capacity. The disposal capacity of the imported grate incinerator is bigger than that of the circulating fluidized bed incinerator which by contrast was at the scale-up stage of technological development at the beginning of the twenty-first century and was not able to deal with large scale disposal as its imported counterpart. Fifth, WIE plants using domestic technologies of different types are mainly invested in and marketed by local private companies, who target and exploit the market of middle scale-cities as a first stage of market entry.

At the beginning of the 1980s, China had experienced a shortage of electricity due to economic development and industrialization. Collective firms and private companies were allowed and encouraged to build thermal power stations. In 1985 a policy was implemented to encourage the comprehensive utilization of the nation's resources. Included in the policy was an 'encouragement list'. The list, issued by the Chinese government is adjusted from time to time. A package of preferential policies, including pre-emptive offers, preferential pricing and the resource being prioritized in government R&D programmes is given to firms. The goal is to re-orientate the nation towards goals of more efficient utilization of those resources which are listed.

Energy recovery from waste was added to the list in 1996 with the need for industry to be at the forefront of WIE developments highlighted. Hence, the market and innovation space for WIE technologies are gradually being created and enlarged, attracting different actors engaged in different relevant activities.

Sixth and fundamentally, the interesting market segmentation phenomenon described in this section, is rooted deeply in the different approaches to waste disposal adopted by different municipalities. WIE technologies grow and develop not in isolation but on the contrary with reference to their insertion into different governance regimes within a society. One particular example of this is provided in the following paragraphs.

According to combustion science, the temperature of the combustion flame correlates with the calorific value of the substance in combustion. Theoretically, substances with a stable calorific value around 3350 KJ/Kg can be combusted stably,[19] for example achieving complete combustion[20] stably, with the temperature of the flame lower than 700–800°C under certain conditions.[21] However, different incinerators provide different fields for combustion, in which the rate of heat loss is different due to the way gas flows in the combustion system, the efficiency of heat exchange in the combustion chamber and so on. Therefore, the combustion temperature in the chamber of a circulating fluidized bed incinerator generally should not be lower than the critical range of 850–950°C in order to achieve stable combustion as well as the control of pollutants produced in the process of combustion, whereas 1000°C and higher is generally required for the grate incinerator.

Due to the heterogeneity of the composition of MSW, the calorific value of waste is generally an average value in the same batch of waste. Therefore, it is generally required that the minimum calorific value of waste should be higher than 6000 KJ/Kg and with this level maintained in order to meet the temperature requirement of combustion in fluidized bed incinerators, and the required minimum calorific value in waste are higher for grate bed incinerators. However, the average calorific value of Chinese MSW is around 4180 KJ/Kg or lower. Moreover, there is considerable fluctuation in the calorific value of Chinese waste over the year. The data from one WIE power plant[22] shows that the calorific value of the waste received in the plant fluctuates in a range from 2500 KJ/Kg to 5200 KJ/Kg in different weather and seasons. The only solution to combusting MSW with unstable calorific value currently is to use auxiliary fuel. Otherwise, incomplete combustion will occur in the system, producing many pollutants and yielding a large amount of residue; or even worse, the combustion process cannot be continued and 'flame out' occurs. Almost all WIE technologies in China, whether imported or native, use some kind of auxiliary fuel,[23] such as coal, certain distillation residues such as heavy fuel oil, or other non-hazardous industrial wastes which have relative high calorific value such as waste rubber tyres. The most appropriate auxiliary fuel varies with the nature of different incineration technologies, for example grate and rotary kiln incinerators generally use oil as auxiliary fuel, and the Chinese fluidized bed incineration system uses coal.

Finally, we would argue that imported technologies cannot compete with the Chinese native fluidized bed incineration system in terms of technology and relevant cost-efficiency.[24] The reasons can be summarized as follows:

1. Imported grate bed and rotary kiln incinerators have to use oil as auxiliary fuel. This design is fatal to the cost-effectiveness in the operation of the technologies. The primary fuel in China is coal, which is abundant and relatively cheap. The price of oil is six times higher than coal. Even though on average twice the weight of coal needs to be added compared with oil[25] to improve the calorific value in the same waste, using coal is much cheaper than oil and the operation fee then can be greatly reduced in a WIE power plant. The choice of oil as auxiliary fuel does not fit the natural resource context of China. Nor does it accommodate the historical trajectory of technological development which is built upon a foundation of a particular natural resource base.

2. The imported facilities are very complicated systems involving sophisticated add-on sub-systems to treat emission and other pollutants produced in the combustion process, and to meet the high pollution control standards in developed countries. These result in a detrimental requirement for a very high initial capital outlay. Another related weakness is high temperature erosion and damage to key components of the system, which leads to high maintenance fees. By and large, we argue, the general engineering concepts and design principles of foreign technologies are over-specified for the Chinese situation and by contrast Chinese technologies reflect greater consideration for simplification of the system to meet high cost-effectiveness.
3. Waste separation at source is conducted routinely for waste destined for WIE in many developed countries. By contrast in China this is just starting to be the case in a very few big cities. The absence of an infrastructure to achieve the objective of separation results in the characteristic unstable quality of Chinese MSW for incineration. This means that imported technologies, where they are installed, malfunction from time to time. Another consequence of the unstable quality of Chinese MSW is that many pollutant substances may be mixed into the MSW. The designs of imported systems give less consideration to this situation because this problem does not generally happen in developed countries where high standard regulation procedures are in place for the collection and separation of MSW. Most imported technologies have been adapted when they are used in China. However, those adaptations are usually add-on sub-systems,[26] which increase both the fixed and variable costs[27] of the plant.

These three points support our conclusion that the design principles of imported technologies are not suited to the Chinese context, an important outcome of this being increased costs associated with the imported systems and technologies. The reason is that these technologies were not originally designed for the Chinese market. When the costs are this high, the wide take-up and diffusion of the imported technologies is difficult. This explains why most of the technology suppliers of the imported WIE products and systems into China find it very difficult to secure a second project in the country during the first stage. They are actually not mature technologies for the Chinese market on first stage of entry before they have been re-developed to fit the Chinese market context.

However, though western suppliers may not be competitive in technological and economic terms; they are competitive on financing methods. Many projects using imported technologies are financed through international aid, or credit facilities, or operate in BOT or BOOT methods,[28] or exploit other new financing ideas and innovations. Furthermore, because the Chinese market is so big and the required information is not well disseminated, local governments may not have sufficient knowledge about different WIE technology options available or entering the market in order to make comparative assessment across competing suppliers. Also importantly, foreign suppliers appear to have better marketing skills, methods and experience, including using specific sales channels and agents. This commercial expertise also stretches across other parts if the sales and project organization process producing a bundle of commercial expertise that is highly persuasive to potential customers in particular local governments making it possible for western players to penetrate and develop the WIE market in China despite the technological and cost issues outlined above. This suggests that non-technological innovations and other social factors are both theoretically relevant and have a considerable substantive impact on the adoption and choice of WIE technologies in China.

CONCLUSION

As we have now seen, through a multi-faceted development process, involving multiple classes of agent and changing relations, structures, patterns and processes encompassing consumption, government regulation, technological development, and market emergence and expansion, WIE is fast becoming a new industry in China. Furthermore it is an emergent industry which has the potential to contribute towards the twin objectives of dealing with China's urgent municipal solid waste problem, and reducing the country's reliance upon, and depletion of coal reserves.[29]

When we think about the ontological issues and theoretical questions revealed by the case, in particular relating to the intersection of innovation studies and industrial ecology, a number of interesting themes come to mind. A first is the theme of transformation. From a macro institutionalist perspective there can be no more dramatic case than China to lend credence to a regulationist perspective on political economy. We see in the overarching and fundamental macro – transformation from Mao era socialism to technology/market liberalism, a process which has involved the re-alignment of a new dominant mode of capital accumulation with new relations of social regulation (albeit still within the context of a strong central State providing the main regulatory authority). Importantly, this re-alignment has occurred in a historically unique and geographically-specific setting. Such attention to history privileges an interest in unique historical/geographical contexts. Importantly, historical analysis appears quite at odds with the generally a-historical method of industrial ecology.

Further, the account illustrates how complex the mix of socio-economic factors of market, consumption, regulation, cost economics, science base and technological development are and how they are dynamically intertwined and co-evolving in a mutually influencing socio-techno-economic milieu. The case, described here in historical terms, suggests that – and this is an important theoretical lesson – these processes are a priori non-deterministic, non-determinable and unpredictable rendering their top-down management by any single interest group very difficult, even within the context of such a historically planned setting as China. Such a priori indeterminism is an essential ontological feature of innovation processes contrasting starkly with the general managerial determinism seemingly taken for granted in industrial ecology.

A related point is that of variety and our appreciation of innovation as a variety generating process. We have seen in the chapter the many permutations of variety, the mechanisms and combinations of which provide locally differentiated city spaces, consumption and material contexts, resource-use contexts and simple economic contexts of cost and affordability. Further as we have seen there is immense variability in the number of technical design solutions which can be mixed and matched at the subsystem level of plant and equipment in settings involving large scale and multiple production operations and processes. By contrast we would argue the underlying assumptions of managerial industrial ecology are fundamentally *variety-reducing*, since industrial ecology strives towards optimal, maximizing positions in terms of 'efficient' resource use. To be fair, it is necessary to distinguish between the technical armoury of managerial industrial ecology and the many metaphors deployed in the theoretical and disciplinary origins of industrial ecology (see Chapter 14, this volume), which in some cases does stress variety, interaction, holism and indeterminism.

It would also be unfair to dismiss the potential that a policy commitment to limiting resource use, backed by regulatory standards and fiscal incentives has, and the affinity that this policy objective has with the normative base of industrial ecology. As we have seen in the case, the ideology of industrial ecology expressed as attention to the technological and market development of solutions which pay greater attention to questions of resource use, re-use, depletion and disposal, have already entered the policy lexicon and regulatory framework of China. This has to be seen as a fundamental achievement for academic and policy sponsors of industrial ecology across the world.

Albeit an essential conundrum remains. All of the policy, technology, and market incentives described operate on the industrial/productive side of the economy. They have so far left the domestic consumption side relatively untouched. And yet the chapter has powerfully illustrated that it is the ratcheting of domestic consumption to mirror the consumption practices of developed liberal market economies of the West that underpins the growing MSW problem in China. How to deal with the ratcheting of consumption under free market capitalism, and its implications and consequences in terms of resource use, will continue to occupy the minds of policy makers of market economies the world over.

NOTES

1. There are several terms used, sometimes interchangeably, such as garbage, refusal, trash and waste. However, solid waste, as we use the term in this chapter is any waste that someone would consider disposing of on the land. It does not include water discharges to surface waters or waste air emissions. Solid waste is generated in offices, factories, through landscaping activities, agriculture and in homes. According to some definitions solid waste can include sludge from sewage treatment, farm manure and industrial pollution. It may contain liquids such as paints, solvents or motor oil. However in professional policy circles, solid waste only refers to waste that is not hazardous or radioactive, both of which need special treatment. In our classification, MSW in our chapter refers to common trash from a residence and from business and industry that has the same characteristics. Garbage, paper and paper products, metal products, packaging, plastic product, appliances and yard waste are all components of MSW. Thus, the composition of MSW is immensely varied and involves more complicated issues than single source or single composition wastes from industry as waste water sludge, or slag from coal mines and so on.
2. Waste-incineration-for-energy (WIE) technologies refer to those used for MSW incineration power plants. Incineration of hazardous waste such as clinical waste and the speciality plants constructed for this purpose are not discussed. The other terms, namely waste-to-energy in the US literature or energy-from-waste in the British literature, include other technologies, such as for generating energy from landfill gas or gas from anaerobic digestion (composting). Those technologies and associated projects are not discussed in this chapter because we take a narrower reading of WIE.
3. In this section, parts of the text are drawn from (Cen 2003) and (Cen *et al.* 2003). Many of the data are from a report of China Engineering Academy (2002).
4. New incentive instruments are added later, such as tax holidays, credit and so on.
5. Refer to the State Environmental Protection Administration of China, National Environment Bulletin, 2001.
6. Fluid phase as termed by (Utterback 1994).
7. Ministry of Construction, 'Technological Policy on Municipal Solid Waste Treatment and Pollution Control and Prevention, 2000.
8. The same situation has happened in Japan and the USA. There was a period when many composting plants were closed and new composting projects are developed slowly for the same reason.
9. According to 1998 data, 1813 tonnes of packaging was produced in China that year. Then 70 per cent or 1269 tonnes became 'waste' which equals 10 kg per person (Chinese Academy of Engineering 2002).
10. The other term of the same meaning used in literature is calorific value.
11. The research system in China involves three components: universities, industrial enterprises and national research institutes. In the case of development of WIE technologies, some university research institutes and national research institutes under the Chinese Academy of Sciences play crucial roles. Because research funds for all the research institutes are not exclusively administered through the government after the reforms in

the 1980s, potential funding sources from industrial enterprises is being sought. Accordingly, institutes now pay much more attention to market needs and the budgets and concerns of industrial enterprises.

12. Later, some other actors have developed native grate technologies based on imported technology or rotary kiln technologies based on imported technology in co-operation with some big boiler works, who have manufactured or assembled WIE systems using imported technologies. Wen Zhou Wei Ming Co., is the one of the first movers.
13. Coal mines in China are geographically concentrated only in a few regions. The bottleneck lies in how to deliver efficiently the coal extracted from these mines to production bases and cities scattered in a country of 9 600 000 km^2. Transportation will take up 80–90 per cent of the cost of the coal that is bought for electricity generation.
14. Coal is generally transported by train in China. Coal dust coal may be emitted during the journeys. Meanwhile, emission and pollutant discharge standard for WIE plants is much stricter than those of normal thermal power plants. For example, the permitted emission of SO_2 and NO_x in a WIE plant is only 50 per cent of that in a normal thermal power plant.
15. Kong and Jiang (2003) give a very conservative estimate of the unit capital investment of a WIE plant. The capital investment of the most competitive native WIE technology/facility is only one third of fully imported WIE technologies and facilities. Taking into account only core equipment and technology imported for a WIE project, and using the average capital cost of WIE project, Kong and Jiang give a moderate unit investment estimate.
16. However, MSW may not always be a fuel free of charge. Some WIE plants have even paid a transportation fee for MSW in the early development stage of WIE industry.
17. Kong and Jiang (2003) assume a 3 per cent increase per year on the price of coal in their calculations.
18. Because there are great differences in economic conditions and other factors between different regions and cities of China, central government issues guidelines and statements relating to the 'spirit' to be adopted in choosing between alternatives, local governments then have the flexibility to enact them according to the local situation. They therefore have decision making authority over details of subsidy or preferential pricing.
19. This has been proved by ITPE in the 1970s in experiments using stone-like coal with a calorific value of 3350 KJ/Kg.
20. Incomplete combustion results in the production of new pollutants and a great deal of substance residue.
21. The other factors of the condition are volume of the air in the combustion chamber, temperature of preheated air, proportion of excess air.
22. The data are offered by ITPE.
23. In developed countries, the calorific value of MSW is generally around 8375 KJ/Kg in average. In this case, auxiliary fuel is generally not required for the WIE power plant. The established waste collection and separation system/infrastructure in those countries guarantee the quality of MSW in terms of removal of the incombustible components around the temperature of 1000°C, such as metals, glasses, and generally improving the calorific value in waste. However, the development of recycling waste paper into paper and recycling waste plastics into plastics pull out the paper and plastic components in MSW, which reduced the calorific value in MSW, which starts to cause problems in WIE industry in some countries.

    Currently, a few native grate WIE plant operators claim that they do not need to add auxiliary fuel during the combustion process because they have used a pre-treatment system to pile up original waste in a large pool for three to seven days so that the calorific value of the accumulated waste can be improved and relatively stabilized when the water contained in the waste is discharged into another leachate pool. However, this needs to be testified through data and inspection. This also leads to a new problem namely how or where to treat the relative large amount of leachate produced?
24. The project investment of the most competitive Chinese WIE technology is only one third of that of an imported technology with most facilities imported from overseas.

25. The data are from the comparative experiments done by ITPE.
26. It is because the design is stereotyped for manufacturing.
27. Some WIE plants have to employ workers to separate and sort waste manually in the plant.
28. In both methods, an enterprise invests in the WIE plant in question with the right granted by the local government, so that the budgetary pressure of local government can be released. BOT (build-operate-transfer) refers to how the firm designs, finances, builds and operates the WIE plants in the franchised period. When the period is ended, the property right of the WIE plant will be transferred to the government without charge. BOOT (build-operate-own-transfer) differs from BOT in that the investing enterprise may own part of the auxiliary facilities of the WIE plant permanently, or may get some payment when the WIE plant ownership is transferred to the government.
29. Though the impact on fossil fuel consumption should not be exaggerated since the 10 million tonnes of coal per annum that estimates suggest MSW could substitute as fuel source is small when set against the 1000 million tonnes per annum currently consumed for electricity generation across China.

REFERENCES

Cen, Y. (2003), 'Technology transfer in the waste incineration for energy industry: a comparison of China and the UK', Masters degree dissertation, prepared for the Institute for Development Policy and Management, University of Manchester, Manchester.

Cen, K., Y. Cen, M. Ni, J. Yan, Y. Chi and X. Li (2003), *The Progress for the Thermal Treatment of Municipal Solid Waste in China*, Sheffield, UK: 4th International Symposium on Waste Treatment Technologies.

Chinese Academy of Engineering (2002), *Report on Policies Recommendation and Technological and Economic Appraisal on Municipal Solid Waste Treatments in China (Chinese)*, Hangzhou: Chinese Academy of Engineering, Institute for Thermal Power Engineering.

Kong, X.-W. and J. Jiang (2003), 'The discussion on economical approaches for solid-waste recycling', *North-West Power Station Technologies (Chinese)*, **10**.

Utterback, J.M. (1994), *Mastering the Dynamics of Innovation*, Boston: Harvard Business School Press.

Zhang, Z.-M., Q.-Y. Wang, X. Zhuang, J. Hamrin and S. Brauch (2002), 'Development of renewable energy in China: potential and challenge (Chinese)', China Sustainable Energy Program, accessed 20 July, 2004 at www.efchina.org/documents/China_RE_Report_ CN.doc.

PART 4

Consumption and intermediation

9. Industrial consumption and innovation

Jeremy Howells

INTRODUCTION

This chapter explores consumption by firms and how it relates to the wider innovation process within the firm and in the formation of a neglected aspect of a firm's capabilities – that of a consumer. Consumption and the way firms consume intermediate goods and services form an important but overlooked part of a firm's capability set. The role of services is highlighted in this process. This is because by introducing a service dimension to the discussion about innovation sheds a new perspective on the process of consumption and its relationship with firm-level innovation. This is for three inter-related reasons. First, it is suggested that services are important in the consumption of new goods (and services). Second, that going back to Edith Penrose's work and earlier, the way (through routines and practices) that firms consume goods yield service-like attributes and these form important and distinctive capabilities for the firm. Lastly, related to this, the process of consumption and the development of routines associated with this process are forms of disembodied, service innovations.

The analysis presented here focuses on the role of industrial consumption of intermediate goods and services and how it influences the innovation process. Although there have been studies that have looked at the demand market factors in innovation and competition, these have been fragmented views which have only considered parts of the wider consumption process. This chapter will explore some of these wider consumption processes and, more fundamentally, seek to define and articulate more clearly consumption by firms.

CONSUMPTION AND THE FIRM: AN OVERVIEW

From a period of relative stasis in terms of academic interest and progress, there have been a number of recent publications (Bianchi 1998; Gualerzi

1998; Loasby 1998; Langlois and Cosgel 1998; Langlois 2001; Metcalfe 2001; Witt 2001) which have sought to review and critique existing consumption theory particularly from a neo-classical perspective (notably Lancaster 1966, 1971, 1991; Stigler and Becker 1977) but also more generally from other reformulated economic studies (Ironmonger 1972; Earl 1986; Scitovsky 1992; Woo 1992). These studies have sought to further develop and integrate consumption theory within a wider socio-economic context, in particular focusing on consumption as a change agent. In particular, they have sought to highlight a number characteristics of consumption building upon the existing body of literature which have not been readily acknowledged in the past (Howells 2004). As such, their work relates to emphasizing the active, rather than passive, nature of consumption where consumers actively seek novelty to satisfy needs and tastes (Hirschman 1980, p. 284; Bianchi 1998, pp. 75–81). Consumers, therefore, act as interactive agents in the wider competitive environment (Gualerzi 1998, p. 59). In addition, effective consumption patterns require time and resources to develop and this also should be seen as forming a key capability of the firm (Langlois and Cosgel 1998, pp. 110–111). This requires a process of learning for consumers (Loasby 1998, p. 98; Robertson and Yu 2001, p. 190; Witt 2001, pp. 28–31), which is also reflected in the development of efficient routines (Langlois and Cosgel 1998, p. 59; Langlois 2001, p. 90) for successful consumption. The recognition of the fact that competences and routines are built up around the consumption process requires a whole set of attributes in investment, knowledge and enterprise in the consumption process. This is associated with the notion of the 'enterprising consumer' (Earl 1986, pp. 53–84), 'skilled consumption' (Scitovsky, 1992, p. 225) and the development of 'market knowledge' (Tiger and Calantone 1998) and 'consumption knowledge' (Metcalfe 2001, p. 38). Most significant here is that in seeking different wants (Ironmonger 1972, p. 13) and novelty, consumption creates incentives for innovation and has an important influence on the selection of new technologies and the new combinatorial use of them.

The role of consumption in shaping industrial change and innovation is therefore starting to be acknowledged. To be an effective consumer involves time and resources. This echoes Lancaster's (1966, 1971, 1991) theory of consumer choice, which highlights the fact that the consumer needs to be an active agent and invest in time and resources to build up capabilities to consume effectively. These studies, in emphasizing the development of consumer competences and routines, also reflect back to Stigler and Becker's (1977, p. 78) neo-classical concept of the accumulation of 'consumption capital'. Just as innovations require considerable investment to produce, so do consumers need to invest in new capabilities and routines to consume them.

Unfortunately, such studies, although significant in the conceptualization of consumption (particularly from a wider economic viewpoint), have two major drawbacks. They, first, only tangentially discuss the role of consumption in the process of innovation. The issue of innovation and technology, if noted at all, is within the broader context of the search for new tastes or in the desire for novelty. The search for novelty and the development of new tastes are important issues which have been neglected in orthodox, neoclassical economics, but their impact on innovation is not pursued. Second, such studies make little or no specific reference to the consumption by firms of (intermediate) goods or services, but instead remained focused on final consumption patterns by the individual or households. Thus the literature seems to stick to the issues of the 'atomistic' individual or (at best) household consumer and their consumption of final goods. Indeed, Earl's (1986, pp. 20–1) otherwise excellent analysis of consumption focuses on the car, using Kay's model of business strategy (Kay 1982, 1984) and readapting it to analyse household consumption patterns (Earl 1986, pp. 63–65 and 74–76) rather than directly exploring how firms may seek to consume. There are a few exceptions to this neglect of intermediate goods and services, most notably by Langlois and Cosgel (1998; Langlois 2001) in their development of a 'capability model' of consumption associated with the development of consumption routines.[1] This capability model refers both to services (Langlois and Cosgel 1998, p. 109) and intermediate goods (Langlois and Cosgel 1998, pp. 89 and 114) but again, although excellent, it does not cover these issues in any detail.

PERSPECTIVES ON INDUSTRIAL CONSUMPTION

If there has been very little discussion about industrial (or intermediate) consumption, with the emphasis on individual or household (final) consumption in the literature; what do we mean by the term industrial consumption? Related to this, can industrial consumption be discussed and analysed in any meaningful sense? And, if so, what processes are involved in industrial consumption?

Firms and other organizations can be seen to 'consume', although clearly it differs from the way that individuals or households consume which is a more voluntaristic and individualistic process (Warde 2005). However, firms still consume in the sense that they (or perhaps more accurately individuals and sets of individuals within them) buy, use, modify and dispose of goods and services and in so doing they yield utility from doing so. Consumption also generates wider benefits ('psychic income') in terms of recognition (or 'classification' as Holt (1995) would put in relation to generating status), the development of key trading relationships (relating in turn to power dependency issues) and indeed articulation of what the organization is.

If firms can therefore be said to consume, in what way does firm consumption differ from individual or household consumption? It would be easy to argue that industrial consumption is a more collective, rational, objective and 'de-personalized' process. However, we should be wary of making such simple comparisons. Let us focus on the initial process of the decision to buy, as the first step in the consumption process. Industrial decision-making is certainly outwardly by its nature a more collective process, but equally purchases of personal and household goods associated with family involvement are often collective in nature. Moreover, the actual decision-making process within the firm is effectively undertaken only by a small group of individuals (rarely larger than a family group). Similarly, the decision-making process in not necessarily any more rational or objective in its decision-making, despite the growth in decision-making models and procedures deployed by companies when they buy or outsource goods and services. Studies have shown that decision-making within firms is just as personal, subjective and irrational as in most personal buying situations (Klein 1989). There is also the supposedly 'de-personal' nature of corporate buying because of its supposed formulaic and objective process. However, paradoxically because industrial 'buyers' and 'sellers' of industrial goods and services form relatively small communities, they often know each other very well and personally. The extreme may be large, specialized industrial equipment such as power plants or aircraft where the effective number of buyers and sellers may be very small indeed and where long term personal relationships may build up. Even with more routine, lower value goods specialist buyer and seller teams still often know the customer/supplier very well. Compare this with personal purchasing. Close personal involvement may still hold for the local corner shop, but for shopping in major multiple chains, such personal relationships and knowledge may be very much more distant.

The most significant difference between individual and corporate consumption may however be in demarcation between buyer and seller and their associated roles. For household durable goods the division of roles and responsibilities between buyer and seller are largely clear-cut, but for complex projects involving the generation of a new good or service the role between producer and consumer may be much more ambiguous and indeterminate (Rosenberg and Stern 1971). This is especially the case where introducing a technological innovation requires close cooperation between consumer and supplier (Vaaland and Håkansson 2003).

In relation to answering what processes are involved in industrial consumption, most studies have focused on what may be viewed as the very 'front end' of consumption, namely the purchasing and buying of (new) goods and services. As such, consumption has been simply equated with the demand process and associated with buying and purchasing. This has particularly centred on the 'buyer-supply' (see, for example, Burt and Soukup 1985; Thomas 1994) or 'user-producer' (see, for example, Lundvall 1992; Von Hippel 1988) relationships. However, even here the focus has been supply-led and discussion has largely centred on implications for production and the producer. Thus, studies examining buyer-supplier relations have focused on: (a) the supplier; (b) the wider supply networks; and (c) how firms have established and integrated their supply networks into their production process (see, for example the Sabel *et al.* (1989) study of the automotive industry).

The focus of these studies has been more on the relations themselves rather than on specific functions. However, there have been a set of studies investigating the activity of buying in terms of the purchasing function within the firm, and more specifically on the role of purchasing on new product development and innovation (see, for example, Burt and Soukup 1985; Thomas 1994; Wynstra *et al.* 2000). However, these studies remain centred on the initial act of buying and how technical inputs can be incorporated into new product developments rather than wider consumption activities.

User-producer relations, have been examined more specifically within the context of innovative activity. Thus, the role of users in the innovation process has also been extensively studied (von Hippel 1976, 1988; Parkinson 1982; Shaw 1985, 1987; Foxall 1987; Holt 1987; Lundvall 1988, 1992). Here the important role that users play with producers has been highlighted, and in particular the links or relationships between them. However the focus has mainly been on the impact of users on producers in terms of what innovations they produce rather than so much on the actual use and shaping of innovations by consumers (however, see Foxall, 1988). Apart from these studies which have a bearing on consumption there have been studies that highlight issues to do with absorption, most notably Cohen and Levinthal's (1990) work on absorptive capacity, and diffusion. Discussion about absorption has been on a rather abstracted level (Zahra and George 2002). In relation to diffusion, studies have sought to map out the diffusion of innovations within the firm (after the initial purchase and adoption) and reveal that this is far from an instantaneous or homogeneous process (Pae *et al.* 2002). Finally here there are more general studies associated with the impacts of buying equipment in terms of their benefits to operational efficiency. This can be related to a whole literature on productivity and improving yields, particularly in relation to materials use.

All these studies, however, provide a rather disparate picture of the 'demand' side of innovation in the context of buying new goods and services. There is a much wider range of functions and activities associated with the consumption of new intermediate goods and services. These relate to activities such as:

(a) the installation and testing of goods and services once purchased;
(b) training in the use of goods and services;
(c) the maintenance and upgrading of them once in use; and
(d) their final disposal.

These activities and relationships associated with consumption within the firm are complex, inter-related, interactive (for example, role of feedback processes) and contested. This moves well beyond the fragmented picture so far described by studies. For example, one common and continuing misconception is the frequent conflation of the role of buyer (or purchaser) with user within the firm. This more fundamentally involves confusing buying with use[2] and the often frequent individualistic and personalized notion of the firm as a single knowing entity. Moreover, this is despite studies highlighting conflicts between the purchasing department and user departments (Howson and Dale 1991), or in the case of new product development between the R&D/product development department and purchasing (Faes 1986).

Between the initial purchase of a new good or service and its disposal there are therefore a series of complex activities and processes within the firm which are associated with the consumption process. It may also involve the sourcing of additional goods, services and knowledge from outside the firm to help consume the initial material or equipment. There will also be feedback between previous rounds of consumption between the initial purchase and its final disposal.

There is a constant round of feedback information loops and iterations over the consumption life cycle between the consumer firm and the producer or supplier firm. This is particularly important in relation to the transfer of information and knowledge by the user firm about the good or service being consumed, to the supplier firm which produced the good. This in turn leads to the creation of complex knowledge structures which co-evolve between producer and consumer. It should be stressed that this is not just about the user firm feeding back information on the consumption process to the supplier; it may involve the user initiating new products and asking a supplier to bid to produce it (Foxall 1988). There is often a complex interplay between producers and users in the innovation process; changes in one sphere often 'sparking off' changes in the other sphere. This can lead to constant rounds of new interaction (von Hippel 1994, pp. 432–434), with the innovation process between producers and users creating its own innovative dynamic. Indeed staff from the user organization may form part of the production function helping to produce the services (O'Farrell and Moffat 1991; see also O'Farrell 1995). Strong behavioural forces also come into play here in terms of how consumers interact with products and how these change with the introduction of something radically new (Cooper 2000, p. 4) and creates the indistinctness of production and consumption noted earlier (Rosenberg and Stern 1971).

Care must also be taken in not assuming there is a single feedback path between disposal and purchasing. Feedback within the consumer firm itself is often likely to be fragmented and sometimes contradictory. Consider the decision to replace a current piece of equipment with a newer model from the same supplier firm. This is often seen as a 'simple' replacement decision by the 'firm', where the supplier and the good are familiar, but is it that simple? The decision to buy a replacement piece of equipment will obviously involve the key user unit or department, a particular factory or manufacturing department, but will also often include many other functions and activities within the firm. Here are just a few main functions:

(a) the purchasing department who handled the order (considerations on how easy it was to find alternative suppliers, processing the previous order with the supplier firm and so on);
(b) the finance department (what finance or leasing arrangements were offered, for example);
(c) the maintenance and engineering department (ease and cost of maintenance, possibility of upgrades or retrofitting);
(d) human resources department (training requirements to use the equipment effectively); and
(e) environmental and disposal issues associated with running and replacing the equipment.

The above are just a few functions and activities that could be involved in making the 'simple' replacement decision. For larger pieces of equipment or for shared facilities there may, in addition, be several user units within the firm that have differing views about whether the upgraded model supplied by the existing supplier firm should be given the 'go ahead' or another model and supplier should be considered. The latter may involve substantial 'switching costs' but may be preferred not for reasons of the quality and attributes of what is being consumed, but rather in terms of strategies over who is supplying (for example, to reduce dependence on a single dominant supplier or to encourage innovation by attracting in new suppliers; Helper 1989). In this case the 'simple' decision may not be about direct consumption now, but about establishing consumption trajectories in the future.

INDUSTRIAL CONSUMPTION: SINGLE EVENT OR COMBINATORIAL STREAM?

Consuming for what?

We have discussed consumption within the firm in terms of the different (and distributed) elements that are associated with it. However, the analysis so far has centred on viewing it largely in terms of the life cycle of a single element; a single good being purchased and consumed. Our conception of consumption should not be restricted to viewing it as a single event or element, but rather as a continuing process associated with combining different elements – other goods, services and resources – to yield a more complete or ultimate consumption experience. This more dynamic view of consumption will be outlined in more detail below. It also links up with more traditional notions about utility derived from consumption. In trying to uncover what the nature of consumption is, it is useful to review the long established concept of utility that has been used by economists and which highlights the distinction between desires and the satisfaction of wants.[3] What discussion there has been about industrial consumption, has been in a very one-dimensional and direct sense associated with the 'act of buying' a specific product, and this is particularly true of industrial services. However, as economists have long recognized in relation to buying goods in general, these goods satisfy certain wants and therefore yield a utility in satisfying these wants. In this context many services may be seen as 'purer' in utilitarian terms as they are often nearer in the spectrum of satisfying ultimate wants, whereas goods may be seen as being more about an interim milestone on the road to such satisfaction (or more specifically provide a solution to a problem; Gadrey *et al.* 1995, p. 5). One may think of purchasing a television set which satisfies an initial desire, but which can only be satisfied (partially or completely depending on what television programme you watch!) by viewing television programmes on the television. This can lead into a deeper metaphysical discussion about what consumption fundamentally is. However, on a more immediate level it also raises the notion that consumption has a strong temporal quality (and with it, dynamic and evolutionary qualities). Consumption is rarely instantaneous (for example, when a good or service is actually purchased) in the sense of satisfying immediate wants. Thus, as Joan Robinson (1962, pp. 122–3) noted 'consumption, in the plain meaning of the term, in the sense that it is connected with the satisfaction of natural wants, does not take place at the moment when goods are handed over the counter, but during longer or shorter periods after that event'.

What is being stressed here is that consumption is a much richer and time-consuming process about satisfying more fundamental wants rather than a simple one-off, one-dimensional experience. This is difficult to illustrate adequately, but even for such a seemingly ephemeral and transient matter of whether watching a film at the cinema is a case in point. Should its 'consumption' solely be seen as watching the film or does it encompass the much wider consumption experience of discussing the film with friends and colleagues afterwards and remembering it in connection with other films and books after the event? With the exception of Greenfield (1966, pp. 8–9) few researchers in the services field have recognized the temporal and durable notions of the consumption process.

However, this leads into a discussion about the similarity and interlinked nature of goods and services (rather than their distinctiveness raised earlier). This can be seen on two levels. First, if one is interested in consumption as satisfying ultimate wants, goods therefore fulfil essentially service-like attributes (Howells 2004). This issue is developed by Saviotti and Metcalfe's (1984) work on attempting to measure technical change of products, for example material artefacts. Saviotti and Metcalfe (1984), building upon Lancaster's (1966) work, stress that a product can have both internal properties, for example those relating to the internal structure of the product, and external properties, relating to wider issues associated with the type of service being offered to users as part of that good.

From a slightly different perspective, if goods are seen as having service qualities in their consumption and utility, it has also been recognized that products and services have long been associated together (De Bretani 1989) and they are often 'co-consumed'. Indeed, this takes up Hill's (1977) notion of services as changing the condition of the consumer. Thus, as far back as 1892, Alfred Marshall highlighted the issue of derived demand and joint demand for goods and services (Marshall 1899, pp. 218–223), exemplified in his notion of 'composite demand'. More specifically this has been explored in more detail by Swann (1999) in his analysis of 'Marshall's consumer' and the increasing levels of sophistication that consumers can present in the consumption process.

Consumption is therefore rarely, if ever, the consumption of a good or service at a single point of time. In the case of the car, consumption has moved from the simple, one-off purchase to the wider process of buying, leasing, using and maintaining a car over the long term (Howells 2001). This shift in focus has major implications for firms that sell such products and services in terms of how they address consumers' needs and satisfy their ultimate demand. Consumption is therefore not a one-off contact, via the sale of a product, but a continuing process involving long-term customer contact. This is to be expected if consumption shifts from a single, one-off act to long-term user support; that is from selling a car at a single point in time to supplying fast/reliable/cheap/flexible/safe transport over a period of time.

INDUSTRIAL CONSUMPTION: COMBINATORIAL AND PROCESS-DRIVEN

If the above has emphasized that industrial consumption should not be seen as a 'one-off' but rather a longer term process, we should also recognize the combinatorial aspects of the consumption process, particularly in relation to the consumption of new goods and services (Howells 2001). The combinatorial aspects of the consumption process can be seen when a firm (or individual) consumes a good and combines it with the consumption and use of other goods and services. In short, services often 'encapsulate', or act as 'wrappers' to, goods and resources. This will not be fully explored here, but there are some aspects which are worth discussing as they help illuminate the consumption process within firms.

In order to consume a certain good, firms frequently consume multiple sets of goods and services to create a consumption experience and to yield some kind of utility. This involves not only purchasing goods and services from other suppliers/producers to help consume the initial good, but also may involve the generation of in-house goods and services from other parts of the firm as well. The consumption of a good therefore can involve multiple sets of other consumption and production activities. The consumption capability of a firm therefore resides not just in consuming the initial product and absorbing it within the firm, but also in combining other goods and services to aid consumption.

This combinatorial aspect in the consumption process is reflected in the way that services can help in the consumption of new goods (and services). This can be seen to operate in a number of different ways (Howells 2004). Thus, existing services can encapsulate new goods (or services) through a number of different means. Existing services can provide a 'familiarizer' effect by providing a familiar, trusted service to a new good, making it more acceptable for consumers to adopt the innovation. They can also offer a 'buffer' effect, so that once a good has been adopted services help enable the consumption of the new good in exactly the same (or very similar) consumption service format as the former, old good (for example earlier vintage) was consumed. Existing services can also provide a facilitator' effect in terms of encouraging and helping consumers to learn new practices and routines through an existing service 'window' or framework to use the new good. As noted earlier, learning plays an important role in the consumption of new innovations.

However, new services can encapsulate existing goods (or services) to form a new innovation. Here services are used to improve the acceptability, flexibility and performance of existing goods and these attributes are outlined in more detail below, using some simple examples. As such, new services encapsulating existing goods can provide a number of revitalizing and innovative roles in the consumption process. This can be as a 'sweetener' effect, by improving the acceptability of a good through a new service format which may overcome obstacles to the adoption or use of a good or service before. This may involve better set-up and operational instructions, which to the consumer may involve simple changes, but to the provider may involve complex, disembodied technical changes to routines and practices in the presentation of the good. Thus, technical documentation, including product configuration data, maintenance and operating instructions for use, may involve changes in highly complex organizational and operating routines both for the service being sold and the purchasing company using such new documentation.

New services associated with an existing good (or service) may also improve flexibility of use; for example, create a 'flexibility' effect. Thus, better maintenance practices and fault diagnostics may, moreover, allow the good to be made available to the user over longer periods of time or during periods when it was previously not possible to use it. A new service may improve the performance of the good, what may be termed as a 'performance' effect. The most obvious example here is a new software programme (with improved performance and functionality) being loaded to run on existing information technology equipment. However, another example is the development of a whole new area of services associated with 'predictive support services' which both improve the efficiency of a good, but also reduces 'downtime' in its use. Lastly, there may a 'functionality' effect, in that new services may allow an existing good to be used in a different way (see, for example, Robertson and Yu 2001, p. 188). A piece of testing equipment, or scientific instrument (von Hippel 1976), may therefore be used to test for things in different environments or conditions it was previously not initially designed for. However, this in turn may involve modifications to the existing good, generating a new round of innovation.

Combining services in this way can be seen as playing an important role in the consumption of innovations by enabling consumers to interact and accommodate these new goods and services more easily. Through mechanisms like branding they can reassure consumers and act as pointers to quality standards which were experienced through previous rounds of consumption. However, services also provide conduits through which innovations are adopted and learning mechanisms through which innovations are used. Adopting a new good therefore requires some of its attributes to be recognized and understood (and in this context, facilitated by services) if the consumption of the innovation is to be successful (Scitovsky 1992, p. 225).

INDUSTRIAL CONSUMPTION AS ROUTINES, PRACTICES AND CAPABILITIES

The shift from a single one-off consumption event to a more long term combinatorial process in turn suggests that firms' develop practices and routines surrounding the process of consumption. This picks up the general theme outlined earlier by Langlois and Cosgel (1998, p. 59, Langlois 2001, p. 90) who noted the need firms have to establish efficient routines for successful consumption. This can be seen to parallel Warde's (2003) outline of consumption practices by individuals, and can also be traced back to earlier work in industrial economics associated with the notion of the development of firm routines and capabilities as they relate to the capacities of firms to change the resources they consume. Thus, Edith Penrose in her book *The Theory of the Growth of the Firm* noted in relation to consumption by firms noted that (Penrose 1995, pp. 78–79) that:

> Physically describable resources are purchased in the market for their known services; but as soon as they become part of a firm the range of services they are capable of yielding starts to change. The services that resources will yield depend on the capacities of the men using them, but the development of the capacities of men is partly shaped by the resources men deal with. The two together create the special productive opportunity of a particular firm.

As such, the distinctive way which firms consume and use physical goods and resources in terms of their service utility, forms a key, but neglected element within a firm's repertoire of capabilities (see, however, Swann 2002). Consumers must learn about and correctly use a product to realize its benefits (Wood and Lynch 2002, p. 425). Indeed, how firms translate goods and resources into services may form an important component and attribute in this whole process. The process of consumption as a capability can form a powerful complementary asset in innovation that consumers can use to exert control over the producer (Foxall 1988, pp. 242–243). This is in turn highlighted by Langlois and Cosgel (1998, p. 110) in terms of consumption requiring a set of routines which can form distinctive capabilities for the firm. These capabilities not only include better communication of a firm's needs to suppliers' capabilities (Langlois and Cosgel 1998, p. 112), but perhaps more fundamentally identify and articulate what these existing and new needs are to itself (Robertson and Wu 2001, p. 190).

Firms need to develop successful routines and procedures in order to successfully compete. However, we need to caution against such routines being undertaken in a necessarily coordinated or harmonized way. As noted earlier, it has been recognized that purchase decision-makers and product users are often not the same group within a firm (Pae *et al.* 2002, p. 720) and that lack of adequate linkages between purchasers and users within an organization can lead to poor purchasing decisions with regard to new goods and services. A fractured and departmentalized process to the buying, use and more general consumption of goods and services can lead to lost opportunities in terms of harnessing a firm's potential capabilities in terms of consumption. Firms need to provide a more integrated consumption knowledge framework from which they can harness in distinctive ways to form a core capability of the firm.

CONCLUSIONS

This chapter has sought to highlight that consumption as a process is not a narrow activity solely focused around simply buying or use, but should be instead considered as a much wider process and activity. As such, consumption does not just stop with purchasing, or even use of the good or service concerned. We should therefore seek to conceive of consumption more as life cycle process, running from expectations before purchasing of the good and service right through to disposal and even the memory or legacy of its use. In this way, consumption should not be seen as a single event but as a long term, dynamic process. The analysis has also sought to stress that many different functions and activities are involved in consumption. It is a combinatorial process not only in the sense that other goods and services used to encapsulate and consume the initial good or service selected for consumption, but also in terms of the different capabilities and routines that the firm has to combine and use if it is to consume something successfully.

An issue remains of how wide the process of consumption should be conceived of. For firms themselves, part of the reason why they have not been able to exploit their position as consumers is that they have considered different elements of the consumption process (buying, adoption internally, use, modification and socialization and so on) as disparate and un-connected elements. As such, they have not been able to leverage more out of their position as consumers, or fully develop their consumption capabilities in relation to the innovation process. For academics the conceptualization of industrial consumption process remains fragmented because of disciplinary or sub-disciplinary interests which have tended to focus on often narrow elements of the wider process. Only by taking a more inter-disciplinary and more holistic view can we gain a real understanding of the process of industrial consumption and how it relates to the innovation process.

ACKNOWLEDGEMENTS

This ongoing research is funded by the UK Economic and Social Research Council. Thanks go to the various members of CRIC's consumption and services teams who have made various useful comments on the article. The views expressed are the author's alone.

NOTES

1. This is echoed in an earlier work by George Katona entitled 'The Powerful Consumer' who talks about habitual problem-solving frameworks which steer decisions in a certain direction (Katona 1960, pp. 58–59).
2. Lundvall (1988, p. 365) uses the term 'final users' to denote the actual users of goods, equipment, and so on that the firm has purchased.
3. Although as Joan Robinson (1962, p. 48) noted in her analysis of neo-classical theory of utility: '*Utility* is a metaphysical concept of impregnable circularity; *utility* is the quality in commodities that makes individuals want to buy them and the fact that individuals want to buy commodities shows that they have *utility*'.

REFERENCES

Bianchi, M. (1998), 'Taste for novelty and novel tastes: the role of human agency in consumption', in M. Bianchi (ed.), *The Active Consumer*, London: Routledge, pp. 64–86.

Burt, D.N. and W.R. Soukup (1985), 'Purchasing's role in new product development', *Harvard Business Review*, (September/October), 90–97.

Cohen, W. and R. Levinthal (1990), 'Absorptive capacity: a new perspective on learning and innovation', *Administrative Science Quarterly*, **35**, 128–52.

Cooper, L.G. (2000), 'Strategic marketing planning for radically new products', *Journal of Marketing*, **64**, 1–16.

De Brentani, U. (1995), 'New industrial service development: scenarios for success and failure', *Journal of Business Research*, **32**, 93–103.

Earl, P.E. (1986), *Lifestyle Economics: Consumer Behavior in a Turbulent World*, Brighton: Wheatsheaf.

Earl, P.E. (1998), 'Consumer goals as journeys into the unknown', in M. Bianchi (ed.), *The Active Consumer*, London: Routledge, pp. 122–39.

Edvinsson, L. and J. Richardson (1989), 'Services and thoughtware: new dimensions in service business development', in A. Bressand and K. Nicolaidis (eds), *Strategic Trends in Services*, New York: Ballinger, pp. 33–49.
Faes, W. (1986), 'Singler chemicals: purchasing's contribution to new product development – a case history in the pharmaceutical industry', in M.J. Baker and S.T. Parkinson (eds), *Organisational Buying Behaviour: Purchasing and Marketing Implications*, London: Macmillan, pp. 59–72.
Foxall, G.R. (1987), 'Strategic implications of user-initiated innovation', in R. Rothwell and J. Bessant (eds), *Innovation: Adaptation and Growth*, Amsterdam: Elsevier, pp. 25–36.
Foxall, G.R. (1988), 'The theory and practice of user-initiated innovation', *Journal of Marketing Management*, **4**, 230–48.
Foxall, G.R., F.S. Murphy and J.D. Tierney (1985), 'Market development in practice: a case study of user-initiated innovation', *Journal of Marketing Management*, **1**(3), 201–11.
Gadrey, J., F. Gallouj and O. Weinstein (1995), 'New modes of innovation: how services benefit industry', *International Journal of Service Industry Management*, **6**(3), 4–16.
Greenfield, H.I. (1966), Manpower and the Growth of Producer Services, New York: Columbia University Press.
Gualerzi, D. (1998), 'Economic change, choice and innovation in consumption', in M. Bianchi (ed.), *The Active Consumer*, London: Routledge, pp. 46–63.
Helper, S. (1989), 'Strategy and irreversibility in supplier relations: the case of the US automobile industry', Boston University School of Management working paper no 89–22, Boston, MA.
Hill, P. (1977), 'On goods and services', *Review of Income and Wealth*, **4**, 315–38.
Hirschman, E.C. (1980), 'Innovativeness, novelty seeking and consumer creativity', *Journal of Consumer Research*, **7**, 283–95.
Holt, D.B. (1995), 'How consumers consume: a typology of consumption practices', *Journal of Consumer Research*, **21**(June), 1–16.
Holt, D.B. (1997), 'Poststructuralist lifestyle analysis: conceptualizing the social patterning of consumption in postmodernity', *Journal of Consumer Research*, **23**(March), 326–50.
Holt, K. (1987), 'The role of the user in product innovation', in R. Rothwell and J. Bessant (eds), *Innovation: Adaptation and Growth*, Amsterdam: Elsevier, pp. 1–12.
Howells, J. (2001), 'The nature of innovation in services', in OECD (ed.), *Innovation and Productivity in Services*, Paris: OECD, pp. 55–79.
Howells, J. (2004), 'Innovation, consumption and services: encapsulation and the combinatorial role of services', *The Services Industries Journal*, **24**, 19–36.
Howson, T.G. and B.G. Dale (1991), 'An example of the purchasing function in a sales-oriented company', *International Journal of Operations and Production Management*, **11**(5), 71–82.
Ironmonger, D.S. (1972), *New Commodities and Consumer Behavior*, Cambridge: Cambridge University Press.
Katona, G. (1960), *The Powerful Consumer: Psychological Studies of the American Economy*, New York: McGraw-Hill.
Kay, N.M. (1982), *The Evolving Firm*, London: Macmillan.
Kay, N.M. (1984), *The Emergent Firm: Knowledge, Ignorance and Surprise in Economic Organisation*, London: Macmillan.

Klein, G.A. (1989), 'Recognition-primed decisions', *Advances in Man-Machine Systems Research*, **5**, 47–92.
Lancaster, K.J. (1966), 'A new approach to consumer theory', *Journal of Political Economy*, **14**, 133–56.
Lancaster, K.J. (1971), *Consumer Demand: A New Approach*, New York: Columbia University Press.
Lancaster, K.J. (1991), *Modern Consumer Theory*, Aldershot, UK and Brookfield, US: Edward Elgar.
Langlois, R.N. (2001), 'Knowledge, consumption, and endogenous growth', *Journal of Evolutionary Economics*, **11**, 77–93.
Langlois, R.N. and M.M. Cosgel (1998), 'The organization of consumption', in M. Bianchi (ed.), *The Active Consumer*, London: Routledge, pp. 107–21.
Loasby, B.J. (1998), 'Cognition and innovation', in M. Bianchi (ed.), *The Active Consumer*, London: Routledge, pp. 89–106.
Lundvall, B-Å. (1992), 'Innovation as an interactive process – from user-producer interaction to the National Systems of Innovation', in G. Dosi, C. Freeman, R. Nelson, G. Silverberg and L. Soete (eds.), *Technical Change and Economic Theory*, London: Pinter, pp. 349–69.
Lundvall, B-Å. (1992), 'User-producer relationships, national systems of innovation and internationalization', in Bengt-Åke Lundvall (ed.), *National Systems of Innovation: Towards a Theory of Innovation and Interactive Learning*, London: Pinter, pp. 45–67.
Marshall, A. (1899), *Elements of Economics of Industry*, 3rd edn (1st edn 1892), London: Macmillan.
Metcalfe, J.S. (2001), 'Consumption, preferences, and the evolutionary agenda', *Journal of Evolutionary Economics*, **11**, 37–58.
O'Farrell, P.N. (1995), 'Manufacturing demand for business services', *Cambridge Journal of Economics*, **19**, 523–43.
O'Farrell, P.N. and L.A.R. Moffat (1991), 'An interaction model of business service production and consumption', *British Journal of Management*, **2**, 205–21.
Pae, J.H., N. Kim, J.K. Han and L. Yip (2002), Managing intraorganizational diffusion of innovations: impact of buying center dynamics and environments', *Industrial Marketing Management*, **31**, 719–26.
Parkinson, S.T. (1982), 'The role of the user in successful new product development', *R&D Management*, **12**, 123–31.
Penrose, E. (1995), *The Theory of the Growth of the Firm*, 3rd edn (1st edn 1959), Oxford: Oxford University Press.
Randles, S. (2001), 'On economic sociology, competition and markets: sketching a neo-Polanyian IEP framework', mimeo, Manchester: CRIC, University of Manchester.
Robertson, P.L. and T.F. Yu (2001), 'Firm strategy, innovation and consumer demand', *Managerial and Decision Economics*, **22**, 183–99.
Robinson, J. (1962), *Economic Philosophy*, London: Penguin.
Rosenberg, L.J. and L.W. Stern (1971), 'Conflict measurement in the distribution channel', *Journal of Marketing*, **8**, 437–42.
Sabel, C.F., H. Kern and G. Heerigel (1989), 'Collaborative manufacturing: new supplier relations in the automobile industry and the redefinition of the industrial corporation', prepared for the MIT International Motor Vehicle Program International Policy Forum, Massachusetts.
Saviotti, P.P. and J.S. Metcalfe (1984), 'A theoretical approach to the construction of technological output indicators', *Research Policy*, **13**, 141–51.

Scitovsky, T. (1976), *The Joyles Economy: An Inquiry into Human Satisfaction and Consumer Dissatisfaction*, Oxford: Oxford University Press.
Shaw, B. (1985), 'The role of the interaction between the user and the manufacturer in medical equipment innovation', *R&D Management*, **15**, 283–92.
Shaw, B. (1987), 'Strategies for user-producer interaction', in R. Rothwell and J. Bessant (eds), *Innovation: Adaptation and Growth*, Amsterdam: Elsevier, pp. 255–66.
Shostack, G.L. (1977), 'Breaking free from product marketing', *Journal of Marketing*, **41**(2), 73–80.
Stigler, G.J. and G.S. Becker (1977), 'De gustibus non est disputandum', *American Economic Review*, **67**, 76–90.
Swann, G.M.P. (1999), 'Marshall's consumer as an innovator', in S.C. Dow and P.E. Earl (eds), *Economic Organization and Economic Knowledge: Essays in Honour of Brian J. Loasby*, Cheltenham, UK and Northampton, MA, USA: Edward Elgar, pp. 98–118.
Swann, G.M.P. (2002), 'There's more to economics of consumption than (almost) unconstrained utility maximisation', in A. McMeekin, K. Green, M. Tomlinson and V. Walsh (eds), *Innovation by Demand: An Interdisciplinary Approach to the Study of Demand and its Role in Innovation*, Manchester: Manchester University Press, pp. 23–40.
Thomas, R. (1994), 'Purchasing and technological change: exploring the links between company technology strategy and supplier relationships', *European Journal of Purchasing and Supply Management*, **1**, 161–68.
Tiger, L. and R.J. Calantone (1998), 'The impact of market knowledge competence on new product advantage: conceptualisation and empirical examination', *Journal of Marketing*, **62**, 13–29.
Vaaland, T.I. and H. Håkansson (2003), 'Exploring interorganizational conflict in complex projects', *Industrial Marketing Management*, **32**, 127–38.
von Hippel, E. (1976), 'The dominant role of users in the scientific instrument innovation process', *Research Policy*, **5**, 212–39.
von Hippel, E. (1978), 'Successful industrial products from customer ideas', *Journal of Marketing*, **42**, 39–49.
von Hippel, E. (1988), *The Sources of Innovation*, Oxford: Oxford University Press.
von Hippel, E. (1994), 'Sticky information' and the locus of problem solving: implications for innovation', *Management Science*, **40**, 429–39.
Warde, A. (2005), 'Consumption and theories of practice', *Journal of Consumer Culture*, **5**(2), 131–53.
Witt, U. (2001), 'Learning to consume – a theory of wants and the growth of demand', *Journal of Evolutionary Economics*, **11**, 23–36.
Woo, H.K.H. (1992), *Cognition, Value, and Price: A General Theory of Value*, Ann Arbor, MI: University of Michigan Press.
Wood, S.L. and J.G. Lynch (2002), 'Prior knowledge and complacency in new product learning', *Journal of Consumer Research*, **29**, 416–26.
Wynstra, F., B. Axelsson and A. van Weele (2000), 'Driving and enabling factors for purchasing involvement in product development', *European Journal of Purchasing and Supply Management*, **6**, 129–41.
Wynstra, F., M. Weggeman and A. van Weele (2003), 'Exploring purchasing integration in product development', *Industrial Marketing Management*, **32**, 69–83.
Zahra, S. and G. George (2002), 'Absorptive capacity: a review, conceptualization, and extension', *Academy of Management Review*, **27**(2), 185–203.

10. Consumption: the view from theories of practice

Sally Randles and Alan Warde

ONTOLOGIES OF CONSUMPTION

Until very recently, problematizing consumption has not been considered an urgent task among industrial ecology (IE) technicians or scholars (see Chapter 1, this volume).[1] An orthodoxy borrowed largely from mainstream economics, seeing consumption as an aggregate outcome of processes of acquisition, of depletion, and importantly of disposal, has often been taken as sufficient to quantify the movement of resources in a highly reduced macro system of resource flows, especially following the lifecycle assessment suite of methodologies (by way of example, Udo de Haes 2002; van der Voet 2002).

It is true that one of the mother disciplines of IE, ecological economics, incorporates the 'problem' of consumption, in two important respects. It takes up the cause of environmentalism, absent from mainstream economics, and constructs a position of critique against the latter (reviewed by Røpke 1999). First, ecological economists argue, the economic system should be conceived as a subsystem of the biosphere such that the environmental 'costs' of resource depletion (consuming resources) and pollution (depleting or 'consuming' the biosphere) should not be externalized as economics traditionally does, but rather they should be internalized to the economic calculation. To do otherwise undermines the functioning of the economic system itself and arguably fails to bring about human welfare or well being. Second, the interests of the environment can only be served by conserving or limiting the rate of use of natural resources, and this is best achieved by driving up the efficiency of resource 'consumption' within the productive system. Both of these arguments draw upon, and simply extend, the macro-scale calculative epistemology of mainstream economics, whilst a familiar parallel argument is made at the micro (firm) level seeking to improve resource efficiency of individual firms (Røpke 1999).

Ecological economics also picks up the baton of other issues and concerns of mainstream economics, such as questions of welfare distribution and of labour. Concern with the relationships between people and work, and people as 'consumers' (which throws the spotlight on 'the consumer' as the main culprit as well as principal lever of change) is quite different to the concern with aggregate (industrial) consumption sketched above. Here, Sanne starts for example by stating: 'consumer's choices are affected by structural factors in society such as working life conditions, urban structure and everyday life' (Sanne 2002, p. 273).

Sanne argues that pathways towards 'sustainable consumption' are severely constrained by working life conditions, such as the demands of childcare and work–time scheduling.

The defining analytical emphasis of IE however is on macro-energy and commodity flows. This extends the work of ecological economics on consumption in two ways. First it provides a more dynamic model than the static calculative snapshot. In so doing it notes that the effects of consumption do not grind to a halt at the point of consumption. Rather consumption provides residual material artefacts, which can be represented either negatively as pollutants, or positively as a 'feedstock' into other production processes. This perspective of consumption as a moment within cyclical processes does indeed open up new and helpful conceptual insights. It means that we can see consumption as related to 'systems of provision' rather than as isolated from them (as much of the sociology of consumption literature before the 1990s was prone to do), pointing to new places for methodological, policy and regulatory intervention not available to those who insist on a more linear and unidirectional view of the production–consumption relation. From the point of view of the individual firm, new directions for technological research and development are also opened up. For example, DuPont offers the idea of the post-consumer in a discussion of how the firm responded to raised consumer environmental concerns about recycling and re-use, by developing a recyclable version of the polyethylene component of envelopes such as are used by the courier FedEx (Sharfman *et al.* 2001).

Similarly, important work on dematerialization and service substitution (for example, Ryan 2000, Guide and van Wassenhove 2002) highlights technical innovations which reduce the impact of consumption on the environment without necessarily requiring a reining in of consumption itself. This provides a rich seam of research that highlights linkages across industry and scholarly IE. For the purpose of this chapter such work also draws attention to the very important point that it is not consumption *per se* which is the *bette noire* of environmentalism, rather it is the particular form, or forms, that consumption takes: sustainable consumption is not consuming less, but consuming differently.

However getting within the 'black box' of consumption as represented in flow diagram analyses, the representational method favored by industrial ecology, has still not been considered an item on the IE research agendas. Of course, questions of human behavior, in this case consumption behaviors, may not be relevant if the object of the science of industrial ecology is the 'aggregate mapping' of resource flows, rather than their explanation. But if asked to provide policy or management advice on the dynamics of consumption (that is, an understanding of how consumption changes), or to recommend policy intervention to effect changes which will halt or slow down the detrimental impacts of consumption processes on the environment, then a better understanding is necessary.

Indeed there have been some recent laudable attempts to critique the absence of agency in IE (Andrews 2001) and this is to be welcomed. However this still leaves the question hanging of what sort of agent (or consumer) – that is, the 'ontology' of the agent (or consumer) – IE envisages. Just what sort of agent is going to underpin the theorization of agency in IE? One risk, for example is that IE adopts the universal micro-agent of economics, also appropriating the rational, optimizing *homo economicus* of free will who has the means to exercise unconstrained choice and who follows rigid pre-set decision making 'rules'. An example of the latter marries the technical preference for sophisticated quantitative modelling in IE with the epistemological preference for macro-quantifiable phenomena, but injects a pre-set, and ontologically dubious rigid rule-driven agent as a proxy for capturing 'human behavior'. Such a development is persuasive for those interested in enabling the technically advanced computational powers of agent-based modelling borrowed from complexity research to play a role in IE. Developments of this epistemology inject a psycho-cognitive 'motivated' agent into the model, and to cope with phenomena of 'emergent' behavior arising from experience and learning, a game-theoretic approach is again borrowed from micro-economics (for example Fischoff and Small 2000).

Our point of entry is not, however, that industrial ecology has conceptualized consumption wrongly, rather that it has barely conceptualized it at all. It has done little more than assume a taken-for-granted popular portrayal of consumption – as brought about the aggregation of the free will of individual agents' in their purchasing decisions, and importantly as a malleable phenomenon open to change by policy makers, producers, and green educational campaigners alike.

By contrast, sociologists of consumption since the mid-1980s have been pre-occupied with the critique and reconceptualization of such notions of consumption. This research effort, we suggest, can assist industrial ecology scholars in their attempts to critically scrutinize the treatment of consumption in their own discipline.

Understanding how people behave in the domain of consumption, persuading others that some aspects of that behavior ought to be altered, and then showing how that behavior can be changed is a complex intellectual endeavor. It has been the task of a large number of scholars over the last few decades, in many different disciplines. How people behave in the domain of consumption, for what reasons they might behave differently, and how they might be persuaded so to do, are areas of controversy and dispute. This chapter can only scratch the surface of some aspects of these questions.

We begin by offering an alternative conceptualization of consumption. We are primarily interested in consumption that takes place within or around the household, and not in industrial or organizational consumption. (By contrast, this volume offers a chapter on industrial consumption. In fact there is no obvious reason why industrial or organizational consumption shouldn't be explored through the practice theory lens outlined below.) We define consumption as a process whereby agents engage in appropriation, whether for utilitarian, expressive or contemplative purposes, of goods, services, performances, information or ambience, whether purchased or not, over which the agent has some degree of discretion. This implies that we should make a radical distinction between purchase and consumption.

The approach that we suggest provides a plausible alternative ontology of consumption to that implied in much of the industrial ecology literature, and moreover one which we believe would helpfully move concern with consumption forwards for those working in industrial ecology. It is based upon a 'theory of practice'.

There are many variants of practice theory the corpus of which fails to provide a cohesive or consistent theoretical base. Indeed contradictions across different positions abound. For example, we can compare the Giddens inspired interpretation of Spaargaren (2004) with our Bourdieusian inspired position (Warde 2005). The former interprets practices as deriving from a duality of structure bridging across from, on the one side, an ontological underpinning situated in discursive and practical consciousness, linked to practices via lifestyle; and from the other, a set of rules and resources which determine a system of provision which in turn informs practices.

Our view is almost the reverse. We see practices as preceding individuals, consumption and lifestyle, on the one hand, and preceding systems of provision on the other. This suggests that the basic ontological unit for analysis is the practice – what do people do when they are shopping, 'householding', playing golf, hiking, eating out and so on? Following Bourdieu (1977, 1990), practices are theorized as being carried out in relatively unreflexive, unconscious ways, as captured by his notion of *habitus*. Notwithstanding there is a very sharp 'feel for the game' – a sense that different activities and behaviors are appropriate and acceptable in different social settings and are rewarded or punished accordingly (different forms of attire, deemed appropriate for different relatively discrete domains of practice, are worn to play golf, tennis, go to the cinema, go to an up-market restaurant for example, and may be punished by ridicule if misjudged).

The characteristics of our preferred theory of practice can only be briefly hinted at in this chapter, but a fundamental implication of all practice theories is that there is no universal solution to the problem of unsustainable consumption, only a set of contingent localized and compartmentalized adaptations of behavior in discrete domains of social life.

The relevance and plausibility of this re-worked conceptualization of consumption will be illustrated with reference to work on waste by Strasser (1999) and by Hand *et al.* (2005) on showering. Finally some thoughts concerning the implications of a practice-theoretic approach to consumption for industrial ecology will be offered. Before proceeding however, we provide a brief and highly selective outline the lineage of consumption research within sociology in order to identify a basis for critical engagement with industrial ecology, as well as the jumping off point within sociology for a revived interest in theories of practice.

PREAMBLE TO PRACTICE THEORY: DISTINCTION AND ORDINARINESS

Perhaps the most significant weakness in the theoretical toolbox of economics, regardless of any heterodox hue, is the commitment to the polarity of macro and micro approaches. Whether we take the macroeconomic calculative method, or the micro-homogeneous agent, in both cases we miss any sense of a meso-level where struggles between variously positioned classes of agent take place. By focusing attention at the meso-level we bring into sharp relief the reality of societies as structurally organized, stratified, and differentiated. Indeed societies universally exhibit a propensity to try and maintain and reproduce structures of differentiation in the face of pressures for change. From the standpoint of consumption, this view understands consumption as one of the arenas and vehicles through which structures (of class, age, gender and space such as national differences, urban versus rural and so on,) are fought over and bounded, shared, mutually understood, and used to self-identify; or by contrast misunderstood, contradicted, or contested from within or outside particular social strata.

In this vein, early institutionalist work was concerned with what Veblen (1953 [1899]) called conspicuous consumption. Individuals and groups mark their social position visibly through possession of items and participation in activities that signify prestige. This idea was developed among others by Pierre Bourdieu who espouses a version of a theory of practice in his seminal work *Distinction* (1984) a classic socio-structural explanation of consumption. Bourdieu argues that struggles between social classes for power and authority are expressed through shared and reproduced understandings of lifestyle and taste – especially 'good taste'. The appropriation of material goods and services – clothes, houses, holidays – corresponds to a (class-ordered) hierarchy of legitimate taste providing the appropriator with identifiable and externally endorsed signs of status, signaling group affiliation – though not in a self-conscious or reflexive (for example, rationally thought out and modified) way. Behaviors and actions recognized as 'inappropriate' to these unspoken understandings produce discomfort: the perpetrator does not 'fit in' with the powerful ordering effects of class distinctions and lifestyles. Bourdieu has been criticized for exaggerating socio-structural rigidity, leaving little room for changing tastes, experimentation or improvization. Since we know that consumption patterns do change, the question arises: How does change occur? This is a question we will return to below.

Subsequent scholars have also drawn attention to the phenomenon of omnivorousness of both activity and taste (Peterson and Kern 1996). Omnivorous consumption is defined not by its exclusiveness but rather by its inclusiveness. Members of upper social classes distinguish themselves by expressing a wider, all-encompassing range of preferences. Peterson and Kern's research, based an empirical study of music preferences in America, found that upper class respondents were knowledgeable and claimed to appreciate a wide range of music types, including country and western and gospel, traditionally the exclusive domain of less educated people with low cultural capital. Translated into implications for practice theory, one explanation for the ratcheting of consumption from the perspective of omnivorous behavior would be the tendency to dip into many more, varied, forms of purchases and experiences than Bourdieu would envisage. For example you would need to buy more CDs and attend more concerts, to give expression to omnivorous taste and behavior in practices associated with the public display of music appreciation. Regardless of these critiques, the power of Bourdieu's classic work still resonates for many sociologists of consumption and many empirical studies still confirm its validity, if in modified form (for example, Southerton's work on the interior design, look, and usage of kitchens and kitchen appliances, Southerton 2001).

Why do notions of socio-structural rigidities and processes of differentiation matter to industrial ecology? They matter because industrial ecology is amiss in assuming a world of standardized products inserted into a world of undifferentiated consumers. Instead it is necessary to understand the close-coupling of structures of differentiated consumption and corporate market segmentation strategies. This is powerfully demonstrated by Bourdieu's (2005) study of private housing in France. He shows how close coupled structures of 'niche' new-build housing provision inserts into the differentiated social, demographic and lifestyles profiles of potential buyers, providing stable structures of economy and economic relations.

It matters also in terms of providing a way into understanding the exponential rise in packaging, which is then jettisoned as 'waste' or 'rubbish'. Take, for example, packaging waste from branded groceries and luxury goods (perfumes are an extreme example). Marketing strategists intentionally accentuate and exaggerate lifestyle and taste distinctions through packaging to appeal to particular social groups (translated in marketing terms to market segments or niches). Packaging becomes less a reflection of function, protecting the product contents, and more a signature of taste, aspiration, relevance, desire (Lury 1996). Interestingly, the more consumers attempt to differentiate themselves through purchasing decisions, by 'choosing' from amongst a plethora of products and product variants, the more, according to structuralist accounts, the opposite occurs. Attempts to reflexively signal individual identity, autonomy and heterogeneity becomes the reverse – a willingness, even desire, to purchase according to advertisers constructions of consumers as grouped and ordered, and thus consuming according to 'type'. Such an understanding demotes and replaces the utility of the material packaging (its role as a container of the product) to a marketing device (for encouraging a less price sensitive response to buying decisions and consumption behaviors). The growth of packaging waste would thus be explained by the continual iteration of producers marketing strategies and consumers wishes to achieve differentiation.

However, as a response to the excessive attention paid to highly visible, spectacular types of consumption on which this literature relies (clothes and cars for example) a counter research agenda emerged stating that it is equally necessary to understand ordinary consumption. (Gronow and Warde 2001; Southerton *et al.* 2004). That is consumption which is unremarkable, everyday, routine and largely pulses away quietly in the background. Examples would be routine shopping (Lai 2001), water usage (Moss 2004), energy (Chappells and Shove 2004), or transport (Cass, Shove and Urry 2004).

Ordinary consumption is best understood in terms of concepts like habit, routine, constraint, and so on and can be summed up as a recognition of the conventional nature of consumption. The notion of routine and conventional consumption can be played out against more prominent ones of individual choice (utilitarian, strategic and rational) and its main contestant (deriving from cultural studies) that consumer behavior was the primary mode of expression of personal identity. By considering consumption behavior as 'conventional', it is implied that people mostly consume in a rather routine, habitual, unreflective way, in accordance with sets of shared aspirations, requirements and expectations. Of course not all consumption is of that kind, but a great deal is. Importantly for this chapter, it claims that unremarkable routine consumption which involves crucially the use of utility services such as water and energy pose at least as great a problem for sustainability as the more conspicuous forms that attract most attention. This is a particularly important bridge between a recent body of sociological research and the substantive concerns it shares with industrial ecology, around the consumption of water and energy resources, for example.

This is also a field which lends itself well to the theoretical lens of practice theory. Practices involve the appropriation of suites of products and services, some provided through the market some not. They incorporate visible as well as less visible items (the designer crockery and utensils alongside the vegetables and washing-up liquid in food preparation). Importantly they do not stand in isolation from provision systems, background institutions and infrastructures but are fundamentally interdependent with them, a point we will return to and illustrate below. For now let us turn to the rudiments of practice theory in order to demonstrate its theoretical power and salience to the study of ordinary consumption.

THEORIES OF PRACTICE

There are many versions of a theory of practice. Briefly, theories of practice begin from the assumption that practices are the core and fundamental unit of social existence and hence social analysis. Practices precede individuals; and practices determine the basic parameters of behavior, because reputation, decency, and so on, requires people to be seen as competent practitioners. The term practice implies modes of conduct collectively shared, historically established, normatively regulated and conventional, but which are imprecisely prescribed, eclectically implemented and gradually improvized upon in the process of people conducting their everyday lives. Participants in a practice exhibit common understanding, know-how, and belief in the value of the practice. People know what to do, even if they could not explain to anyone else how to do it, and act in accordance with a set of conventions which seem appropriate given their resources, dispositions and previous experience.

As regards consumption, we propose that items consumed are put to use in the course of engaging in particular practices (for example, eating or traveling) and that being a competent practitioner requires appropriate consumption of goods and services. The practice, so to speak, requires that competent practitioners will avail themselves of the requisite services, possess the appropriate tools, and devote a suitable level of attention to the conduct of the practice. Such a view is consistent with an approach to consumption which stresses the routine, ordinary, collective, conventional nature of much (final) consumption. It focuses attention first and foremost on practices – on what people do and think they are doing at the point of consumption. It is also consistent with the view that practices are internally differentiated, such that persons in different positions in social space do the same activity differently.

The importance of theories of practice for most of their adherents is that they start from distinctive presuppositions which are not explained on the basis of individual decision making, as with rational action theory, nor on the basis of functioning systems (where the operation of the society or the organization accounts for the behavior of its members). Particularly important in consumption studies is the distance it establishes from orthodox models of individual action – whether *homo economicus, homo sociologicus* or *homo aestheticus*. These all tend to start with explanations from the point of view of individuals undertaking many separate acts voluntarily – though subject to certain constraints – calculating self-interest, internalized social norms or rules, or considerations about presentation of self. Instead, analysis begins from understanding the history and development of the practice itself, the internal differentiation of roles and positions within practices, with the consequences for people of being positioned when participating. As Reckwitz puts it, what is distinctive about individuals is, so to speak, the sum of their positions:

> As there are diverse social practices and as every agent carries out a multitude of different social practices, the individual is the unique crossing point of practices. (Reckwitz 2002, p. 256)

So put another way, the uniqueness of an individual person is to be located as the point of intersection of all the practices within which he or she is positioned.

So, a myriad of established practices populate the social universe. Practices precede individuals; and practices determine, because reputation, decency, and so on, requires people to be seen as competent practitioners, the basic parameters of consumption behavior. The practice selects its items, the state of the practice explains the material conditions of its performance.

INSTITUTIONAL AND INFRASTRUCTURAL SETTINGS

Practices do not float free of technological, institutional and infrastructural contexts. Sporting and professional associations set the rules that govern the practice of their members, providing stability but also the means of acknowledging pressures to modify practice, alter rules, or institute change. The practice of driving a car is governed by the driving licence, technical capabilities of the car, laws of the road, and the state of the road network. Car driving is interdependent with each of these, but in a non-essentialist or deterministic way. The contingent way in which practice interrelates with the evolution of technologies, institutional context and physical infrastructure therefore cannot be attributed to a single universal causal mechanism, but rather to the specificity of local histories and conditions. Practice theory thus rejects deterministic accounts with a single factor responsible for causally directing events. Equally, in terms of policy response it rejects single source solutions, whether directed at institutional, infrastructure, or technological change. Practice is integrative across this trilogy of technology, institutional backdrop, and physical infrastructures and is only likely to be successfully modified through a combination of interventions which impact on each of these in a cohesive way.

It is the interdependence of practice and changing infrastructures of provision which is a key object of analysis in the book collection edited by Southerton, Chappells and Van Vliet (2004). For example, Tim Moss investigates the problems of infrastructural overcapacity that occurred when an infrastructure built to physically distribute and supply water was no longer synchronized with levels of water uptake and use in East Germany. His piece highlights the problems not of overconsumption, as popularly depicted, but of underconsumption. He illustrates powerfully that stagnant or declining consumption can be as much a problem for sustainability as overconsumption (Moss 2004). The editors summarize by discussing issues of infrastructural legacies and inflexibility, the role and organization of multiple supply side agencies in provision systems, and the role of intermediaries as change agents (see Chapter 11, this volume). The editors further draw attention to the differentiated nature of consumption and its social and collective nature, and the multidimensional modes and scales of organization that can configure different opportunities for environmental action (Chappells *et al.* 2004, pp. 144–9).

HOW DO PRACTICES (AND THEREFORE CONSUMPTION) CHANGE?

Among those who seek strategies for counteracting some of the alleged damaging consequences of mass consumption, three ways to change behavior are most commonly advocated. The first, the strategy of education, is to pass on information about the harmfulness of particular behaviors, on the presumption that if people know what is good for them they will, simply by consulting their own self-interest, alter their behavior. The second, the strategy of market regulation, usually entails the state altering the conditions of sale, prohibiting, restricting access to, or imposing punitive taxes upon a particular good or service to discourage bad behavior and, vice versa, encourage a more beneficial pattern. A third strategy is conversion, explicitly trying to persuade people by appeal to the moral, ethical or political superiority of reformed behavior. Green consumerism, the movement for fair trade, and so on, rely on such a strategy. All three have a part to play in changing behavior, but none has been resoundingly successful on their own in the past, and this is in part because they give undue primacy to a model of consumption as personal choice, failing to devise strategies for dealing with the collective and conventional grounding of consumption patterns.

The principal implication of a theory of practice is that the sources of changed behavior lie in the processes whereby practices themselves are reproduced or altered. Practices have some considerable inertia – and one can point to a series of features of everyday life which sustain that inertia, namely:

- prevalence of habits and habitual behavior;
- embodied know-how, which maybe costly to abandon and, if new forms of behavior are required, difficult to relearn;
- the community of other practitioners, who offer support, may also want to keep the structure of the practice much as before;
- infrastructure, which constrains behavior, steers it in the manner in which it is accustomed, especially if new forms of behavior require considerable investment; and
- excellence is conventionally established and any threat of reduced recognition is likely to be resisted.

People are not 'available' for change for those reasons. However, practices are also in continual evolution, usually as part of a process of incremental, step-wise, path-dependant development. The sources of dynamism include:

- pursuit of excellence and a degree of competition in all practices – this leads practitioners to want state-of-the-art equipment and experience which will enhance performance;
- the need to keep up and maintain standards of performance – and since this is relational there will be constant upgrading of material items (to the extent that there is invention of the same or its adoption by other and leading members);
- capitalist economic growth orientation – this panders to the insatiable wants of consumers, persuading people to adopt new things;
- market segmentation – the targeting of different sections of the market with particular items which are appropriate to people in that sector (for example, Amazon sends you a list of books that other people like you have also ordered) raises levels of consumption.

The collective development of practices transforms patterns of consumption and is in turn a primary source of innovation and expansion of demand. As Alfred Marshall observed (see Swann 2002), the expansion of demand is a process whereby activities generate wants, rather than vice versa. One point, or rather the hundreds of points, at which leverage to change in consumer behavior might be exerted is in the organization of common, frequent everyday practices. Effective strategy (one of contingent adjustment) entails concentration on the internal dynamics of particular practices, where all the items and resources required for the practice are scrutinized for their appropriateness and their necessity.

A HISTORIC EXAMPLE – WASTE IN PRACTICE

It may be helpful to illustrate empirically how things look different from the perspective of a theory of practice. We use domestic waste as an example. Waste is, of course, a primary environmental problem, one for which a great many people would express a diverse set of concerns about the profligate and inefficient use of nonrenewable natural resources, the throw-away attitude towards objects or the impact of waste disposal strategies. There is also an associated moral and distributional question, for long associated with notions of luxuries but now perhaps transposed into an issue of discarding usable materials, the issue being that some people lack the means to meet basic needs while others possess in superabundance items which they make little use of and which contribute little to their well-being.

There is a small but significant literature on the sociology of waste. Frequently it begins by citing Mary Douglas's anthropological account of pollution and taboo (Douglas 1996 [1966]). Douglas traces the social construction of dirt across societies and histories, and its corollaries, cleanliness and hygiene. Dirt is depicted as offending disorder onto which order is imposed through a range of strategies: tidying, disposal, and removing offending items to an invisible place in unrelenting attempts to organize the environment according to contemporary social norms and moral pressures. Bound up in belief systems expressed as righteousness and good citizenship, and translated into forms of coercion, the organization and elimination of dirt has both instrumental and representational dimensions. A second classic text is Michael Thompson's (1978) *Rubbish Theory*. Thompson again points to the changing meanings of what constitutes rubbish. He notes that material objects are overlain with social meaning which determines their value as either 'transient' (objective and sentimental value decreasing), 'rubbish' (no sentimental value), or 'durable' (objective and sentimental value increasing for example, antiques, family heirlooms, collectables, or yesterday's kitsch). More recently Chappells and Shove (1999) provide a clever analysis of the role of the changing technology of the dustbin as an entry to the same point about the historically shifting meaning of, and social value of, rubbish. Unlike others however the authors note the connections between domestic practices of managing waste and disposal, and their necessary insertion into physical systems and infrastructures of waste collection at the level of civic authorities. Noting shifts in attitudes (and civic infrastructures) towards presorting rubbish for recycling, Recycling bins, (see Withers 1996, p. 4), are described as 'conspicuous icons of our environmental responsibility'.

Of particular relevance to a practice perspective is the work of Strasser (1999) where the historical shift from the stewardship of objects to the throwaway society is charted. Most people for most of history threw very little away and cultivated the skills to make use of trash. There was a recycling industry (and an informal sector too) through the nineteenth century in the USA, fascinating in its organization for delivering items and collecting rags and pots and so on. One indication of the shift of attitudes was that obsolescence came to be seen as a good thing in the 1930s, a strange sort of Keynesian-inspired notion that it would be best for the economy and culture if things did not last long. The Second World War produced a quite different response, however, when saving things, giving things back to be turned into war materials, recycling and repair became a patriotic duty, though Strasser notes ironically that many relevant skills had been lost so that some items could no longer be made use of. The shift in culture, according to most historians is seen as a consequence of abundance, which changed the character of everyday life. Abundance caused a shift from thrift and saving to buying things new and rejecting things before they were worn out. Contemporary society is characterized by its partial utilization of items and its lack of the means to utilize waste products.

The force of Strasser's account is twofold. First, she shows that even as recently as the early twentieth century in America there were infrastructural systems in place for the re-cycling of most items. When new pans were sold, old ones were taken away by the salesman for reprocessing. Traveling salesmen for various products would make up loads of used goods and despatch them by railroad back to factories so that the used raw materials could be recycled. Clothes were reprocessed, sometimes domestically, to make dishcloths and rugs, not to mention in the repair and re-making of worn clothing, and commercially for the making of paper, stuffing furniture, and so on. This was a highly organized set of systems, where it was not the frugality of the consumer that accounted for the outcome (though in poorer times the small economic incentives involved may have encouraged participation) but the existence of an infrastructure for making recycling plausible, simple and costless. What appeared as the activities of individuals reducing waste was the result of a systematic set of arrangements for making that possible. Second, she shows that these arrangements harbored understandings and know-how with respect to the reuse of worn out objects. As she argues, during the Second World War when there was a major, widely supported, drive to economize on the use of raw materials, many of the previous means of conservation were no longer feasible either because the technological and organizational systems which facilitated recycling had been disbanded or because the practical skills involved no longer existed. The skills of dressmaking, sewing, darning, and so on, need to be learned and practiced if they are to make any contribution to the furnishing of suitable materials for the reproduction of households and their apparel. The passing of the rag and bone merchant, who came round the street with a horse and cart on a weekly basis symbolizes the end of an era of industrial organization in which recycling was an automatic and normalized component.

A CONTEMPORARY EXAMPLE: SHOWERING

A further explicit contemporary attempt to illustrate the effectiveness of the approach is the paper of our colleagues (Hand *et al.* 2005) which is concerned with one particular domestic practice, showering, now a very widespread and mundane activity in the UK. The rapid increase in the frequency of showering in the home might be explained in terms of changing household technological apparatus, changing standards of body maintenance or new pressures on personal and household schedules. Not only are all clearly relevant to the shift from the weekly bath to the (almost) twice daily shower, but are probably better understood conjuncturally, as a configuration, than as the outcome of some linear causal sequence. That is, though a domestic infrastructural system which delivers universally electricity and water, environmentally sensitive products is indeed a precondition of the normalization of showering, it is not its cause. Equally essential is both a form of understanding of the body and of conventional standards of cleanliness and sensibility, and a conventional concern with the temporal sequences of daily life. Instead Hand *et al.* (2005) conceptualize the process in terms of the dynamics of the organization of a particular practice – the practice of showering – in which the arrangement of elements of technology and infrastructure, cultural ideas about the body and the social ordering of time coalesce. There are pressures towards fixing and reproducing the mode of activity, but also crises within and disjunctures between the elements which see them co-evolve. The practice is, in effect, configured by the arrangement of the relations between these three elements and can be described in terms of understandings, know how and values which constitute competence in that practice.

This is, altogether, a way of trying to articulate an account of an activity which is subject to quickly moving alterations and development of organization and convention. Articulation is difficult because it is necessary to decenter the account from that of individual intentional action, to dispense with the narrative form which encourages a causal and sequential account; to refuse to succumb to the temptation to sharpen the account by saying that one factor is the major one in explaining outcomes; and to express and account for the interdependence (and connections) of three diverse institutional complexes (technology, body maintenance and the scheduling of daily life).

IMPLICATIONS OF PRACTICE THEORIES OF CONSUMPTION FOR INDUSTRIAL ECOLOGY

There has recently been a resurgence of interest in practice theory to explain and understand consumption from within sociology. However this is itself a divided literature, and we presented our preferred approach which is one inspired by the work of Pierre Bourdieu. For industrial ecology two questions remain. First, is it necessary or appropriate for IE to concern itself with the inner workings of consumption at all? Second, if the answer is yes, what is the theoretical value and policy relevance of practice theory?

We acknowledge that becoming acquainted with sociological theories of consumption is an onerous task for a discipline as far away epistemologically as industrial ecology and IE scholars may find the task both uninteresting and irrelevant. But if IE is genuinely interested in engaging with explanations of consumption behaviors or, in particular, in contributing to the policy debates concerned to understand how, if at all and when, current consumption patterns and the practices underpinning them can or 'should' be changed, then, engaging with such a literature is not only relevant, but imperative.

Above all, a practice theory of consumption suggests consumption is 'sticky' and is not readily amenable to education campaigns appealing to the norm-driven rational-choice consumers who are able and willing to respond as individuals by changing their behaviors as a result of education campaigns pointing out the errors of their ways. There are plenty of examples attesting to the stickiness of consumption. One is the eating habits which underpin obesity. It appears that the trend to consume excessive amounts of certain kinds of food continues unabated despite numerous sources of nutritional advice and educational campaigns aimed at curbing overeating. The phenomenon is deeply embedded in habituated practice, institutions of eating (quantities and frequencies of food consumed), technologies which change the composition of the food (adding fat, sugar and salt) and a provision system which rewards increased consumption via increased profits accruing to producers regardless of the health effects. This is just one of many examples which demonstrate that consumption is crucially situated in interdependent relations with and within physical and institutional infrastructures, such that consumption and infrastructure can only be nudged in new directions together, not independently. This does not mean, however, that practices are immune to change, and as Strasser's account on the history of waste demonstrates, changing practices can have far reaching implications for the environment.

The contingent nature of the formation and evolution of practice emphasizes specific historical settings, and would methodologically advocate comparative studies to highlight parameters of difference. We are talking about a unit of analysis which sits above micro-analysis, whether individual goods, services or commodities, or individual agents, but below macro-quantification of resource flows. We noted that practices utilize suites of technologies situated within infrastructures and institutions governing both provision and consumption. As such, practice theories reject universal or generalized accounts of consumption, and emphasize historical and spatial contingency. The approach equally rejects both an ontology of the reflexive, voluntaristic, individual actor and of the 'dupe', the easily led and easily manipulated consumer. Such a reworked conceptualization of consumption provides, we suggest, a useful complement to resource flow analyses, providing explanatory leverage on phenomena which are crucial to the research programs of industrial ecology but which to date have apparently been left in a black box.

NOTE

1. For notable recent exceptions see the special edition of the *Journal of Industrial Ecology*, **9**(1–2), edited by Hertwich (2005).

REFERENCES

Andrews, C. (2001), 'Building 2 micro foundation for industrial ecology', *Journal of Industrial Ecology*, 4(3), 35–50.

Bourdieu, P. (1977), *Outline of a Theory of Practice*, Cambridge: Cambridge University Press.

Bourdieu, P. (1984), *Distinction: A Critique of the Judgment of Taste*, London: Routledge.

Bourdieu, P. (1990), *The Logic of Practice*, Cambridge: Polity Press.

Bourdieu, P. (2005), *The Social Structures of the Economy*, Cambridge: Polity Press.

Cass, N., E. Shove and J. Urry (2004), 'Transport infrastructures: a social–spatial–temporal model', in D. Southerton, H. Chappells and B. Van Vliet, *Sustainable Consumption: The Implications of Changing Infrastructures of Provision*, Cheltenham, UK and Northampton, MA, USA: Edward Elgar.

Chappells, H. and E. Shove (1999), 'The dustbin: a study of domestic waste, household practices and utility services', *International Planning Studies*, **4**(2), 267–80.

Chappells, H., B. Van Vliet and D. Southerton (2004), 'Conclusions', in D. Southerton, H. Chappells and B. Van Vliet (eds) (2004), *Sustainable Consumption: The Implications of Changing Infrastructures of Provision*, Cheltenham, UK and Northampton, MA, USA: Edward Elgar, Chapter 10.

Douglas, M. (1996), *Purity and Danger: An Analysis of the Concepts of Pollution and Taboo*, first printed 1966, London: Routledge.

Fischhoff, B. and M.J. Small (2000), 'Human behavior in industrial ecology modelling', *Journal of Industrial Ecology*, **3**(2/3), 4–7.

Gronow, J. and A. Warde (eds) (2001), *Ordinary Consumption*, London: Routledge.

Guide, V.D.R. Jr and L.N. van Wassenhove (2002), 'Closed-loop supply chains', in R. Ayres and L. Ayres (eds), *A Handbook of Industrial Ecology*, Cheltenham, UK and Northampton, MA, USA: Edward Elgar, Chapter 40.

Hand, M., E. Shove and D. Southerton (2005), 'Explaining showering: a discussion of the material, conventional, and temporal dimensions of practice', *Sociological Research Online*, **10**(2), accessed at www.socresonline.org.uk/10/2/hand.html.

Hertwich, E. (ed.) (2005), 'Consumption and industrial ecology', *Journal of Industrial Ecology*, **9**(1/2), guest edited edition on consumption.

Hetherington, K. (2004), 'Secondhandednesss: consumption, disposal and absent presence', *Environment and Planning D: Society and Space*, **22**, 157–73.

Lai, Shou-Cheng (2001), 'Extra-ordinary and ordinary consumption: making sense of acquisition in modern Taiwan', in J. Gronow and A. Warde (eds), *Ordinary Consumption*, London: Routledge.

Lury, C. (1996), *Consumer Culture*, Cambridge: Polity Press.

Moss, T. (2004), 'Institutional restructuring, entrenched infrastructures and the dilemma of overcapacity', in D. Southerton, H. Chappells and B. Van Vliet (eds), *Sustainable Consumption: The Implications of Changing Infrastructures of*

Provision, Cheltenham, UK and Northampton, MA, USA: Edward Elgar, Chapter 7.
O'Brien, M. (1999), 'Rubbish-power: towards a sociology of the rubbish society', in J. Hearn and S. Roseneil (eds), *Consuming Cultures: Power and Resistance*, Basingstoke: Macmillan, pp. 262–76.
Peterson R.A. and R. Kern (1996), 'Changing highbrow taste: from snob to omnivore', *American Sociological Review*, **61**, 900–909.
Reckwitz, A. (2002), 'Toward a theory of social practices: a development in culturalist theorizing', *European Journal of Social Theory*, **5**(2), 243–63.
Røpke, I. (1999), 'The dynamics of willingness to consume', *Ecological Economics*, **28**, 399–420.
Ryan, C. (2000), 'Dematerializing consumption through service substitution is a *design* challenge', *Journal of Industrial Ecology*, **4**(1), 3–6.
Sanne, C. (2002), 'Willing consumers – or locked in? Policies for sustainable consumption', *Ecological Economics*, **42**, 273–87.
Sharfman, M., R. Ellington and Meo (2001), 'The introduction of postconsumer recycled material into TYVEK®: production, marketing and organizational challenges', *Journal of Industrial Ecology*, **5**(1).
Southerton, D. (2001), 'Ordinary and distinctive kitchens; or a kitchen is a kitchen is a kitchen', in J. Gronow and A. Warde (eds), *Ordinary Consumption*, London: Harwood Press, pp. 159–78.
Southerton, D., H. Chappells and B. Van Vliet (2004), *Sustainable Consumption: The Implications of Changing Infrastructures of Provision*, Cheltenham, UK and Northampton, MA, USA: Edward Elgar.
Spaargaren, G. (2004), 'Sustainable consumption: a theoretical and environmental policy perspective', in D. Southerton, H. Chappells and B. Van Vliet (eds), *Sustainable Consumption: The Implications of Changing Infrastructures of Provision*, Cheltenham, UK and Northampton, MA, USA: Edward Elgar, Chapter 2.
Strasser, S. (1999), *Waste and Want: A Social History of Trash*, New York: Metropolitan Books.
Swann, G.M.P. (2002), 'There's more to the economics of consumption than (almost) unconstrained utility maximisation', in A. McMeekin, K. Green, M. Tomlinson, V. Walsh (eds), *Innovation by Demand*, Manchester: Manchester University Press, Chapter 3.
Thompson, M. (1979), *Rubbish Theory*, Oxford: Oxford University Press.
Udo de Haes, H.A. (2002), 'Industrial ecology and life cycle assessment', in R. Ayres and L. Ayres (eds), *A Handbook of Industrial Ecology*, Cheltenham, UK and Northampton, MA, USA: Edward Elgar, Chapter 12.
Van der Voet, E. (2002), 'Substance flow analysis methodology', in R. Ayres and L. Ayres (eds), *A Handbook of Industrial Ecology*, Cheltenham, UK and Northampton, MA, USA: Edward Elgar, Chapter 9.
Veblen, T. (1953), *The Theory of the Leisure Class: An Economic Study of Institutions*, first published 1899, New York: Mentor.
Warde, A. (2005), 'Consumption and theories of practice', *Journal of Consumer Culture*, **5**(2), 131–53.
Withers, J. (1996), 'Designs on your dustbin', *The Independent*, 10 April.

11. Ecology of intermediation

Will Medd and Simon Marvin[1]

INTRODUCTION

This chapter explores how the concept of intermediary organizations can contribute to the development of industrial ecology and our understanding of the relationships between material flows, technologies and social logics. We argue that intermediaries play an important role in understanding transformation in the flows of resources by enabling the introduction of new technologies or changed social practices into different contexts. In their work across different social logics, intermediaries also work across different spatialities through which they enable the introduction of new social practices or technologies that implicate resource flows. In line with the principles of industrial ecology (following Erkman and Ramaswamy 2000) intermediaries make visible the ways in which value can be extracted from resources that may be under-utilized or seen as waste, and make linkages of innovation between firms.

We develop our focus on the concept of intermediaries by drawing upon research about sustainable water management as an example of the infrastructural challenges of developing more sustainable material flows. One of the challenges of understanding sustainable water management is that there is no clearly defined water sector. Indeed, the proliferation of intermediaries within the water sector highlights the hidden role that different organizations play in effecting the uses of water. In England and Wales regional water companies are the dominant water supplier (in addition are some private companies abstracting water on specific sites) and developing strategies at this level that can be operationalized in the daily practices and routines of different types of users has proved problematic. Increasingly water companies and regulators are looking to the work of intermediaries in translating those strategies into local contexts.

The chapter is set out as follows. First, we show how changes to the structuring of the water sector have been the proliferation of intermediary organizations. Second, we overview some of the characteristics of these organizations and the work they do. Third we argue that the significance of intermediaries is in the work they do across in translating technologies into new social contexts by working across different social logics and spatialities. Finally we conclude with some reflections on the implications of the concept of intermediaries for the industrial ecology metaphor.

A note of caution: these new intermediaries are not always successful or necessarily inherently positive for either the environment or all users. For example intermediaries tend to be highly selective and may focus on premium users who are offered additional services and savings, while low-income or small users may not be so attractive to intermediaries driven by a commercial logic. Although intermediaries may actively promote resource-saving practices and activities to reduce the costs of utility services to large users, they can also help stimulate additional demands by lowering costs of service or promoting technologies, such as power showers intermediary action is not therefore inevitably resource-saving.

TRANSFORMING INFRASTRUCTURE: THE EMERGENCE OF INTERMEDIARIES

Transformations in the social organization of infrastructural provision create a new context that has stimulated the development of intermediaries within an enlarged space between production and consumption. More specifically, intermediaries bring together new technologies and changed social practices that offer important possibilities for reshaping production, consumption and the relations between them. But what are intermediaries?

If we take intermediaries as people and objects that act 'in-between' different processes, then everything within an industrial ecology can be considered an intermediary. Every moment through which different resources are transported and transformed requires a form of intermediation. In the case of water, for example, a whole host of intermediaries bring potable water to our taps, from the tap itself through to treatment works and reservoirs, and the complex of pipes, laboratories, institutions and people in between. In other words, assemblages of technological, social and hydrological intermediaries combine at various points to form what has become termed the water cycle. Indeed we could trace the 'biography' of water to explore the role of intermediaries throughout its history, from 'the first flood' to the 'water wars' (Ball 1999). An enlarged approach to intermediaries builds awareness of the significant role that a variety of phenomena play in the production and consumption of an apparently simple resource. This view focuses attention on the heterogeneity of infrastructural provision and to the importance of 'heterogeneous engineers' in creating the complex assemblages we call water infrastructure (Latour 1996; Law 1987). But if everything that sits between a supplier and a user is an intermediary what is the explanatory value of the concept?

In its everyday use the term 'intermediary' tends to have a more specific meaning that incorporates the intentionality of the intermediary. Intermediaries are deliberately positioned to act in between by bringing together and mediating between different interests. There are for instance spiritual intermediaries, financial intermediaries, health intermediaries, peacekeeping intermediaries and web-based intermediaries. These intermediaries position themselves to have a particular role in literally intermediating between sets of different social interests, to produce an outcome that would not have been possible, or as effective, without their involvement. An intermediary is not necessarily either a producer or consumer but they may well work between these interests to achieve an objective. Such intermediaries are, then, strategic.

Bringing the discussion back to water as an example of a material resource, we can argue that under classic monopoly conditions water networks were configured around a powerful supply logic designed to provide clean water to domestic users to ensure wider public health, and to industry to support economic development. This 'state hydraulic' logic (Bakker 2003) aimed to ensure that the production and use of water was uninterrupted so that there was no need for intermediaries to intervene between producers and users unless they accelerated the provision of water. Water supply passed through a complex range of social, natural and technical intermediaries and these were assembled to narrow the gap between production and consumption interests to ensure the rapid, reliable and continuous flow of water to users. Domestic water metering, for instance, has been rare in the UK because of the desire to guarantee that water was not restricted in households and to ensure collective public health. Only at times of water shortage was the relationship between supply and use brought into view and subject to scrutiny as public water providers exhorted users to temporarily reduce demand during a drought, before re-establishing normal relations.

Hence while traditional actors such as consultants or public health professionals might be considered as intermediaries, 'new intermediaries' emerged in the water sector when the context of producer-consumer relations changed. Organizing infrastructure under classic monopoly conditions meant that the distance between production and consumption interests was deliberately short, as the public producer of water was responsible for assembling the complex socio-technical-natural relations required to deliver water to mass markets. Public health intermediaries had a role in ensuring that households used water to ensure cleanliness and disease prevention. Users were assigned a relatively passive role that was to consumer sufficient clean water to ensure that wider public health conditions were met. Within the context of privatization in 1989, there have been shifts in the social organization and priorities of the water sector in a move to what Bakker (2003) has termed a 'market environmentalist' logic. This new logic combines market-based instruments for economic regulation (by comparative competition) with scarcity-responsible demand side management. Consequently it has significantly reshaped the relations between production and consumption interests, creating an enlarged context for intermediaries and intermediation. The key to understanding this context is the dual process of the unbundling of integrated infrastructure networks and their selection re-bundling by intermediaries (see Graham and Marvin 2001).

Commercialization, privatization and liberalization of networked infrastructure effectively unbundled an integrated network into competitive and monopolistic segments. There has been separation of roles vertically and horizontally, of spatial areas through which services are delivered, and of types of consumers whom the market-oriented companies target. At the same time new concerns have arisen around particular consumer interests, regulation and sustainability that, coupled with processes of fragmentation, have consequences for the coordination of consumption and provision. It is within the space of infrastructural fragmentation that new possibilities for the role of intermediaries are emerging. These 'intermediaries' are deliberately positioned to have mediating roles in relation to sustainability.

THE DIVERSITY AND WORK OF INTERMEDIARIES

Within the context of infrastructural transformation we are interested in the emergence of strategic intermediaries that have a role in reshaping the flows of water in relation to production and consumption practices. These intermediaries always appear in hybrid forms, combining technology, organizations, networks as well as texts, money and people (see Callon 1997), and constitute new forms of interdependencies and socio-technical assemblages, as with infrastructure itself (Graham and Marvin 2001, pp. 30–31). As such the legitimacy and expertise of intermediaries do not emerge within a space between distinct spheres of production and consumption, but gain their legitimacy by working across the boundaries of a range of interests not normally considered as significant to strategies of sustainability. Importantly, and in contrast to other work that has highlighted the systemic characteristics of intermediaries in relation to innovation systems (van Lente 2003) our understanding of strategic intermediaries points to the diversity of forms and selective, rather than network wide, inter-linkages they can take.

There are different ways in which we could draw distinctions about intermediaries. For example intermediaries may be small scale consultancies perhaps with only one person offering advice on water saving, or they may be large scale companies offering a range of infrastructural services. They may be commercial, non-profit, forms of government organization or network organizations. They may gain their status through their expertise in a particular area, the technology they have developed, the voice they represent or the areas of influence they have. Due to limits of space, instead of describing more fully these characteristics, more relevant to the argument of this chapter is the work that intermediaries do. In particular we want to show how intermediaries work in different ways with technological development, work across different social logics and work across different spatialities.

The move from a 'state hydraulic' to a 'market environmental' has involved an opening of technological innovations relevant to water management. Technological considerations of the water sector now need to consider the large scale technical infrastructure such as the pipes that, with the help of gravity, carry water over 100 miles from the Lake District to the city of Manchester, as well as the small scale technologies such as power showers, water saving taps and flow-metering equipment. This technology matrix involves a hybrid set of old and new, large and small scale, low and high flow water usage, cheap and expensive, low tech and high tech. Technological solutions are one thing, putting them into context is another and here we are interested in the role that has come to be played by intermediaries in working in-between technologies and particular social contexts.

The work that intermediaries undertake in relation to making technologies work varies in relation to their different sets of focus. Four aspects in particular can be identified. First intermediaries may promote existing technologies, for example demonstrating to potential users the benefits of an existing product. This might mean promoting water saving devices, water recycling or sustainable drainage systems. Second, developing new technologies often results from having identified a problem that is locally very specific and requires a particular design system. Third, the development of a technology market might involve the promotion of existing markets, enabling the development of prototypes or finding new applications for existing bespoke technologies. Finally the challenge can be in making better use of existing technologies or exploring the possible linkages a technology may offer.

Particularly important to the work of intermediaries is how their differing work with technologies involves working across different sets of logics. We use the term logics to refer to different types of interests into which 'water' is put to work. The term logic is used to denote a dynamic that is beyond specific individual organizational motives and which extends into a wider of set of relationships, in turn embedded within other logics. This is not to deny the ways in which a logic is always located locally, but it is to show how such localization is also connected to other local events. Our interest then is how intermediaries play a key role in translating water – symbolically and materially – into different logics, how water is put to work in and across different social logics. Water clearly holds many values, embedded within cultural and institutional dynamics (see Strang 2001; Swyngedouw 2004). Our intention here is not to identify all the different dimensions of values that water holds and the contradictions between them. Instead, our concern is to illustrate the ways in which intermediaries play a role in translating across different logics and how in doing so make water work with differing sets of interests at the same time. There are three sets of logics we want to outline: environmental logics, economic logics, and social/health logics.

The environmental logics are made visible by intermediaries in different ways. First, the environmental may appear in the background, for example with the focus being on how the sustainability of a sector may require environmental considerations in terms of regulation, or even seeing the environment as a market opportunity. A specific resource, like water, would then appear as a focus within the context of such regulation or opportunity. Second, the focus of the intermediary may be part of a wider environmental strategy. In this case, a resource like water may be explored alongside a wider set of environmental concerns, perhaps with climate change or regeneration. In the case of climate change this means water problems appear more as a consequence of environmental degeneration – in terms of flooding and drought – and the solutions lie elsewhere, for example in energy use, emissions and so on. Third, water itself can be the specific environmental focus, for example aiming to reduce levels of water abstraction or pollution. In practice the ways in which these environmental logics are constituted is more complex, for example the ways in which the environment is positioned may vary and involve, for example, combinations of political will, economic value, symbolic concern or policy driven concerns.

The economic logics that intermediaries work with are also manifest in three ways. First water can be translated into a direct commodity with economic value. Hence a range of commercial intermediaries pursue the direct sale of water competing on economic grounds. Second, water might be translated into business efficiency, for example recycling water to reduce abstract and disposal charges or reducing insurance bills by developing a sustainable drainage system. Third, water may be seen as being carried for other markets, for example raising property values, increasing tourism and attracting investment.

Water is also translated by intermediaries into different social and health logics. While the two could be analysed separately they come together in the sense that they are often tied to a concern about the loss of democratic accountability of the water sector and the role of water as a public good. First intermediaries might be concerned with the health implications of water ensuring for example that public health is not jeopardized in the context of commercial pressures. Second are concerns about the implications for equity of water charging, for example weighing up the extent to which environmental improvement should be paid through water bills. Third are increased concerns about engaging the public in water decision making.

INTERMEDIARY SPACE

Spatial considerations are crucial to water management and to industrial ecology. This is not surprising when we consider how the hydrological cycle 'carries water on an unending journey through streams, rivers and oceans, the atmosphere, the ice sheets, living systems and the deep earth' (Ball 1999, p. 24). This cycle is only further complicated when one adds the reservoirs, treatment works, pipelines, drains, sewers and taps of infrastructures guiding water that have become so taken for granted in the modern world (Illich 1986). The ways in which socio-technical systems have come to involve new forms of spatial connectivity have been well documented. Formulations such as 'time-space distantiation' (Giddens 1984) or 'time-space compression' (Harvey 1989) point to the ways in which space is not an abstract container within which events happen, but time-space relationships are constituted through the orderings of social-technical relations. Here we draw upon work by Mol and Law (1994) who make the distinction between regions, networks and fluids:

> First, there are *regions* in which objects are clustered together and boundaries are drawn around each cluster. Second, there are *networks* in which distance is a function of the relations between the elements and difference a matter of relational variety . . . [and fluids, where] boundaries come and go, allow leakage or disappear altogether, while relations transform themselves without fracture. (Mol and Law 1994, p. 643)

Regions

For sustainable water management there are problems of defining water by a regional space. This has been highlighted by considerations of the potential impact of the European Union Water Framework directive (European

Community 2000), the implementation of which will involve a nested character of multiple spaces that we can characterize as 'regional', for example the water companies management boundaries, local government boundaries, regional government, national government boundaries, non-governmental organizational boundaries, local communities, urban boundaries, rural boundaries and so on. Thus, while strategies for sustainable water management tend to represent particular regions (Europe, nation state, region, sub-region, river basin and so on), difficulties soon emerge about the interconnections and nestedness between regional spaces. Indeed, as institutions adapt towards a river basin model as the territorial unit of water management, 'problems of interplay between water and other relevant institutions – such as for spatial planning, agriculture or nature conservation – may be exacerbated' (Moss 2003, p. 86; White and Howe 2003).

Networks

A focus on the multiple and nested regional boundaries itself raises important questions about the translation of material resources across different spatial scales. This is important to understand. However, what is missing if we remain within a regional imagery alone, are the ways in which other spaces are also important and through which regional entities are constituted. Much work, in particular, has argued for the importance of network spaces in the constitution of regional spaces, for example through the road networks, telecommunications and, of course, water (Graham and Marvin 2001).

Mol and Law argue, 'the space in which regions can be drawn and differentiated exists . . . is an effect or a product that depends on another quite different kind of space, the space of networks' (1994, pp. 648–649). The formation of the North West Regional Water management was premised on the notion of generating an 'integrated zone' (Chappells 2003). However, if we take the perspective of network space we see the distribution of water throughout the North West through a vast array of all-channel networks incorporating a complex assemblage travelling in all directions, more or less tightly coupled at different points. While such a network may appear, or indeed be represented by corporate bodies as an integrated whole, it is also characterized by fracture, disruption and differentiations. The people in the City of Manchester, for example, use a large supply of drinking water supplied from the Lake District which travels through large scale pipelines running over 60 miles. Yet as the water also passes through, that water supply remains untouchable for people in between whom the pipeline runs nearby (see Graham and Marvin 2001 on these 'by-pass' issues). Importantly, the network space of the water infrastructure is one that disconnects as much as it connects.

Networks then cut across regional spaces and complicate our picture of the North West regional water sector. Network space is a space that splinters the apparent integrated regional space (Graham and Marvin 2001). The infrastructure of water networks connects and disconnects asymmetrically different locations, differentiating, for example, between industrial sites and domestic homes, supplying water from parts of the region to other parts of the region, discharging waste water in some areas and not others. Such splintering is multiplied. Just as there are different and overlapping regional spaces, so too there are networks among networks. There are networks of local water supply, for example, but also interlinkages between water supply and electricity (to drive pumps and treatment works), between waste water and rivers. These spaces are significant for they point to the limitations of a perspective that seeks to integrate water management at a river basin level, adding to the problem of overlapping regional boundaries the fragmentation of linkages across, through and in-between different networked spaces.

For sustainable water management this interdependence of regional and network spaces is made further problematic by the wide spread distribution of different types of water users. Within a regional spatial framework patterns of water consumption and waste-water production can be aggregated. Such aggregation can include some differentiation of types of users, for example households, industrial users, small to medium enterprises and public sector organizations. However, within these groups we find their geographical dispersal involves quite different linkages to the networks of infrastructure enabling the supply of water or taking away waste-water. As we start to look to the networked sets of relationships that connect different users then we move down in scale to the intricacies of local networks that reveals sets of inter-dependencies otherwise not visible. Here we want to focus on one particular example, the position of small to medium enterprises (SMEs). In contrast to large-scale industrial users whose activities are easily identifiable by the water companies and the regulators, the activities of SME organizations often remain harder to identify. Individually their water consumption is marginal such that changes in individual water usage would not have a huge impact on water resources. Collectively, however, their individual iterations through the networks lead to a significant impact both in terms of water usage and waste-water disposal. For the environmental regulator the geographical distribution of SMEs across the region poses particular problems for introducing sustainable water management. Here we want to turn to an example where an intermediary took on an important role for the regulator and in doing so we see a further space, a fluid space, in-between networks and regions that enables the translation of sustainable water management from the strategic level into these specific contexts (see Hayes 2002).

Fluid space

Introducing fluid space in a chapter about water is not intended as a pun. Mol and Law's (1994) formulation of the distinction between region, network and fluid space draws upon the metaphor of the body as region, the veins as network and the blood as fluid. They draw attention to the ways in which the boundaries of a fluid 'come and go, allow leakage or disappear altogether' (Mol and Law 1994, p. 643). The boundaries of water dynamics do indeed come and go. Times of flooding make this only too apparent, as do times of drought. The imagery of a fluid is also useful, however, for understanding the types of dynamics involved in translating sustainable water management across different contexts. It offers a way of understanding how the very specific contexts of local practices and technology introduction are enabled and reconnected into the water networks and the regional water spaces.

We want to illustrate our argument through the example of a small project that we consider to be an intermediary organization that works in between the regional and network spaces by working as a more fluid space that allows adaptable and mutation across different boundaries. The organization we will call the SME Water Advice Project (SMEWAP). SMEWAP emerged from a problem faced by the environmental regulator, the Environment Agency (EA) and the constitution of SMEWAP involves the enrolment of yet more networks into the problems of water management. The problem faced by the EA was the need to promote sustainable water practices while at the same time being the enforcer of environmental standards, imposing fines for breach of waste water, for example. Of particular concern to the EA was the role of SMEs both in terms of water use and waste-water disposal. While the practice of SMEs might become visible to the EA when an infringement takes place – for example when a company releases waste water into a river this is reported by the public – it did not have the resources to regularly check and monitor the activities of SMEs distributed around the region. Further still, SMEs were reluctant to approach the EA for advice because of the risk of subjecting their practices to investigation and risking high penalties or investment costs. The solution was for the Environment Agency to establish a project that would act as an advisory service for SMEs. But such a service, if it was to be acceptable to SMEs, would need to link into SME networks and to do this could not be based within the EA. A pilot project was established within a local area and was set up within an existing business advisory centre. In this way, the project could work through the existing local business networks offering free advice to business on sustainable water management issues. At first sight then we see the opening up of another regional and network space for water management. The project is established to cover a particular geographical area (a regional space) and to link into existing business networks (a network space). A look more closely at how SMEWAP works reveals the constitution of a fluid space that enables it to work.

To illustrate this we can turn Law's (2002, pp. 99–100) more detailed specification of fluid space in relation to objects. He identifies four characteristics. First, there are 'no particular structure of relations is privileged'. In the case of SMEWAP, its ability to present itself in different forms is crucial.

'If they need to be greener and they know they need to be greener I keep mentioning the agency. If they need to make a business improvement because they're losing money we keep mentioning the BEA because it sounds more business. So you're sort of, you choose the slant to take with them . . .' (interview with SMEWAP adviser). While some companies are motivated by money saving, others might, however be motivated by achieved ISI 14001 status. Hence SMEWAP might present to a company in relation to how it can avoid high fines for waste-water discharge; it may present to a company in terms of the potential costs savings of reduced water use or of reduced insurance through a sustainable drainage system (reducing flooding risk); or it may present to a company in terms of the symbolic value a company might gain by being seen to contribute to sustainability. In each of these not only does SMEWAP present a different form (for example sometimes it makes explicit its work for the EA, sometimes it presents instead in relation to the local business network) but is also shows how the issues of sustainable water management become transformed into different contexts: as legal practice, a commercial practice, a symbolic practice. And in doing this through the fluidity of SMEWAP is translating sustainable water practices and technologies into the localized regional space of the SME that is in turn connected to wider networks of water supply and waste-water disposal.

However, thought change and adaptability is important, those 'relations need to change a bit rather than all at once', otherwise the work of SMEWAP loses identity. SMEWAP, while presenting itself in different forms does nonetheless maintain a focus, namely on SMEs. It always presents itself as SMEWAP even if the emphasis of what the organization is changes, for example emphasizing the regulator's funding or emphasizing the business network links. Third, 'no particular boundary around an object is privileged', SWEWAP depends on a range of funding for the project to work. The organization as a whole works with multiple funding sources and will be audited by different funding bodies, each drawing different distinctions to measure the work SMEWAP has done. Sometimes this is the same work, each body counting the same work, and sometimes the work is divided up, some parts attributed to a particular funding programme and other parts to other funding programmes. Funding sources come and go and the work remains. In this sense the funding stream from the regulator is alongside a range of funding streams, increasing and decreasing in value and intensity while the work of SMEWAP continues. Again, SMEWAP presents itself to different funders in different ways. Finally, 'mobile boundaries are needed for objects to exist in fluid space' and hence SMEWAP does remain as SMEWAP though it adapts to particular circumstances, for example charging consultancy fees to companies outside of its boundaries.

The important point is that SMEWAP occupies a more fluid space that is neither defined by clear a cut regional space not fully integrated into a particular network, rather it works across these spaces and by that enables regional level concerns to be translated into local practices.

REFLECTIONS: ECOLOGY OF INTERMEDIARIES

The challenge faced by industrial ecology of understanding the relationships of material flows, technologies and social logics is clearly highlighted by the work of intermediaries. Such is the problem of translating across different social logics that our work suggests increasingly utility companies and regulators are looking to the work of intermediary organizations to aid in the implementation of sustainable strategies. In this chapter we have outlined the ways in which intermediaries have become significant to the water sector, highlighted some of the key aspects of different environmental, economic and social/health logics through which value is attributed to water and shown how different spatialities are brought to light in the work of intermediary organizations. The important point is that they challenge the discipline of industrial ecology faces – the relationship between material flows, technologies and social logics – is the every challenge that intermediaries themselves seek to address. What then can industrial ecology learn from intermediaries?

The industrial ecology approach draws upon the imagery of ecology to denote the complex interdependencies resources flows and production/consumption processes. Interestingly, while the work of industrial ecology tends to be one of trying to increase our understanding of the overall resource flows of a given region, the work of intermediaries works in a different direction, usually pointing to the very specific practices within particular localities. And they do this without full understanding of the flows and patterns of resource use. Arguably then, the concept of ecology within industrial ecology needs incorporate the hidden work that is less understood and that will often remain invisible because of its fluid nature, and yet which is crucial for resource flows. Rather than romantic conceptions of looking up towards a more integrated whole, the localized and fluid work of intermediaries is suggestive of the need to look more locally at the ways in which global processes are manifest in local activities and how through local interventions resource flows can be effected (see Kwa 2002).

NOTE

1. This chapter is derived from work undertaken by Will Medd and Simon Marvin as part of the EU F5 Intermediaries Project (EVK1-CT-2002-00115) which examines the rise of new intermediaries in the water sector. See www.irs.net.de/intermediaries.

REFERENCES

Bakker, K. (2003), *An Uncooperative Commodity: Privatizing Water in England and Wales*, Oxford: Oxford University Press.

Ball, P. (1999), *H_2O: A Biography of Water*, London: Phoenix.

Callon, M. (1997), 'Techno-economic networks and irreversibility', in J. Law (ed.), *A Sociology of Monsters: Essays on Power, Technology and Domination: Sociological Review Monograph*, London: Routledge.

Chappells, H. (2003), 'Reconceptualising electricity and water: institutions, infrastructures and the constitution of demand', PhD thesis for the Lancaster University Department of Sociology.

Erkman, S. and R. Ramaswamy (2000), *Applied Industrial Ecology: A New Platform for Planning Sustainable Societies*, India: AICRA.

European Community (2000), 'Directive 2000/60/EC of the European Parliament and of the Council of 23 October 2000, establishing a framework for community action in the field of water policy', *Official Journal of the European Communities*, **327**(1), 1–72.

Giddens, A. (1984), *The Constitution of Society: Outline of the Theory of Structuration*, Cambridge: Polity Press.

Graham, S. and S. Marvin (2001), *Splintering Urbanism: Networked Infrastructures, Technological Mobilities and the Urban Condition*, London: Routledge.

Harvey, D. (1989), *The Condition of Postmodernity*, Oxford: Blackwell.

Hayes, E. (2002), *Mountains, Sheep and Fences: A Study of the Network of Reconciliation within the UK Lake District National Park*, Lancaster: Lancaster University Centre for Science Studies, p. 230.

Illich, I. (1986), *H_2O and the Waters of Forgetfulness*, London: Marion Boyars.

Kwa, C. (2002), 'Romantic and baroque conceptions of complex wholes in the sciences', in J. Law and A. Mol (eds), *Complexities: Social Studies of Knowledge Practices*, Durham, NC and London: Duke University Press.

Latour, B. (1996), *Aramis, or the Love of Technology*, Cambridge, MA: MIT Press.

Law, J. (1987), 'Technology and heterogeneous engineering: the case of the Portuguese expansion', in W.E. Bijker and T.P. Hughes (eds), *The Social Construction of Technical Systems: New Direction in the Sociology and History of Technology*, Cambridge, MA: MIT Press.

Law, J. (2002), 'Objects and Spaces', *Theory, Culture and Society*, **19**(5/6), 91–105.
Mol, A. and J. Law (1994), 'Regions, networks and fluids: anaemia and social topology, *Social Studies of Science*, **24**, 641–71.
Moss, T. (2003), 'Solving problems of "fit" at the expense of problems of "interplay"? The spatial reorganization of water management following the EU water framework directive', in H. Briet, E. Engels, T. Moss and M. Troja (eds), *How Institutions Change: Perspective on Social Learning in Global and Local Environmental Concerns*, Opladen: Leske and Budrich.
Strang, V. (2001), *Evaluating Water: Cultural Beliefs and Values About Water Quality Use and Conservation*, Ipswich: Water, UK.
Swyngedouw, E. (2004), *Social Power and the Urbanization of Water: Flows of Power*, Oxford: Oxford University Press.
van Lente, H., M. Hekkert, R. Smits and B. van Waveren (2003), 'Roles of systemic intermediaries in transition processes', *International Journal of Innovation Management*, **7**(3), 1–33.
White, I. and J. Howe (2003), 'Planning and the European Union water framework directive', *Environment Planning and Management*, **46**(4), 621–31.

PART 5

Governance and values

12. Enabling redesign for deep industrial ecology and personal values transformation: a social ecology perspective

Stuart B. Hill

Most new initiatives start with a 'planning' process; and the outcomes are frequently disappointing. Underneath planning lies 'imagination and creativity', and underneath this lies 'passion and feelings' – all within an internal context of 'values and worldviews', and a specific external context. Engaging first with these latter areas generally leads to innovative plans and programs that are genuinely progressive and transformative.

Similarly, most initiatives focus on 'efficiency' and 'substitution' strategies. These predictably fail to address the causes of problems. What is needed is a 'whole system design/redesign' approach that aims to make systems problem-proof and that enable sustainability and wellbeing.

Furthermore, problems tend to be addressed in fragmented ways, and within the confines of disciplines and specialities. Again, what is needed is a holistic, integrated, whole system approach . . . To be able to do this external redesign it is usually necessary to also engage in some liberating internal redesign – in terms of our understandings and ways of working and collaborating.'

(Statement by the author for a proposed position of 'Provocateur' with the Department of Primary Industries, Government of Victoria, Australia; 13 October 2004)

INTRODUCTION

What I am arguing above, and in this chapter, is that the redesign, design and innovation that is needed at the industrial and business level needs to be 'enabled' by supportive changes in our institutional structures and processes (at the political and socio-cultural level), and that changes at both of these levels can, in turn, only be effectively 'enabled' by radical (deep, root level) transformation at the personal level. Such personal change usually involves healing and liberational processes that result in empowerment, expanded awareness and visioning, clarification and transformation of values and worldviews, and an ability to live more fully and more relationally in place and in the present, while also having much greater concern for other humans, other species, ecological processes, and the long-term wellbeing of all life. To put it negatively, psychologically wounded individuals will always tend to design and manage structures and processes that will, sooner or later, result in problems. Such personal change can be enabled by psychotherapeutic processes that support natural recovery and healing from past psychological wounding (from which we have all suffered, despite our tendency towards adaptive denial), and through the provision of supportive present environments. Without this necessary internal level of transformation and redesign, all external innovation and change is likely to be compromised, adaptive of the status quo, and consequently 'shallow' (versus the much needed genuine 'deep' ecological transformation). Far from being depressing, I find this perception incredibly hopeful in that it opens up numerous as yet untapped opportunities as we learn our way into the future. I should also add that my assumption is that at every moment all of us are doing the best we can (which may include rejecting much of what I am arguing here), given our natural potential, the 'positive' and 'negative' effects on us of our past experiences and the nature of and level of support within our present environment.

To introduce this personal level of understanding I will first share some of my own experiences that led me to taking this multi-layered approach to enabling innovative and effective redesign for a 'deep' industrial ecology.

MY PERSONAL JOURNEY FROM SCIENCE AND TECHNOLOGY TO PSYCHOLOGY AND BEYOND

My earliest experience of working in industry was in the late 1950s as a chemist and trouble-shooter in an electroplating and light engineering company. In an early effort to improve efficiency and reduce resource consumption and pollution I investigated the use of ultrasonics in enabling improved deposition of protective coverings. Although the concept was good, and has subsequently been further developed, at that time I was confronted by the common challenge of costs, various issues relating to practicality, and to largely unknown health and other side effects of the new technology. These are common experiences facing innovators. I eventually had to settle for less radical innovations and focus on improving the efficiency of current systems.

In the early 1960s I went to University to study marine biology and – through learning how to effectively 'farm' the sea – save the world from starvation (having been regularly told as a child about the 'starving Chinese', mostly as a way to persuade me to eat my food). This led to my second major industrial experience when, as a summer student, I went to work in Germany testing pesticides. Although the company's scientists were extremely thorough and efficient in their testing methodologies, on reflection I quickly became aware of the conceptual flaws and of the numerous problems associated with such curative, and what I have subsequently come to refer to as 'back-end', approaches to problem solving. This eventually led to the development of my 'efficiency-substitution-redesign' model for evaluating problem-solving initiatives (Hill 1984, 1985, 1998; Hill and MacRae 1995). The essence of this insight was that the most effective way to solve problems is to redesign the systems involved to make them, as far as possible, problem proof. This design approach is ideally done proactively rather as a reactive response to problems and crises. Although 'efficiency' and 'substitution' strategies may reduce resource dependence and environmental impact, by not addressing the causal design flaws they can, usually unintentionally, protect and perpetuate the very design features that are responsible for the problems. Because of this, 'efficiency' and 'substitution' initiatives are best conceived as transition strategies toward whole system 'redesign', or as second choice and emergency strategies. It should be noted that redesign/design initiatives often paradoxically result in much greater gains in efficiency than when efficiency is the limited focus of an innovation (Fletcher and Olwyler 1997). I have subsequently further developed these ideas and applied them not only to pest management (Hill 2004), but also to soil, landscape and natural resource management (Hill 2003a), as well as to numerous other areas including learning and education (Hill *et al.* 2004), health and wellbeing, peace, community and organizational development (Hill 2005), and now industrial ecology (Hill 2006).

What also emerged from these experiences was a realization of the importance of gaining a better understanding of bio-ecological processes, which I argue comprise the real bottom-line of our survival and wellbeing over the long term (Hill 2005; Mulligan and Hill 2001). I was able to considerably progress this understanding through an opportunity in 1965 to go to Trinidad to do a PhD on the total ecology of a bat-inhabited cave (not because I was particularly interested in bats and caves, but because the cave could serve as a 'model' ecological system in which it would be easier than in more open systems to measure and track ecological processes). This particularly expanded my appreciation of the complexity of bio-ecological processes, and of the need to always take into account implications for the functioning of whole systems over the long term, and also of distant, indirect effects of even apparently minor interventions. With this whole systems understanding, I was developing my competence to approach design in a much more holistic and holographic way than was common at the time (Wilber 1982).

My first academic appointment was in 1969 as a Research Associate with the outstanding soil zoologist Professor Keith Kevan, who was Chair of the

Department of Entomology in the Faculty of Agriculture of McGill University in Quebec. There I became increasingly horrified by the way agriculture students were being taught – with little recognition of agriculture as a bio-ecological system, and little included on design or on system maintenance – the focus was on management for maximizing productivity and profit.

I started collecting critical literature relating to the design and implementation of an ecological approach to agriculture, and in 1974, with the support of a benefactor, David Stewart of the Macdonald-Stewart Foundation, Ecological Agriculture Projects was established. This quickly became Canada's, and possibly the world's, most comprehensive resource centre for information on ecological and sustainable approaches in agriculture. During my 20-year Directorship of this centre, as well as producing numerous papers (www.eap.mcgill.ca), our group obtained a contract with the Department of Agriculture in Quebec to service and support extension agents in that Province in their efforts to enable producers to become more ecologically sustainable (Hill and MacRae 1992). This subsequently led to my doing similar work throughout North America and in many other parts of the world. This extensive experience, and access to numerous case studies, repeatedly confirmed my earlier insight, that it is possible to design systems that are both ecologically and economically sustainable. It also became clear, however, that flaws in our economic system – particularly the biased rewarding of marketable yield, and lack of rewards for the rehabilitation, construction and maintenance of 'healthy' systems; and growing economic globalization, with its bias towards cheapness, the short term, and single commodities (versus whole systems) – put ecological producers at an economic disadvantage (Hill 2001a, 2006; MacRae *et al.* 1989a; MacRae *et al.* 1993). This became particularly evident in the first major comparative study of organic farming in North America, which found that although both organic and conventional farms achieved roughly the same levels of profit and productivity (organic having an advantage in wet years – because herbicides don't work well when it is dry), the organic producers were able to achieve this on 20 per cent of the amount of energy required by the conventional group. Thus they were not rewarded by the market or by government for the 80 per cent saving in fossil fuel consumption (Lockeretz *et al.* 1984). Clearly, if we are to design and manage ecologically sustainable industries, such market-based inequities must be addressed, through appropriate political and economic instruments (MacRae *et al.* 1990). Furthermore, it should be noted that these farmers achieved these remarkable outcomes with virtually no research and extension support. This highlighted the need for the funding of more appropriate research and extension (MacRae *et al.* 1989b).

The other thing that became clear was that psychological factors, which are commonly largely neglected in most redesign and social change initiatives, must be addressed to achieve significant sustainable progress (Hill 2001b). These insights were deepened through my own personal 'healing work', and subsequent training as a psychotherapist. This is the most challenging area to discuss, and the one most subject to rejection and denial, particularly because denial is a primary strategy for surviving trauma in the absence of support for healing (through discharge and recovery). The late Scottish psychiatrist R.D. Laing (1969) characterized this as an adaptive double hypnosis in which a constructed pseudoreality replaces the reality, and then we deny that this has happened. So, it is very difficult to engage in meaningful dialogue about this, because its very mention commonly triggers a, largely subconscious, retreat into denial. To challenge such denial, so that we can move on to discuss the topic on hand, I sometimes encourage workshop participants to engage in a two-minute exercise in which pairs are asked to face each other, hold hands, make eye contact, and take turns to talk only in the present. For most people this is virtually impossible (especially for deeply wounded individuals, for whom it may be perfectly sensible not to participate), yet for a psychologically 'well' (unwounded or healed) person, being fully aware in the present, and able to clearly communicate experiences gained through our sensory systems from outside and inside, would be easy to do. To some extent this simple exercise provides us with an indication of the extent of our woundedness, and of our subconscious preoccupation with negative past influences. Conversely, this also gives us some indication of our untapped potential and reason to be optimistic about the future. To be fully available to design and redesign systems for ecological sustainability (and all other noble goals), we must either first recover from these hidden undermining, limiting and distracting influences, or be provided with such powerfully supportive environments that there is no chance of any of these potential influences from being reawakened and restimulated (Hill 2003b). Although my particular pathway to improved clarity was primarily through radical psychotherapy (Hill 2003b) and 'co-counselling' (Jackins 1978), there are examples of the enormous power of having access to a benign and supportive environment. The most impressive case of this that I know of is the Peckham Experiment, in which the provision of such an environment – essentially a community centre in which the locals in that part of London were free to pursue their own learning and activity agendas – enabled the participants (over 1000 families over 12 years) to behave in ways that had both personal and social benefits that were unprecedented, and that included an enhanced innovative capacity (Stallibrass 1989; Williamson and Pearce 1980) [http://www.thephf.org.uk]. Clearly there are lots of implications here for the redesign/design and management of the workplace (and also all centres of learning, and our homes); and of the potential of improved access to, and use of, appropriate and diverse psychotherapeutic services.

These, and other related, experiences enabled me to be appointed to my present position as Foundation Chair of Social Ecology at the University of Western Sydney in Australia (I currently define social ecology as: 'the study and practice of personal, social and ecological sustainability and progressive change based on the critical application and integration of ecological, humanistic, relational, community and "spiritual" values' (Hill 1999). This also enabled me to renew my interest in agriculture, particularly in the extremely innovative work of the late P.A. Yeomans, who developed the 'Keyline' system for landscape management, as well as an award-winning chisel-type plough, an improved method for making farm ponds (called dams in Australia), and a system of livestock management that dramatically increases soil organic matter, soil formation, soil fertility and productivity (Yeomans, K. 2002; Yeomans, P.A. 1958, 1971, 1978). For me, Yeomans embodied and exemplified most of what I have been arguing for in my approach to industrial ecology. What he lacked was particularly the psychological component, and this has, I believe, subsequently limited the more widespread adoption of his brilliant innovations (Hill 2003a, 2006). This will be discussed further below. A recent book by his middle son, Allan Yeomans, may help remedy this (Yeomans, A. 2005). Allan has further developed the Yeomans plough and has shown how its widespread use may enable us to capture as much carbon dioxide, and store it in the soil as humus, as the amount that is being released into the atmosphere as a result of burning fossil fuels (see www.amazingcarbon.com). Although this would not provide a permanent solution to the 'global warming' problem, it could buy us time to develop non-fossil fuel based technologies, while addressing this potentially devastating challenge. The contributions of P.A. Yeomans will be discussed in more detail as a case study below.

Based on the above, other experiences and the extensive literature relating to ecological sustainability and the process of change, I have compiled a set of assumptions that, I consider, should be taken into account when developing and implementing ecological initiatives, including those in industry.

Some Assumptions

The following assumptions are discussed in more detail in Hill (2006).

Nature

Nature functions according to ecological 'laws' and processes that involve limits and opportunities, cycles, non-linear and threshold relationships, complexity and high functional biodiversity, widespread mutualism, with competition usually being a last resort, and most resources being used for maintenance (sustainability) and regulatory processes, with 'production' being a by-product of this (Commoner 1970; Hill 1991). There can never be a non-ecological long-term future for our species, including our industries. Because we are products of nature we are all subject to nature's limits and opportunities.

Industry

Industry, like economics, politics and religion, is a social construct. Designed and used appropriately, industry can serve us in supporting the wellbeing of both people and the planet. Conversely, with personal disempowerment, lack of awareness and vision, undeveloped worldviews and confused values, we are susceptible to being enslaved by industry (as we are by any of our other social constructions). The more powerful the social construction, the more powerful and clear we need to be to not become victims of such enslavement. In this regard, for industrial ecology initiatives to achieve their full potential they must focus on fundamental whole-system eco-design and redesign, and not be regarded as add-on or fine tuning activities.

Sustainability

Sustainability is concerned with the long-term regeneration and maintenance of living systems. It has a paradoxical relationship with progressive change and personal and ecosystem development, for which it is a co-requisite. Ecological sustainability affects the survival and wellbeing of all life. Social and cultural sustainability relates only to human groups, and personal sustainability to individual wellbeing. Because money and economic systems, like politics, technology and even religion, are human constructions (in a sense, merely 'tools') that enable us to act on our values, they should not be accorded similar status to the environment or personal wellbeing when considering sustainability. Like all tools, they must be regarded as subject to being changed as needed, and their appropriateness must be judged against a broad range of life affirming values. To allow any of them to assume the role of a higher value, as we have for growth, wealth, ownership and global trade, is paradoxically an indicator of our collective disempowerment (it is a predictable, associated compensatory behaviour) and of the loss of our humanness, and/or of its untapped potential. Consequently, for me, any triple bottom line must relate to ecological, personal (including 'spiritual') and social (including economics, politics and so on) sustainability. If we are to survive, then economics and money must eventually be put in its place, and not allowed to dictate our values or be the sole factor in determining our decisions.

Change

Change is a natural whole-system process that in nature mostly occurs gradually (with occasional bursts) in a highly integrated way that is adaptive over the short term and co-evolutionary over the long term (Norgaard 1994). Effective sustainable and psychosocial evolutionary change in human societies is supported by being based on this awareness, by having shared emergent and contextually appropriate goals and agendas, being clear (not naïve) about the contexts within which one is operating, and having the knowledge, skills and psychosocial maturity to collaboratively implement our visions and bring about progressive changes (deMause 1982, 2002). We must constantly be open to change in direction (including paradigm shifts; Kuhn 1970) as we sensitively and imaginatively learn our way into the future. One key to effective change is to focus on small meaningful initiatives that can be accomplished with the widest possible sense of ownership, and to publicly celebrate progress (to acknowledge achievements and facilitate copying by others). The importance of this approach to change cannot be overemphasized. Mega-projects 'owned' by experts and those with positional power are the least likely to succeed, and the most likely to experience low compliance and, over time, lead to unexpected negative outcomes, and be ultimately unsustainable (Hill 2001b).

Redesign

All existing systems can benefit from fundamental redesign based on the creative application of our understandings of life, particularly in relation to ecology and psychology. An initial list of such understandings in ecology with some of their social implications is provided in Table 12.1. This deep approach to industrial ecology, natural resource management and change is profoundly different from the more usual tinkering approaches that aim to improve efficiency within flawed designs (such as monocultures in agriculture, forestry and fisheries), substitute inputs (such as renewables and biologicals, now including genetically modified organisms, for non-renewables and synthetics), and that focus on problem solving and symptoms (usually regarded as 'enemies' instead of feedback from poor designs and mismanagement). Instead, deep redesign initiatives aim to use bio-ecological and psychosocial insights to create self-maintaining and self-regulating, optimally productive, sustainable, healthy systems.

Knowledge

Despite the extent of our accumulated knowledge and technological power, our species has still only scratched the surface of its potential in terms of personal and cultural development, and of our understanding of the workings of nature. Most of what is remains unknown (Voisin 1959), see

Table 12.1 Comparison between prevailing assumptions and practices and ecological understandings within industrial societies

Prevailing assumptions and practices	Ecological understandings and biases
• Wait for crises	• Responsive to early indicators
• Linear material flows	• Cyclical, regenerative relationships
• Unlimited growth (unsustainable)	• Growth subject to limiting factors
• Production overemphasized	• Most resources used for maintenance
• Reliant on fossil fuels and nuclear power	• Based on solar and renewable energy
• Competition emphasized	• Mutualism favoured
• Simplified, highly controlled systems (dependent and unstable)	• Functional diversity and complexity confer stability
• Few specialists and roles valued	• Rich diversity of specialists, generalists, roles and niches within communities
• Structures and processes universalized (everything the same, everywhere, all the time)	• Uniqueness of time and place (reflected in all structures and processes)
• Rapid, forced change with few beneficiaries and many 'casualties'	• Gradual co-evolutionary structural change, with occasional bursts of creativity
Cultural and personal imperatives	
• Inequitable and accumulating personal wealth (unsatisfiable and unsustainable); living off the capital	• Building personal, social and ecological capital and wellbeing, and a sense of enough; living off the interest
• Growing consumption (increasingly emphasizing compensatory wants)	• 'Conserver Society' (equitably meeting basic and aesthetic needs)
• Mega, powerful, resource consuming structures process and technologies that are waste producing and impacting	• Appropriate scale, resource efficient (solar renewable) structures, processes and technologies that minimise waste and impact
• Market forces political and consumer manipulation through advertising and exclusion; short-term narrow focus, with neglect of externalities – monetary system of values (economic rationalism)	• Values-based decision making by an informed, participatory population (public education, access, transparency and inclusion) – for the greatest good (social justice)
• Transglobal corporate managerialism and hierarchical control; homogenized designs, products and services	• Regional self-reliance, shared leadership and responsibility; and context sensitive and specific designs, products and services

Table 12.1 (continued)

Prevailing assumptions and practices	Ecological understandings and biases
• Mobile, disposable workforce (loss of sense of purpose, meaning, connection to place and community)	• Right to meaningful work (sense of purpose, place and valued roles within vibrant communities)
• Controlling and problem solving, specialized science and technology (understanding science and arts as disposable luxuries)	• 'Understanding', creative, and design-focused science, technology and arts, and their integration

especially his Figure 1, p. 3), and, in any event, because all knowledge is constructed, it can never be absolute or complete and must always be regarded as provisional, open to revision, refutation and elaboration. Paradoxically this is cause for hope, because the opportunities for improvement and progress are enormous. This will be realized, however, only if we are willing to become much less arrogant about our 'knowing', and much more imaginatively proactive in our psychosocial and cultural evolution, and in our learning from and working with nature. In particular, this will require us to courageously let go of dysfunctional and life-threatening assumptions, biases, visions, preoccupations, designs and practices.

Humans

Humans are not 'good' or 'evil'; rather they are potentially both. However, the life force within each of us, together with our social nature, biases us towards the benign and relational end of the spectrum, as evidenced by our passion for learning and improving, and caring and collaborating (Hill 2003b; Josselson 1996; Shem and Surrey 1998). Contextual factors, particularly busyness, inappropriate reward systems, and lack of supports and regulations can be major barriers to the expression of these qualities. Spontaneity and being in the present are the most reliable indicators of psychological and emotional wellbeing (Williamson and Pearce 1980), which are prerequisites to genuine progressive change.

Communication

Because of the factors referred to above under 'Humans', most communication about change is predictably relatively shallow and ineffective. Feelings of really being listened to are rare and misunderstanding is widespread. Effective communication is made particularly challenging as a result of our enormous individual variability. This may be related to differences in personality preference (Keirsey 1997), gender (Tannen 1986), age, cultural background, lifestyle preference, knowledge, skills and psychosocial development (Beck and Cowan 1996; Josselson 1996; Lauer 1983; Wilber 1998), our past experiences, including both those that were liberating and developmental, and those that were wounding (Hill 1991, 2003b; Jackins 1978), as well as our substantial biological differences. As a consequence, much communication is adaptive and is concerned with negativity (or, conversely, with ungrounded and 'patterned' positiveness), trivia, tiptoeing around issues, rather than dealing with them (and other postponing strategies), and reactive defensiveness and power games. Dysfunctional communication is, in my experience, a much more common barrier to making progress in most areas of industrial ecology than is the need for technological innovation. Because of this, it is imperative that much greater attention be paid to improving communication.

To complement the many examples and case studies from manufacturing industries, discussed in other chapters, I have chosen to briefly analyse the ecological initiatives of P.A. Yeomans, the innovative Australian farm landscape redesigner referred to above.

P.A. YEOMANS, THE PROTOTYPE ECO-REDESIGNER AND INNOVATOR

In the 1940s when virtually all agricultural experts and producers were busy finding ways to control and manipulate farm landscapes to make them immune from the vagaries of nature – through clearing of the land, the use of agricultural chemicals, invasive cultivation, 'improved' plant varieties and irrigation, P.A. Yeomans, with a background in mine engineering and earth moving, was boldly experimenting on his farm in NSW with ways to work with and effectively use nature's physical structures, rich biodiversity and ecological processes to develop a farming system that would not only be sustainable, but also build natural capital. P.A., as he was usually known, was a world leader in the application of ecology to the design of managed ecosystems. His story is illustrative of the complexities involved in the origin of great ideas, their development and application and the attitudes of others, particularly those with threatened positional power, to such challenging ideas and their originators. Yeomans was not only ahead of his time, but willing to work with complex systems in holistic ways using the energies of nature when the dominant focus was to simplify and control systems with powerful machinery and synthetic chemicals. His ideas are even more important today as we witness the results of what he referred to as the 'bastardization of agriculture', which has resulted in widespread resource degradation, desertification, salinization, pest and disease outbreaks and dependence on curative interventions, farm bankruptcies, the decay of rural communities, fights over access to water, increased dependence on subsidies, the slow death of the family farm, and a growing gap of misunderstanding between rural and urban communities. The widespread application of Yeomans' ideas since the mid-1950s could have prevented some of these outcomes – yet this was not to be – and still his ideas remain unknown or only superficially known by most agriculturists in Australia and elsewhere. Because Australia's future will be increasingly limited by access to water – for drinking, industry and irrigation – and because Yeomans discovered how to most efficiently manage our water, as his ideas are implemented his name will likely become known to all. But this could still take a long time as we continue to trundle down the various paths of magic bullet curative solutions, the latest and most potentially dangerous being the narrowly conceived biotechnology path.

Yeomans was driven to find ways to design and manage landscapes to make optimal use of water

> I was always interested in water control, and whether experimenting with 'wild flood' or contour furrow irrigation or getting oneself saturated watching run-off in heavy rainfall, the flowing water seemed to hold many of the answers to the questions of land. (1958, p. 262)

This led him to design a new type of plough, now called the Yeomans Keyline Plow, a pattern of ploughing that optimally retains and distributes rainfall and irrigation water within the soil and across the landscape, an integrated series of farm dams, which he used for sheet irrigation (we could improve on this today), and a systematic way for planning the design of the farm and its operation. Later in his life he applied his water management plans to the design of cities and towns (Yeomans 1971). Put simply, his approach was to get the most out of the water that falls onto the land by making it travel the greatest distance across the landscape and do the maximum work on its journey to the sea.

> The floodwaters from prolonged heavy rains, which now go to sea within a few days, would still be in the soil and in the farm dams months later. Some of the water would remain there for a year or more. During this time the increased soil moisture would be feeding ground water supplies which flow as springs to feed creeks and rivers. Therefore, river flow would be more constant. Then the continuous but slow seepage from farm dams would be adding to these underground supplies. This would be clean and clear, as well as constant. The present silting

> up of rivers would cease and the constant flow of silt-free water would speedily regenerate them. (1958, pp. 9–10)

A fuller version of Yeomans' story has been published earlier (Mulligan and Hill 2001), as has an analysis of the lessons that may be learned from the process of his innovations and their adoption (Hill 2003a, 2006). Below is a summary of some of those findings.

Yeomans' personal qualities that enabled him to be so innovative included the following:

- Exceptional powers of observation and creativity.
- Deep and broad interests, commitment, rebelliousness, 'drivenness' and 'stickability'.
- Diverse complementary enabling experiences and competencies, and extensive reading, international networking and travel.
- Cross-boundary, integrative, lateral and paradoxical thinking.
- Ongoing experimentation and careful record keeping.
- Implementation of small, meaningful initiatives (including small risks) that could contribute to larger, longer-term plans.
- Passion about communicating his ideas, through books, a magazine he established, letters, farm open days and talks.

He was also limited by the following personality and psychological characteristics:

- Somewhat intolerant, low level of patience, isolated in some ways.
- Difficulties with collaboration, and a challenging writing style.

In addition to this, the usual range of social factors limited Yeomans:

- Most of society was in a relatively uncritical phase of fascination with deceptively simple 'magic bullet', technocentric solutions to complex ecological, social and personal problems.
- Unavailability of affordable enabling technologies (for example, electric fencing).
- Lack of access to funding for research and development (this needs to be long term and include transdisciplinarity).
- Lack of supportive government policies and programmes and interest by researchers in universities and government laboratories (and even ridicule by some of these individuals).
- Lack of consumer demand and markets for his 'green' products, and low public awareness of ecological imperatives.

Possible strategies for addressing these limitations might include the following:

- Personal development work (recovery, therapy, self-knowledge, relationship counselling, group support).
- Collaborating more widely to achieve shared ownership and enrichment of the project (with those in the region and beyond, those in university and government, public interest and consumer groups).
- Linking his radical innovation to one(s) that has (have) already achieved some level of acceptance (capitalizing on the existing trends).
- Working with a smaller part of the enterprise as a more intensively managed experiment (with controls for comparison), and so generate better data, and an operation that can be maintained over the long term.
- Working with others with better communication skills (possible use of signage, well-written pamphlets, articles and books, grant proposals and submissions to government).
- Seeking access to all of the resources listed above as limiting factors.
- Greater effort to form alliances and linkages with others to achieve a shared sense of ownership, and greater collaboration in achievement of aims.
- Greater use of the media for public education and for influencing political and cultural change.
- Going further in mimicking and working with nature.
- Being willing to 'become the other' as a strategy for deepening one's understanding of limiting factors, influencing variables, relationships and opportunities.

My hope in relation to the above is that others concerned with landscape design and management will now investigate and further develop Yeomans' innovative approaches.

CONCLUSION

The central message here for those involved in industrial ecology initiatives is that to achieve sustainable progress we must pay much more attention to the factors discussed above, which are commonly neglected when working with change. Key among these are the broad range of personal and psychosocial limiting factors, whole-system design/redesign approaches, cross-boundary and transdisciplinary thinking, being more open to working with the unknown, and with the full spectrum of co-factors involved in change. This includes, in addition to focusing on innovations, to be also simultaneously working with others to facilitate enabling structural and institutional transformation, based on the kinds of assumptions discussed above. If we are willing to risk doing this (and I acknowledge that for many it will involve a significant challenge and risk), then I believe that significant progress can be made. If we persist in denial, postponement, and in focusing on reactive and limited approaches (for example, just efficiency and substitution strategies), rather than on broad, integrated, whole-system, deep design/redesign approaches, grounded in our understanding of nature, ecology, psychology and culture, then progress will remain slow and much of the change will be counter-productive. The choice is ours. Because effective change is limited by our awareness, empowerment, vision, values and worldviews, and by the contexts within which we are operating, these are the areas where I believe that most attention will need to be applied.

REFERENCES

Beck, D.E. and C.C. Cowan (1996), *Spiral Dynamics: Mastering Values, Leadership, and Change: Exploring the New Science of Memetics*, Cambridge, MA: Blackwell.

Commoner, B. (1970), 'The ecological facts of life', in Johnson, H.D. (ed.), *No Deposit – No Return: Man and His Environment, A View Toward Survival*, Don Mills, ON: Addison-Wesley, pp. 18–35.

deMause, L. (1982), *Foundations of Psychohistory*, New York: Creative Roots.

deMause, L. (2002), *The Emotional Life of Nations*, New York: Other Press (see also: www.psychohistory.com).

Fletcher, J. and Olwyler, K. (1997), *Paradoxical Thinking: How to Profit from Your Contradictions*, San Francisco, CA: Barrett-Koehler.

Hill, S.B. (1984), 'Ecological pest control: confronting the causes', *International Journal of Biosocial Research*, **6**, 1–3.

Hill, S.B. (1985), 'Redesigning the food system for sustainability', *Alternatives*, **12**(3/4), 32–6.

Hill, S.B. (1991), 'Ecological and psychological pre-requisites for the establishment of sustainable prairie agricultural communities', in J. Martin (ed.), *Alternative Futures for Prairie Agricultural Communities*, Edmonton, AB: University of Alberta Faculty of Extension, pp. 197–229.

Hill, S.B. (1998), 'Redesigning agroecosystems for environmental sustainability: a deep systems approach, *Systems Research and Behavioral Science*, **15**, 391–402.

Hill, S.B. (1999), 'Social ecology as future stories', *A Social Ecology Journal*, **1**, 197–208.

Hill, S.B. (2001a), 'Health, food and the right to choose', in *Inaugural OFA National Organics Conference 2001, Record of Proceedings*, publication no. 01/121, Barton, ACT: RIRDC, pp. 160–64.

Hill, S.B. (2001b), 'Working with processes of change, particularly psychological processes, when implementing sustainable agriculture', in H. Haidn (ed.), *The*

Best of . . . Exploring Sustainable Alternatives: An Introduction to Sustainable Agriculture, Saskatoon, SK: Canadian Centre for Sustainable Agriculture, pp. 125–34.

Hill, S.B. (2003a), 'Yeomans' Keyline design for sustainable soil, water, agroecosystem and biodiversity conservation: a personal social ecology analysis', in B.P. Wilson and A. Curtis (eds), *Agriculture for the Australian Environment. Proceedings of the 2002 Fenner Conference*, Albury, VIC: Johnstone Centre, Charles Sturt University, pp. 34–48.

Hill, S.B. (2003b), 'Autonomy, mutualistic relationships, sense of place, and conscious caring: a hopeful view of the present and future', in J.I. Cameron (ed.), *Changing Places: Re-imagining Australia*, Sydney, NSW: Longueville, pp. 180–96.

Hill, S.B. (2004), 'Redesigning pest management: a social ecology approach', in D. Clements and A. Shrestha (eds), *New Dimensions in Agroecology*, Binghamton, NY: Haworth, pp. 491–510.

Hill, S.B. (2005), 'Social ecology as a framework for understanding and working with social capital and sustainability within rural communities', in: A. Dale and J. Onyx (eds), *A Dynamic Balance: Social Capital and Sustainable Community Development*, Vancouver, BC: University of British Columbia, pp. 48–68.

Hill, S.B. (2006), 'Redesign as deep industrial ecology: lessons from ecological agriculture and social ecology', in R. Cote, J. Tansey and A. Dale (eds), *Industrial Ecology: A Question of Design*?, Vancouver, BC: University of British Columbia.

Hill, S.B. and R. MacRae (1992), 'Organic farming in Canada', *Agriculture, Ecosystems and Environment*, **39**, 71–84.

Hill, S.B. and R. MacRae (1995), 'Conceptual frameworks for the transition from conventional to sustainable agriculture', *Journal of Sustainable Agriculture*, **7**, 81–7.

Hill, S.B., S. Wilson and K. Watson (2004), 'Learning ecology: a new approach to learning and transforming ecological consciousness: experiences from social ecology in Australia', in E.V. O'Sullivan and M. Taylor (eds), *Learning Toward an Ecological Consciousness: Selected Transformative Practices*, New York: Palgrave Macmillan, pp. 47–64.

Jackins, H. (1978), *The Human Side of Human Beings*, 2nd edn, Seattle, WA: Rational Island.

Josselson, R. (1996), *The Space Between Us: Exploring the Dimensions of Human Relationships*, Thousand Oaks, CA: Sage.

Keirsey, D. (1997), '*Please understand me II: temperament character and intelligence*', Amherst, NY: Prometheus.

Kuhn, T.S. (1970), *The Structure of Scientific Revolutions*, 2nd edn, Chicago, IL: University of Chicago.

Laing, R.D. (1969), *The Politics of the Family*, Ottawa, ON: CBC Publications.

Lauer, R.M. (1983), 'An introduction to the theory of adult or after Piaget what?', in M. Levy (ed.), *Research and Theory in Developmental Psychology*, Lovington, NY: NY State Psychological Association, pp. 195–219.

Lockeretz, W., G. Shearer and D.H. Kohl (1984), 'Organic farming in the corn belt', *Science*, **211**, 540–47.

MacRae, R.J., J. Henning and S.B. Hill (1993), 'Strategies to overcome barriers to the development of sustainable agriculture in Canada: the role of agribusiness', *Journal of Agriculture and Environmental Ethics*, **6**(1), 21–51.

MacRae, R.J., S.B. Hill, J. Henning and A.J. Bentley (1990), 'Policies, programs and regulations to support the transition to sustainable agriculture in Canada', *American Journal of Alternative Agriculture*, **5**(2), 76–92.

MacRae, R.J., S.B. Hill, J. Henning and G.R. Mehuys (1989a), 'Farm-scale agronomic and economic conversion from conventional to sustainable agriculture', *Advances in Agronomy*, **43**, 155–98.

MacRae, R.J., S.B. Hill, J. Henning and G.R. Mehuys (1989b), 'Agricultural science and sustainable agriculture: a review of the existing scientific barriers to sustainable food production and potential solutions', *Biological Agriculture and Horticulture*, **6**(3), 173–219.

Mulligan, M. and S.B. Hill (2001), *Ecological Pioneers: A Social History of Australian Ecological Thought and Action*, Melbourne, VIC: Cambridge University.

Norgaard, R. (1994), *Development Betrayed: The End of Progress and a Coevolutionary Revisioning of the Future*, New York: Routledge.

Shem, S. and J. Surrey (1998), *We Have to Talk: Healing Dialogues Between Women and Men*, New York: Basic Books, see also: www.wcwonline.org.

Stallibrass, A. (1989), *Being Me and Also Us: Lessons from the Peckham Experiment*, Edinburgh: Scottish Academic.

Tannen, D. (1986), *That's Not What I Meant!*, New York: Ballantine Books.

Voisin, A. (1959), *Soil, Grass and Cancer*, London: Crosby Lockwood.

Wilber, K. (ed.) (1982), *The Holographic Paradigm and Other Paradoxes*, Boston, MA: Shambala.

Wilber, K. (1998), *The Marriage of Sense and Soul*, New York: St Martin's.

Williamson, G.S. and I.H. Pearse (1980), *Science, Synthesis and Sanity*, Edinburgh: Scottish Academic.

Yeomans, A.J. (2005), *Priority One: Together We Can Beat Global Warming*, Arundel, QLD: Keyline, see also: www.yeomansplow.com.au.

Yeomans, K. (2002), *Water for Every Farm: Yeomans Keyline Plan*, 2nd edn, Southport, QLD: Keyline Designs, see also: www.keyline.com.au.

Yeomans, P.A. (1958), *The Challenge of Landscape: The Development and Practices of Keyline*, Sydney, NSW: Keyline.

Yeomans, P.A. (1971), *The City Forest: The Keyline Plan for the Human Environment Revolution*, Sydney, NSW: Keyline.

Yeomans, P.A. (1978), *Water for Every Farm Using the Keyline Plan*, Ultimo, NSW: Murray Books.

13. The social and political ecology of industrial ecology

Kieron Flanagan, Ian Miles and Matthias Weber

INTRODUCTION

This chapter argues that a multidimensional examination of governance issues is an essential part of strategic analysis of the process of achieving more sustainable industrial systems. It begins with some clarification of terminology – first of governance itself, and then of the 'knowledge-based economy'. The latter term points to some major social developments – also signified with different patterns of emphasis, by such terms as service economy, information society, learning organizations, and so on – that have substantial implications for governance. Challenging issues are also associated with other (purported) developments that are rarely included within the 'knowledge economy' literature, but are part and parcel of the complex of changes influencing governance and the future of manufacturing. Shifting cultural assumptions and social values are highlighted by such concepts as risk society, post-industrial value structures, and so on. The chapter reviews some major themes in this literature as they bear on industrial transformation, and its concluding sections address the policy and research issues that are raised.

THE SIGNIFICANCE OF 'GOVERNANCE' AND WHY IS IT IMPORTANT?

Concerns about the issues of governance raised by developments in science and technology (S&T) are manifold and in some cases massive. Perhaps the most extreme example of the latter is the concern about the access of individuals and small groups to 'knowledge enabled weapons of mass destruction', raised by Bill Joy in his widely discussed 'Why the Future Doesn't Need Us' (in the April 2000 issue of *Wired*).[1] This also raised the spectre of threats from self-replicating nanobots and artificial intelligence. But even without this futuristic perspective, there are major challenges posed to contemporary governance systems by developments in S&T. Consider the following formulation:

> In the early part of the 21st century, the technologies emerging from the information technology and biotechnology revolutions will present unprecedented governance challenges to national and international political systems. These technologies are now shifting and will continue to affect the organization of society and the ways in which norms emerge and governance structures operate. How policymakers respond to the challenges these technologies present, including the extent to which developments are supported by public research funds and whether they are regulated, will be of increasing concern among citizens and for governing bodies. New governance mechanisms, particularly on an international level, may be needed to address these emerging issues. (From the Summary of Francis Fukuyama and Caroline S. Wagner 2000)

While Fukuyama and Wagner are very much concerned with the impacts of technological revolutions, the points made might equally well apply to efforts to change technological regimes around environmental objectives. The question of what might constitute effective and appropriate modes of governance is being widely raised in connection with changes in technology and industrial organization. But what do we mean by 'governance'?

From Government to Governance?

> In the last decade of the twentieth century the concept of 'governance' has emerged from virtual obscurity to take a central place in contemporary debates in the social sciences. The concept has come to be used frequently, but often with quite different meanings and implications. (Pierre and Peters 2000)

The term 'governance' has numerous connotations, and is applied across numerous contexts. Common to many of these is the reference to the institutions, frameworks, procedures and principles whereby an organization – or a looser system of related entities – is managed or governed. This management in question can refer to the organizational structures, resource allocation, performance reporting and assessment, and stakeholder relations.

Why has the term 'governance' come to prominence over that of 'government'? It reflects an increasing recognition of the dispersion of power and authority across societies, beyond the apparatus of the state. It does not presuppose that political action is the only, or even the principal, means of achieving shared economic and social goals. Pierre and Peters (2000) see governance as involving 'moving up'; 'moving down'; and 'moving out'.

Moving up This phrase denotes the dispersal of power upwards to international organizations, in which nations choose to surrender part of their sovereignty in order to achieve a wider policy goal. Examples include such regulatory bodies as the World Trade Organization, but the international organization which most encapsulates the potential for the upward transfer of sovereignty and power is, of course, the European Union.

Moving down This concept captures the growing importance of subnational levels of social and political organization at the regional, local or even community level. In many cases the transfer of power to these levels is the result of intentional decentralization (for instance, changes in France and the UK, both until recently known for being highly centralized states). Such decentralization changes the nature of policy networks in the geographical areas involved, by encouraging stakeholders to increase their engagement with regional or local, as opposed to national, policy-making institutions (Pierre and Peters 2000).

Moving out This final concept encapsulates the movement of power and capabilities traditionally held by the state into (at the very least) a more arm's length relationship to political actors. This encompasses developments such as the movement of large parts of government bureaucracy into 'executive agencies' run by professional managers on a quasi-commercial basis and with commercial-style incentives and practices; the contracting-out of formerly publicly-performed services to private (or third) sector suppliers; and the movement of organizations out of public sector ownership (often termed 'privatization'). It is important to recognize that this 'moving out' does not always imply the transfer of power to the private sector, and also encompasses the greater involvement of NGOs and not-for-profit organizations in policy networks.

In his textbook on the emergence of 'policy networks' of governmental and non-governmental actors in which power and political resources are dispersed, Rhodes (1997, pp. 46–47) identifies 'at least' six separate uses of the term 'governance', some of which really represent the transformation in thinking about the role of the state which occurred during the 1980s.[2] But some ideas that are relevant here are briefly discussed in the following subsections.

'Good' governance

This phrase encapsulates the concept that there is a best practice pattern of governance which can and should be widely adopted. It is strongly associated with the 'institution-shaping' agenda of international organizations such as the World Bank, which actively promotes a model of good governance based around liberal democracy flavoured with the tenets of the 'New Public Management' philosophy which espouses greater competition, contractorization and marketization in governance (Rhodes 1997).

Multi-level governance (MLG)
Rhodes sees this concept as having grown out of political science discussions of the relationship between EU level policy making, national and regional or sub-regional policy making. It reflects two of Pierre and Peters' (2000) three dimensions of governance, namely 'moving up' and 'moving down'.

Corporate governance
Corporate governance – the generation and exercise of authority within firms – has attracted much public attention recently, not least in the wake of the Enron debacle. Tylecote and Conesa (1999, p. 25) define the term as pertaining to 'the system by which companies are controlled, directly or indirectly, by shareholders and other stakeholders'. It thus implies a rather broader influence over the direction of a company than that implied by the term 'management'. Contrasts are drawn between different styles of governance, for example the shareholder-dominated model of corporate governance typified by the US and UK as compared with the more stakeholder-dominated model exemplified by Germany and Japan. As evidence has accumulated about variations in governance styles among countries, researchers have been interested in exploring how these may be linked to innovation and other aspects of performance and practice. This has been a focus of considerable policy attention in recent years, but we should not assume that one model is superior to another in all circumstances. Indeed, Tylecote and Conesa, rather than simply evaluating the strengths of one model over another, suggest that the different models of corporate governance may be appropriate to different industrial sectors. This is clearly relevant to a discussion of the development of manufacturing in Europe and we return to these points on p. 277.

THE CHALLENGES FOR GOVERNANCE

Governance in the Knowledge-based Economy

The rise of 'governance', as a necessary complement to 'government', is intimately associated with broader socioeconomic developments. In particular we can point to the development of what is widely known as the knowledge-based economy. Admittedly, this term is problematic. All human societies intrinsically rely upon knowledge – what is new about the current epoch? If we are thinking of scientific and technological knowledge as the basis for economic activity, then this is surely something that dates back to the early industrial revolution? The DG Enterprise report, *Innovation Tomorrow*, (DG Enterprise (2003)) argued that the intersection of three trends characterises our contemporary epoch:

- the 'service economy';
- the 'Information Society'; and
- the rise of 'learning organizations'.

These three features are also intimately connected with such developments as globalization and a new stress on technological (and organizational) innovation.

Service economy

The bulk of economic activity, employment, and output is taking place in service sectors of the economy. This is the case across industrialized countries in general, and reflects the growth of marketed services as well as public services. Service-type work is prevalent in all sectors. White-collar work (and higher skill work in general) has grown as a share of employment compared with blue-collar (and low-skill) work within practically all sectors, as well as in the whole economy. More knowledge-intensive work characterises most sectors. The population contains many more people with requisite skills and experience.

The notion of service extends to all sectors – including manufacturing. Firms are oriented to providing services – whether their products are raw materials, goods or intangible products – focus increasingly on what their users are achieving. Their commercial strategies are oriented to achieving markets and customer loyalty by responding to user requirements – which means understanding of these requirements, for example knowledge. Customers require service from firms and public organizations.

Specialized services provide critical inputs to organizations in all sectors on a vastly increased scale. Some Knowledge-Intensive Business Services (KIBS) play important roles in facilitating technology choice, diffusion and implementation; others support organizational innovation and adaptation to changing market and regulatory circumstances. Technology-based KIBS, such as computer and engineering services, technological training and consultancy services, and R&D services, play important roles in generating innovations, and in improving the quality of innovation-relevant knowledge around the economy, as they grapple with the problems of their clients.

The information society

Information society rests upon the large-scale diffusion and utilization of new Information and Communications Technology (ICT). ICT allows for unprecedented capabilities in data capture and information production, and in the processing, storing, and communicating of data and information, to be used across the economy. They allow for near-instantaneous communication on a global scale; much greater access to people in previously unreachable locations and circumstances; copying and sharing of information at very low cost; ability to process huge amounts of information in little time, and so on. This allows for transformation of established business processes, and the development of quite new products and business models. The need for tacit knowledge and expertise has meant that the Information Society changes the significance of spatial location, but has not rendered space irrelevant.

ICTs have diffused increasingly widely, from back-office applications in large organizations and process control in some areas of large-scale manufacturing, to being used in practically all business units in firms of all sizes. Their use involves substantial learning, and this is reflected in the evolving organizational strategies and government policies for them. Mobile and networked communications – voice and data – are moving Information Society on from a phase dominated by personal computing to one where networked computing is evermore central. The characteristic ways in which ICT is used now are quite different from those prevalent a decade ago, and continuing change is likely.

The globalization of economies is facilitated by new ICT. The technology allows more co-ordination of economic activities on a wide geographic scale. It also increases the tradability of many services – or elements of services that are informational ones, at any rate. (Much of the globalization of services takes place not through conventional exports, but through a variety of investment-related methods. Facilitating these, ICT can enable management control of far-flung branches.) Many firms and sectors that have so far been relatively sheltered from international competition are now confronting it.

Learning organizations

Organizations are confronted with an increase in the volume and variety of information, and of the knowledge with which to effectively use this information. More sorts of knowledge are required, as well as deeper knowledge of traditional areas of business. New products and processes often draw on very diverse bodies of knowledge. Some authors claim that a new mode of knowledge production has emerged (Gibbons *et al.* 1994). Here, there is a closer connection between knowledge and application, with traditional distinctions between pure and applied research breaking down. The problems that drive research, and the theories that guide it, are increasingly derived from practical situations (for example, in microelectronics, genomics . . .), and increasingly solved 'in the context of application'. Many scientists adopt a strategic approach to their own careers – they become 'scientific entrepreneurs'. Furthermore, this analysis also indicates that knowledge production, informed by a context of application rather by the concerns of traditional disciplinary communities, is forced to become more transdisciplinary.

Organizational knowledge is more than just a matter of scientific and technical knowledge. Knowledge of markets and user requirements, of regulatory systems and trends, is vital for business practices in general. Globalization promotes demand for better understanding of diverse cultures and regulatory systems, and allows for new avenues of learning from the experience of other organizations and countries. Governments also find themselves dealing with increasingly complex (and uncertain) knowledge, and governance reform is one element of their response. Another is the effort to work much more as a facilitator rather than controller of change, and to bring together different sets of knowledge – for example, through Foresight programmes.

The growing complexity of knowledge means, among other things, that companies have to collaborate to access the knowledge required to enter new markets and to confront new challenges. This applies to innovations too, where collaborative R&D has become more important. Another result of the increased complexity of knowledge is that interdisciplinarity, and the capacity to manage multidisciplinary teams and dialogues, are highly sought after capacities.

Governance and the Knowledge-based Economy

The emergence of networked companies and company networks raises concerns about the 'governability' of these global structures. In manufacturing industries, this trend can be observed most notably in car manufacturing, ICT industries and chemicals.

The rise of the globalized network company has not really found a response from the side of governments. Castells (1996) notes that how to govern the network society is a contentious issue. But decisions do have to be made here and now. In order to establish reliable institutions in particular fields or networks, new (and often largely private) forms of coordination and governance have been established. This is especially visible in the generation of standards, but applies to many other efforts to coordinate practices. These have either been led by individual companies (for example, car manufacturers or large retailers dominating their supply chains) or in a co-operative mode between companies and/or governments (for example, internet domains, codes of conduct). In those cases where government agents are involved (for example, the well-known case of GSM standards for mobile telephony), they tend to take the role of process managers or moderators, thus reflecting a network model also with respect to governance.

The changes in occupational and educational levels, together with the widespread adoption of ICT, lead to changing social demands and expectations of business. Citizens and consumers are more sceptical about expertise and authority, and may be placing more value on environmental and other issues. This is associated with the development of what has been labelled the 'Risk Society' (Beck 1992). Beyond the conventional welfare and growth objectives that traditionally underpinned policy on science, technology and innovation, new societal objectives are raised with respect to what S&T should contribute. Most notably sustainability has turned into a major public policy objective raising major challenges due to the long time horizons it requires to look at and the multi-dimensional and often ill-defined character of the objective itself. For manufacturing it could be translated as fostering 'innovations that de-couple the environmental impacts of products from their functional performance and value-added' (CEC 2001). But considerations of sustainability are not easily incorporated into the operating procedures of market-based companies – nor of public policy-makers.

One of the major challenges for governance that also has a high relevance to manufacturing is therefore this question of how societal objectives such as sustainability could be incorporated in processes and criteria leading to both private and public decisions. In the public policy domain, increasing attempts are made to cross-check all decisions and initiatives with respect to their impacts on sustainability, based often on qualitative assessments (Coenen *et al.* 2001). How satisfactory such methods are remains debatable. As regards decisions of private agents such as firms and individual consumers, for several years there have been debates about how to achieve the incorporation of environmental and/or sustainability considerations. Examples include methods of ensuring the internalization of external costs, procedural requirements labelling, and voluntary agreements (for example, EMAS for environmental management standards), and so on.

The growing complexity, uncertainty and ambiguity of S&T-related decisions affect governance in the political realm (Renn 2002) as well as in economics. At the same time, policy choices inevitably exert a major influence on the trajectories that certain industries will take in the future – though this influence is very difficult to ascertain in advance. Current S&T

policy decisions have an influence on the future competitive position of Europe in manufacturing. Regulatory and taxation measures have the potential to help incorporate sustainability considerations more neatly into private sector decisions regarding the production processes of their products and services. For example, this might be by 'going with the flow' and supporting a service- rather than a product-based approach to manufacturing. Policies could also encourage closed material cycles, though this would represent more of a step-change.

These key characteristics make the limitations of quantitative risk assessment all the more apparent. Given that we can rarely if ever achieve scientific certainty about the impact of technologies that are still in the making, the 'assessment of risks in a quantitative, technical style needs to be complemented by attention to the contextual aspects of the complex systems in which hazards arise and within which social significance and acceptability must be appraised' (Funtowicz *et al.* 2000). Greater attention to the application of the precautionary principle has been called for, also in EU policy, but it is also recognized that scientific evaluation and an analysis of the associated uncertainties will continue to be crucial in the future (CEC 2000).

The management of dispersed governance processes leading to societal choices on science and technology is arguably likely to be far more difficult to put into practice than top-down steering. Ways of anticipating emerging opportunities and challenges, and developing strategies to cope with them in a participatory mode, need to be devised. A growing need for strategic intelligence has been identified (Tübke *et al.* 2002); Foresight is a prominent policy response, but far from the only one.

COMPETING MODELS OF CORPORATE GOVERNANCE

Tylecote and Conesa (1999) consider the relationships between innovation, finance and corporate governance, and, as noted earlier, to argue that different models of corporate governance are likely to favour success in different industrial sectors. They distinguish 'insider-dominated' (roughly equivalent to the more common term 'stakeholder dominated') systems of finance and corporate governance from 'outsider-dominated' (roughly equivalent to shareholder-dominant) ones. In the former, stakeholders such as employee groups or major institutional or private shareholders (as typified by the German system), or the State (still typified by the French system despite the onset of privatization), exist in a long-term relationship with the management of the organization. In the latter systems, exemplified by the UK and US systems, influence is limited to shareholders who are generally 'outsiders' to the firm and industry concerned, and who may not exist in a long-term relationship to that firm, and who are therefore in a weaker position to exert influence on the firm's operation.

Crucially, in a move away from traditional, well-rehearsed arguments about the primacy of one model over the other, the authors suggest that the innovation dynamics of certain sectors may be suited to certain models. The prevailing corporate governance climate may therefore be one (though not the only, or even necessarily the most dominant) factor in determining the success or failure of domestic industrial sectors. Tylecote and Conesa argue that the outsider-dominated model, exemplified by the UK system, may be most appropriate to sectors in which returns are highly appropriable (typified by industries in which patenting is a central part of IP protection), where technological change is rapid, and where technologies are function-oriented rather than object-oriented (such as automobile production). In contrast, they suggest that the insider-dominated model may be most appropriate for sectors in which user-supplier relations are more significant to innovation, where technological change is more incremental and design- or engineering-based, and where technologies are object-oriented. They suggest that sectors such as pharmaceuticals, general chemicals, and perhaps software industries, are more favoured by the outsider model. Sectors based on engineering technology and perhaps speciality chemicals, are more suited to the insider model. While empirical testing of these hypotheses against the success of various sectors in the UK, US, French and German economies yielded mixed results, the key implication of their conceptualization is a telling one. Governance regimes in general, and corporate governance and financial systems in particular, are not simply 'good' or 'bad' (in terms of generating innovation, or particular styles and trajectories of innovation) but are liable to encourage some sectors whilst discouraging others. Such a statement provides much food for policy analysis in relation to 'good governance' and harmonization of regulatory frameworks at the European level.

GOVERNANCE AND SOCIAL CHANGE

The 'Risk Society'

Ulrich Beck (1992) defines risk as 'a systematic way of dealing with hazards and insecurities induced and introduced by modernisation itself.' (p. 21.) He argues that in contemporary societies, risk is increasingly created and managed – and (critically) seen as such.

> In contrast to all earlier epochs (including industrial society), the risk society is characterised essentially by a lack: the impossibility of an external attribution of hazards. In other words, risks depend on decisions, they are industrially produced and in this sense politically reflexive. (p. 183)

Modern risks are created by decisions and choices – by agency rather than by natural hazard or chance (disease, flood, famine, and so on, or war, invasion, crime, and so on). Increasingly major risks are deliberately shouldered when we undertake decisions to utilize our technological control of nature: the aim of course is to achieve benefits from so doing, and risks are a 'by product' of this. Natural hazards can of course put our projects at risk. But modern publics expect our decision-makers to have taken these threats into account. When problems develop, someone is held responsible.

This is a persuasive argument. It chimes with our perceptions of the rise of a 'blame culture' (where someone is held responsible when things go wrong – they should have been able to foresee dangerous consequences, it is claimed. It chimes also with the growing importance attached to the creation of audit trails and of risk management approaches in industry and elsewhere. In our everyday lives, too, many people assiduously seek to identify the individual consequences of their consumption decisions (food, narcotics, travel, sport . . .) and a proportion are also concerned about broader social and environmental consequences (fair trade and ecological labelling reflect this, for instance).[3]

There are probably several related trends here, which are hard to disentangle (and Beck's analysis does not always help here). There are likely to be trends in the incidence of consequences of different kinds of risk (for example, deaths from untreatable infectious disease as opposed to deaths in wars, as opposed to deaths from road and industrial accidents and from major pollution disasters). There are likely to be trends in the perceived salience of different kinds of risk. And there are probably trends in the extent to which (particular kinds of) human agency associated with their production and outcomes.

Risk and Motivation

Much of the 'Risk Society' literature is sociological. However, a more social psychological approach – though pursued by quite different scholars – tends towards rather similar conclusions. This is probably the most influential of a fairly large number of efforts to examine trends in values and motivations in industrial countries (and beyond), though other approaches have been developed in the context of international economic development. For an early review and critique of both strands of work, see Miles (1975).

The work of Ronald Inglehart (1971, 1977, 1990, 1995) has been extensively adopted by social analysts. It follows a Maslow-type analysis of motivational hierarchies, in which (to put it in a language echoing Beck) the sorts of risk experienced in childhood help to shape the sorts of adult we become. To summarize rather simplistically: if our childhoods are marked by fear about the satisfaction of basic needs for food, shelter, and so on, we are liable to carry high concern for these issues through our adult lives; if these risks are not prominent, we may be more concerned about satisfaction of interpersonal needs or self-realization. As societies have become more free from the threats to basic welfare, so a gradual intergenerational value shift has been underway – Inglehart's 'Silent Revolution'.

Economic deprivation or worse has been a central issue for most of the population, of most societies, for most of human history. Following the Second World War, however, Western countries experienced the 'long post-war boom', with the establishment of welfare states to alleviate the worse extremes of poverty, and a steady increase in average incomes (and such other indicators as life expectancy). Inglehart argued that this change of experience was affecting the cohorts who grew up in the more secure periods, leading to a gradual shift from 'materialist' values towards 'post-materialist' ones – from emphasis on economic and physical security to emphasis on self-expression and the quality of life.

Beck's arguments are largely based on observation of political movements, media concerns, and the like. Inglehart in contrast undertook a series of large-scale surveys seeking to measure the preponderance of various value clusters, with cross-national surveys (initially involving six West European countries) going back to 1970. The early studies found the anticipated differences between the value priorities of younger and older generations: later surveys found ongoing trends towards postmaterialism, and evidence for the stability of perspectives forged in childhood. Eurobarometer surveys have incorporated simple Inglehart indicators for surveys across the European Union from the early 1970s, and US, Japan, and other areas have also been included in numerous studies. As younger birth cohorts replaced older ones, the adult population has shifted towards postmaterialist values. In current European surveys, postmaterialists and materialists are roughly equally balanced in Western countries, while in the early 1970s the ratio was 3 to 1 in favour of materialists. (Inglehart and Abramson 1994.)

Environmental attitudes are tied to postmaterialism, though it is unwise to assume that so-called postmaterialists no longer require material goods. (If anything, their material aspirations are particularly high – it is what else they desire, and thus how they may require their material needs to be met, that is changing – see Miles 1975.)[4] More recent birth cohorts emerge as generally more environmentalist than cohorts born earlier – and the

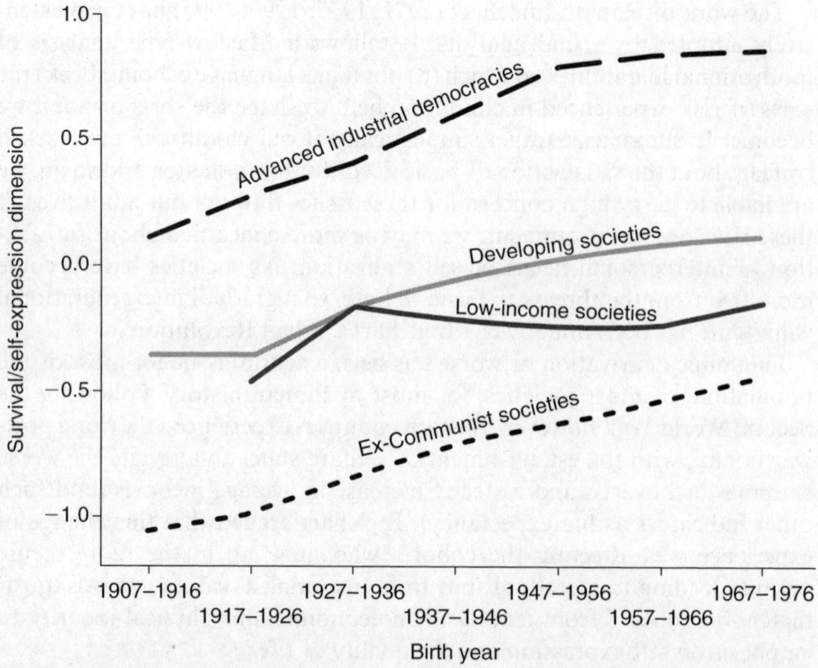

Note: High values on the survival/self-expression dimension indicate high self-expression. Data are weighted to give each society equal weight.

Source: *Figure 8* in Inglehart, R. and Baker, W.E. (2000), 'Modernization, Cultural Change, and the Persistence of Traditional Values', *American Sociological Review*, 65, 19–51, available at: *http://wvs.isr.umich.edu/papers/19-51_in.pdf*

Figure 13.1 An example of data on postmaterialist shift: survival/self-expression values by year of birth for four types of societies

argument is that this is a more stable and consistent relationship than that between environmentalism and such variables as gender and occupational status. Postmaterialists are likely to give higher priority to protecting the environment and to be more prone to joining environmentalist groups than are materialists. Interestingly, the correlation between materialist-postmaterialist values and environmental activism (joining such a group) is higher than that with expressions of sympathy for environmental concerns – the argument being that activism is more likely to reflect deep-seated commitments, whereas verbal concerns may reflect conformity with fashion. (Inglehart 1990, Kanagy *et al.* 1994). These social scientists see the shift towards postmaterialism as a major driver of the environmental concerns and movements of recent years.[5]

Of course, there are many other factors involved in the emergence of environmentalist attitudes – Inglehart himself finds both environmental problems and cultural propensities to be important influences. But the existence of an intergenerational trend throughout industrial societies is a striking claim. Even if we have reservations about some features of Inglehart's analysis, the trends appear to be fairly well substantiated by a series of studies, and require taking seriously. If in industrial or knowledge-based societies economic behaviour in general is becoming less motivated by survival concerns and more a matter of creating a higher quality of life, this has major implications for concerns with the environmental dimensions of life. There would be evident implications for attitudes to decisions about and strategies for industrial development.

Social Change and Industrial Futures

Neither Beck nor Inglehart can (as yet, at least) tell us much directly about how the changing values that they depict are liable to bear upon moves towards more sustainable industrial ecology. But their analyses do have a number of implications for this topic that will bear further examination.

But first, a few cautions concerning the ramifications of these analyses.

First, nature can strike back. Major catastrophes can occur, and even our most advanced science can probably alert us only to a few of the threats that are not just possible, but practically certain on a timescale of a few centuries. (Among these are asteroid impacts, mega-tsunamis, supervolcanoes, new diseases, and so on.) Other threats emerge as nature rebounds from human instrumentality (for example, the development of antibiotic-resistant infections) or responds to our interventions (for example, global warming-related disasters). These may even more readily provide opportunities for others to be blamed. (Though to the faithful any disaster can be seen as a Newtonian god's response to human frailty and error.) But the outcome of unmanageable 'nature' may be a reversal of Beckian or Inglehartian trends.

Second, disruptive human action may undermine expectations of progress, security and manageable risk. From economic crises to 9/11, events may provoke widespread and potentially long-lasting insecurity. The draconian responses to 9/11, in particular, have not just come out of the blue. The security apparatus in many countries has been promoting its safeguards steadily and insistently around a whole succession of perceived threats (internet crime, for instance – or the Beslan school hostage massacre in Russia in 2004, which was used immediately by the highest organs of state to intensify their grasp). Perhaps this is the last gasp of an eroding culture. Perhaps there are more deeply-rooted concerns with security than at least the Inglehart analysis would suggest. (Beck's thesis is sufficiently flexible to allow such developments – indeed, it anticipates them in many ways.)

The upshot of both of the first two points is that we should not necessarily expect trends to persist, especially where the underlying drivers are vulnerable to disruption.

Third, social affairs are highly mediated, and social action is reflexive. The perceived risks, and the perceived causal mechanisms which lead to responsibility for outcomes being attributed to various parties and actions, are not exactly simple derivatives of the social trends that have been alluded to. Whether the removal of mosquitoes (or sharks!) from an area is seen as an improvement in the quality of life is liable to be contentious even among postmaterialists. Whether the solution to environmental problems is to reduce demand drastically, to re-engineer production in the direction of clean technology, or to apply advanced techniques of clean-up and environmental management is even more challenging. The conclusions that individuals are liable to draw, and the actions that they will take and the policies that they will support, are products of argument and experience as well as deep-seated views about risks and social values.

This being said, some lines of argument can be developed.

- To the extent that there is a shift towards risk society and postmaterialism, we would expect an increasing recognition, at both a political and a corporate level, to the view that manufacturing processes should be governed so as to take a much fuller account of their environmental consequences.
- This receptiveness is liable to vary in fairly predictable ways across societies and age cohorts, though a great deal of 'noise' is liable to be introduced by cultural factors and by experiences of disruptive events (for example, the economic crisis in Argentina).
- Environmental concerns are liable to interact in complex ways with other 'new social movement' issues, for example those concerned with identity politics, with sex and sexuality, lifestyle and ethnicity. For example, the technical means of self-expression for some subcultures are very material-intensive – and even apparently sustainable activities such as cycling can threaten natural environments. Thus 'sustainability' can operate in different, sometimes contradictory directions: whether policy goals such as human health, employment, or environmental protection are emphasized by 'sustainability' will depend on the interaction between networks of governance and prevailing social values.
- It is important to articulate arguments about the effectiveness of different strategies for moving towards sustainability; and quite

possibly these arguments will need to be tailored to the wider experiences of the social groups to whom they are directed, and to be substantiated through experiential learning of one sort or another.

- Industrial and other actors will need to accept that they are seen as responsible for various things that were traditionally regarded as externalities, and will need to account for these effects of their actions (in accounting terms among others) and to account for their responses in terms of justification and demonstration of, for example, ameliorative actions and strategies for improvement).
- Such lines of action will quite possibly mean, moving beyond the traditional deployment of authoritative expertise, to engaging more in dialogue with representatives of new social movements. Wider participation in decision-making – a phenomenon emphasized in several official pronouncements concerning governance and a trend clearly observable in many areas of policy in which scientific and technological factors are significant – is often liable to be demanded, though the forms that this takes are extremely variable. It is likely that manufacturing firms will have to accept more transparency of and access to decision-making processes which affect their businesses (for instance, regulatory processes – or even their own internal decision-making process) and this implies also a better documentation and justification of the societal and environmental risks they intend to accept in their manufacturing activities, as well as of their benefits and potentials. Intellectual capital reporting is just a first step on the way towards more transparent accountability.[6]
- Proponents of 'industrial ecology' concepts will need to evaluate the risks and perceived risks of their strategies. What can go wrong, where are the critical dependencies, how can risks be managed, where do responsibilities lie, who will take the blame for problems, who will carry the can? These questions will concern political as well as economic actors.

It is often suggested that better public education will make the governance of scientific and technological issues more effective. A technocratic rationale is often the underpinning for this: if people are better informed, they will recognize the wisdom of expert advice. This particular technocratic rationale, dubbed by its critics the 'deficit model' (for the deficit in knowledge which the public supposedly exhibits) has been progressively discredited. It seems more likely that a scientifically aware public may become more sensitized to scientific disputes, to the uncertainties that are inevitably associated with the application of knowledge, to the new uncertainties that are liable to be associated with the application of new knowledge. Whether this leads to more, rather than less, perception of and caution about risks depends on more than simply scientific literacy.

A rather different technocratic argument is that a better-informed public will be more able to feed its improved awareness into policymaking structures: public consultations, for example, will evoke more and higher quality participation. Early recognition of the problems associated with attempts to bring about structural change (for example, in manufacturing regimes) may save considerable expenses before these are incurred in the wake of substantial investments. (There are plenty of examples of 'grass roots' or 'shop floor' participants being aware of the impending problems of new industrial systems long before senior management are alerted to the issue.) Better inputs of knowledge may enable early modification and better organizational learning. This argument about the mobilization of different knowledge bases[7] evolves into an argument about increasing democracy – more open and participatory governance, in the current jargon – where a more informed citizenry takes a more active role in decision making. The familiar problems of democratic governance – how to articulate and consolidate interests (corporatist, parliamentary, or other types of structures?) and how to deal with dissensus and strongly-held minority views, for example – are well known.

Recent years have seen numerous (and diverse) practical efforts to create new fora for public consultation and dialogue about major innovations and directions of technological change – ranging from consensus conferences to Foresight programmes, from web-based consultations to experiments in deliberative democracy.

Current and Emerging Responses in Governance and Politics

The reform of governance systems has arisen as a political concern in part because of strong evidence of a growing loss of confidence in policy institutions. Poorly understood and complex systems of policymaking are not trusted to deliver the policies that citizens want, or to produce them in the way that they want. Reform of governance recognizes the need to treat citizens as (actually or potentially) knowledgeable and informed participants in policy processes. Their participation and consent is required for regulatory (and other) policies to be effective and robust. Whether this is a sufficient condition, as opposed to a necessary one, is less clear.

The EC identified the reform of European governance as one of its four strategic objectives in early 2000. The White Paper on European Governance [COM (2001)428] proposes:

> opening up the policy-making process to get more people and organizations involved in shaping and delivering EU policy. It promotes greater openness,

accountability and responsibility for all those involved . . . The quality, relevance and effectiveness of EU policies depend on ensuring wide participation through the policy chain: from conception to implementation.

The *White Paper* recommends:

- Less of a top-down approach, with EC policy tools more effectively complemented by non-legislative instruments.
- Better involvement and more openness – with up-to-date, on-line information on preparation of policy through all stages of decision-making.
- Stronger interaction with regional and local governments and civil society. Member States bear the principal responsibility for achieving this, but the Commission has a role to play.
- Subsidiarity: to clarify and simplify proposed regulations and support schemes and determine if support can be decentralized, with consequences for empowerment at national, regional, sectoral and other levels. This should strengthen 'local' infrastructures where necessary. The importance given to industry and technology clusters in recent innovation management thinking might be considered alongside these developments.
- The right mix between imposing a uniform approach when and where it is needed, and allowing greater flexibility in the way that rules are implemented on the ground. This should encourage the diversity of European culture and systems, an important strength of Europe's knowledge-based society.

These considerations clearly respond to widespread expressions of dissatisfaction with remote and nontransparent policy institutions – and can be seen as a manifestation of the emergence of the knowledge-based economy and society, as well as of the recognition of the limitations of political steering.[8] What results from these developments is a shift in the governance of major societal choices regarding technology, with public policies turning more and more in only some of the contributors to these choices, be they very influential ones.

First developments in this direction can also be observed in political practice. As already noted, there is a clear, if uneven, trend towards more participation in any decision leading to major technological decisions, in particular in several fields related to science and technology, most notably large-scale infrastructures, genetic engineering, biotechnology, but also in relation to the information society. With respect to manufacturing this means for instance that large-scale manufacturing plants are subject to complex permit procedures involving the public, in particular if they are based on the use of hazardous chemicals or conflict-laden technologies (nuclear, genetically modified products, and so on). But already in the early phases of research, debates and decisions about political and regulatory framework conditions involve participatory processes and can affect the perspectives for future manufacturing (Glynn *et al.* 2001). The recent debates about the use of stem cells in research give evidence of this. Indeed the specific question of the governance of scientific research which may underpin future industrial developments is an important question which has not been addressed in detail in the project, but which is attracting the attention of European policy-makers.[9]

GOVERNANCE AND THE POLITICS OF MANUFACTURING

Given these pressures on governance that have been described above, this section asks: What sorts of actors are required to effect changes in manufacturing regimes? What sorts of strategies may be pursued?

Who?

There are numerous lines of sociopolitical enquiry that may be pursued here. However, particularly useful for the analysis of sociotechnical transitions is that developed by Alfonso Molina (1990), in the context of examining the origins of large-scale (and more modest) research and development programmes.

In Molina's view, the development of socio-technical constituencies – ensembles of institutions and entities that interact with one another through and within the development of a technology – is the key to understanding the relationship between technological and industrial change, the accumulation of knowledge-based skills and capabilities and regulation. The resultant technology can be thought of as a physical manifestation of the workings of the constituency that shaped its development. Constituents may include technical knowledge and technological artefacts as well as people, interest groups, and so on. All these factors are intertwined, changing their interactions dynamically in ways that result in the strengthening or weakening of the constituency. Like the actor-network approach developed by Michel Callon and colleagues (for example, Callon 1986), the idea of a socio-technical constituency recognizes that technology can only be shaped within limits imposed by the physical world, but that – within these limits – technological development is usually the result of the combination of human, material, financial and time and space resources. The combination

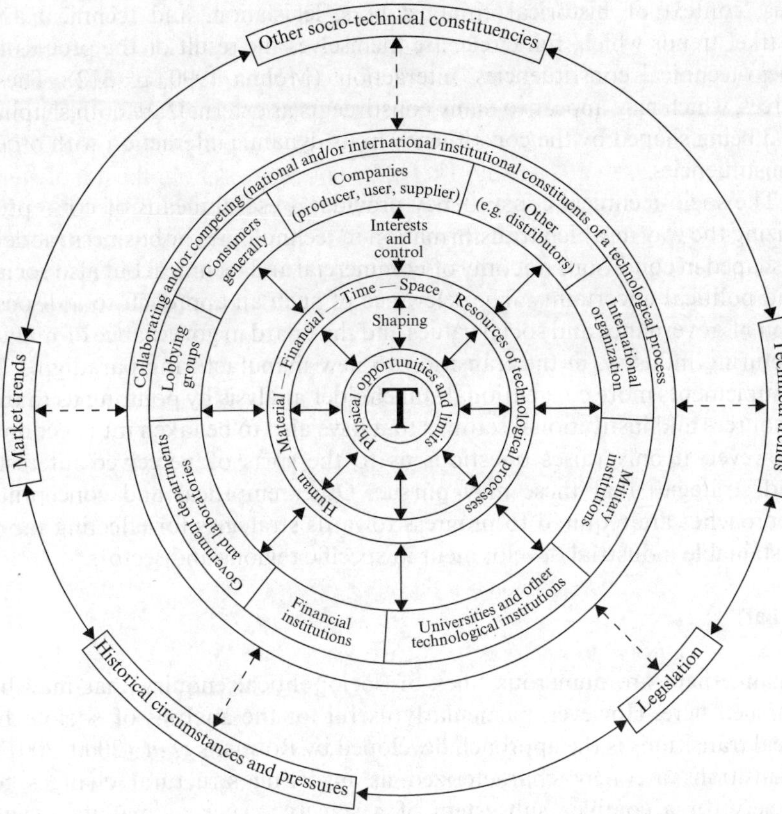

Figure 13.2 Institutional representation of a possible socio-technical constituency

of these elements depends upon interactions between the people and institutions that control these resources, thus shaping the development of the technology. These latter interactions may be national or international, competitive or collaborative in nature. In the model of a socio-technical constituency, these three levels – the technical level, the resources level and the social/institutional level – form three concentric circles linked by two-way flows of influence (see Figure 13.2). The constituency itself is further influenced by (and will in turn influence) technical and market trends, regulation, and historical pressures all of which are the result of the effects of interactions with other constituencies.

Thus, technology, according to Molina, is simultaneously shaped by intra-constituency interactions and by interactions with other constituencies in this 'context of historical circumstances, legislation, and technical and market trends which, full circle, are themselves the result of the process of socio-technical constituencies' interaction' (Molina 1990, p. 312). These forces, which may appear to many constituents as external, are both shaping and being shaped by the constituency in its dynamic interaction with other constituencies.

The socio-technical constituency provides a useful means of conceptualizing the way in which transformation in technologies or business models is shaped in conditions not only of commercial and technical but also social and political uncertainty. The relevance of such an approach to a discussion of governance and social values and the board in governance of manufacturing industry, in the transition to new manufacturing paradigms. It complements more conventional stakeholder analysis by pointing us to the resources and institutional settings that have also to be taken into account. However, it only raises questions as to the sorts of perceived interests and strategies that these may pursue. Other empirical and conceptual approaches are required to progress towards strategies for effecting more sustainable industrial development in specific regions and sectors.

What?

Again, there are numerous lines of sociopolitical enquiry that may be pursued here. However, particularly useful for the analysis of sociotechnical transitions is the approach developed by Rotmans *et al.* (2000, 2001). Transitions are here characterized as involving structural changes to society (or a complex subsystem of a society – such as manufacturing industry), which typically unfold in a gradual way. This gradual systemic change involves both slow changes (developments in stocks) and more rapid dynamics (flows). But it is liable to span at least one generation (for example, several decades). The complexity of the system means that technological, economic, ecological, socio-cultural and institutional developments interact; and these are at different scale levels.

Rotmans *et al.* introduce the notion of transition management as a process aimed at exploring, guiding and fostering such transitions, aiming to help movement towards more desirable outcomes. This approach echoes concerns for more participatory and open governance – and the hope is that greater participation will help build legitimacy and support for the policies adopted. The potential conflict between long and short-term policy thinking is to be tackled by situating short-term policy development in the context of longer term ambitions. Intermediate aims are to be formulated on the basis of the longer-term perspectives. The challenge is to find structured ways of doing this.

One element of this is the interactive development of a transition goal that 'sketches the ambitions in terms of quality images', providing a vision of the 'corridor' of development. The policy corridor for key variables indicates the margins within which the associated risks are considered acceptable. The commitment of different actors (in the sociotechnical constituency – and possibly in its opponents and bystanders) is thus critical. Government perspectives at any one point in time are only one set of such goals. A collective transition goal will thus not be determined on the basis of government fiat. Transition management needs approaches that can leave different development paths open, and span different goals and ambitions.

The transition goal needs to be flexible. Key actors will need to be involved in ongoing evaluation of the process and its goals, which raises further design issues. Government has a key role to play in determining these key actors. The transition goal comprises a multitude of policy objectives and actors' aims: it is not the traditionally quantitative risk-based target-setting approach. The objectives are more flexible, and at best semi-quantitative; sustainability issues pose complex, multi-scale problems whose risks cannot easily be expressed in the traditional ways. Transition management aims at an integrated risk analysis which can involve setting minimum levels for certain stocks (where unacceptable consequences, irreversibilities, or rapid deterioration is likely) and for aspiration levels (of goals that must be achieved). The risk estimates are subjective to the extent at least that the system's structural uncertainties are matters of speculation and disagreement.

If transition goals are flexible and development paths to be kept open, monitoring and assessing of progress along the transition path needs to be conducted continuously, as well as looking for alternatives that could turn into new promising pathways. In other words, what is needed is a system of distributed and strategic intelligence (Kuhlmann 2001) that allows to gather this information, and interpret the findings in terms of alternative transition scenarios. This is an important issue because contingencies may arise that put into question the transition path taken. The oil crisis or 9/11 took most actors by surprise, and there was a real need to be able to react promptly and consistently. This kind of robust strategy development is an important complementary element to the transition management approach.

Rotman *et al.* believe that transition management offers

> a basis for achieving more coherence and consistency in public policy and societal initiatives towards sustainability and will increase the chances for an actual transition to a more sustainable future . . . It offers a framework for the choice of instruments and institutional arrangements. Transition management does not exclude the use of control policies, such as the use of emission trading and standards, and hence is not an alternative for global climate change policies.

> Transition management helps to find additional instruments and arrangements that contribute to the transition process. It thus offers a framework for policy deliberation and the choice of instruments and societal action. Policies are evaluated against two criteria: first, the immediate . . ., and second, the contribution of the policies to the overall transition process. Learning, maintaining variety and institutional change are thus important.

The evolution of this approach – mainly through studies of, and policy consultation about, transitions in energy and transport regimes – is liable to have lessons for industrial ecology. The types of instrument proposed seem in accord with those discussed earlier, but the specific instruments of transition goals and corridors, together with the methods employed to identify these, look promising tools.

Summary and Synthesis of Findings

From government to governance

The term 'governance' has numerous connotations, and is applied across numerous contexts. In political science, the term implies the increasing dispersion of power and authority beyond the apparatus of the state. Questions of governance are thus questions of how a wider network of actors, within and without government, steer (each applying their own social, economic and political resources) economy and society towards shared goals. Crucially, the term governance does not presuppose that political action need be, or should be, the only means of achieving those goals.

Political governance

This can be considered to have three dimensions: the dispersal of power upwards from nation states to international organizations, in which nations choose to surrender part of their sovereignty in order to achieve a wider policy goal; the dispersal of power downwards towards sub-national (regional, local or even community) levels of political organization; and the dispersal of power outwards through reforms which place activities traditionally carried out within the public sector at arm's length to political actors, or into the private or third sectors. The term 'multi-level governance' has been coined to characterize the relationship between international (EU), national (member state) and regional/sub-regional policy-making.

Corporate Governance

This is usually taken to mean the system by which companies are controlled by shareholders and other stakeholders, thus implying a rather broader influence over the direction of a company than that implied by the term 'management'. Many authors contrast the shareholder dominated model of corporate governance typified by the US and UK with the supposedly more stakeholder dominated model exemplified by Germany and Japan. The extent to which different styles of governance may influence innovation has become a focus of considerable policy attention, though some argue that the evidence suggests that different models may appropriate to different industrial sectors. Such a finding, moving away from long-standing arguments over the primacy of one or the other model, is clearly highly relevant to the EU context.

Governance in the knowledge-based economy

The rise of governance is intimately associated with broader socio-economic developments. In particular, policy problems have become more complex, and the resources needed to tackle them have become more dispersed, in the context of an increasingly knowledge-based economy. This development is itself driven by the growth in economic and social importance of the service sectors of the economy; by the diffusion and large-scale take-up of new technologies, particularly new ICTs (which facilitate economic globalization); and by the growing knowledge-intensity of all kinds of organizations (which is driving the concern to build learning organizations), both public and private. In the political realm, scientific and technological knowledge is ever more pervasive (and ever more necessary in dealing with policy problems), but the growing complexity, uncertainty and ambiguity of science and technology related decisions has affected the reputation of traditional governance mechanisms to the extent that many member states have introduced substantive changes. The development of new networked forms of organization, and dispersal of knowledge and resources across supply chains and other collaborative relationships (including public-private supply relationships) has also raised new governance questions. At the same time, consumer attitudes are changing, with more scepticism about expertise and authority, and more value placed on environmental and health issues (see the massive growth in the organic food market in the UK, which has been driven not by the major retailers but by consumer demand). A major governance challenge then is how societal objectives such as sustainability can be incorporated into public and private decision-making processes. As part of this, a growing need for strategic intelligence has been identified – methods of Foresight, technology assessment, transition management come to the fore.

Governance and social values

The 'risk society' hypothesis holds that modern society is characterized by the extent to which the risks it faces are the result of its own actions and decisions, rather than natural hazards or chance occurrences. This, it is argued, constitutes a fundamental transformation in the social understanding, and perception, of risk (and in expectations concerning risk reduction or avoidance). Risk is one driver in changing social values (a shift which can be understood more broadly as one from materialist to post-materialist values), characterized by a higher priority for sustainability, environment, health and quality of life over more traditional wealth-driven consumer values. Added to this must be a growing disenchantment with traditional representative democracy on the party politics model, and a corresponding increase in single-issue politics which in itself may be a response to the 'hollowing out' of government implied by the 'governance' model and which might ultimately lead to new political alignments.

Current and emerging responses

For all these reasons, the reform of political and corporate governance systems has become a major policy concern, with the EC identifying reform as one of its four strategic objectives in early 2000. Reform initiatives tend to encourage greater openness, improved accountability and wider participation in decision-making processes, less of a top-down approach to policy-making and implementation, and striking a better balance between diversity and uniformity. New policy networks (or socio-technical constituencies) need to be built that bridge the gaps that have resulted from the progressive dispersal of power, resources and knowledge throughout economy and society in order to allow for the efficient and effective regulation of manufacturing in the context of new social values. New constituencies (combinations of knowledge, actors and resources) are also required within and across industrial sectors, taking in suppliers, end-users and other stakeholders, if the development of more sustainable products, processes and services is to be encouraged. Fundamentally, governance may be about making choices between the different and sometimes divergent 'pillars' of sustainability – with sustainability in employment or economic terms often pulling developments in a different direction from sustainability in resource or environmental terms.

New Modes of Manufacturing and Governance: Lines of Enquiry

This chapter may have posed some challenging questions about the strategy for industrial transitions, but at least it suggests a rich agenda for research. Greater participation and openness in decision-making challenges traditional bureaucratic – and technocratic – approaches to policy-making. In the context of industrial ecology and new modes of manufacturing, particularly challenging sets of issues arise.

First, central to many of these is the role of expert knowledge. Industrial and trans-industrial design requires detailed knowledge of many matters that are poorly understood by the general public (and often by many more informed people, too). But there is evidence that public distrust of scientific and other authoritative advice is growing, as is unease about the sourcing and utilization of such advice by policymakers. Public trust in the integrity of regulatory institutions needs to be maintained (or regained where it has been eroded). This applies especially to those regulatory institutions which represent public interests and air concerns in respect of environmental issues and where major transformative technologies are concerned. These institutions may need to be designed and revitalized to create and maintain trust.

Second, there are bound to be many uncertainties and requirements for ongoing learning in the transition to new industrial systems. Advice can only be provisional, experimentation will be required. Risks will remain incompletely ascertained, while some people and places may be in the position of being guinea pigs.

Third, political democracy confronts the problems of economic feudalism in new guises in a globalizing economy. Industries are increasingly footloose and may refuse or cease to play ball with an industrial ecology regime for a variety of reasons. These may involve discontent with the environmental regulations themselves, or extend into broader motives to do with social regulations, labour markets, or economic incentives.

This analysis points to some key research topics some of which could be further addressed in future research:

- Much better knowledge is required about the relations between regulations, governance and innovations of the sort involved in the transformation of manufacturing. For instance, the findings of Morgan and Morley (2003), on the role of EU public procurement regulations in inhibiting the formation of local production/consumption networks for food in Wales, stress both a strength and a weakness of the EU system of multi-level governance. This dialectic concerns the tension between uniformity of regulation across the member states, and diversity in practices and traditions, and in the application and interpretation of EU regulations from member state to member state. The study *Innovation Tomorrow*[10] was only a first step at analysing the relationships between regulation, governance and innovation. Much more work needs to be done in order to underpin the development of intelligent policy for a more sustainable manufacturing future.
- Approaches to the transformation of manufacturing systems informed by the concepts of transition management should be

tested, but informed by a solid understanding of the underlying 'socio-technical constituencies'. New modes of interest articulation in relation to the above phenomena need to be tested, in order to explore the scope for utilization of new mechanisms of deliberative democracy in the course of transition management.

- More analysis is required of the links between risk society (and other macrosocial analyses) and the sorts of value changes examined in the analyses of postmaterialism – and how these relate not just to environmentalist sympathies, but how the discourses and experiences of environmental problems and solutions are intertwined with these elements. Public appreciation of different types of risk and uncertainty is a key topic for analysis. Some work has been done in this area – much remains to be done.
- The educational repercussions of the above debate need to be explored more deeply. Probably, additional interventions will be required to align education systems to the transformation in manufacturing.

ACKNOWLEDGEMENTS

This chapter derives from the report 'The Future of Manufacturing in Europe 2015–2020, the Challenge for Sustainability', produced under the FuTMan project for DG Research funded by the EU Framework Programme.

NOTES

1. Issue 8.04, April 2000, available at http://www.wired.com/wired/archive/8.04/joy_pr.html for some of the subsequent reactions, see http://www.wired.com/wired.archive/8.07/rants_pr.html.
2. We should also note that the term carries ideological baggage for some people: take for instance the sentiment expressed by Stoker that 'Governance is the acceptable face of spending cuts' (Stoker 1997) or the implication that the governance, by stressing the distributed nature of power in relation to policy action, is little more than an apologia for government inaction.
3. See the volume by Lash *et al.* (1996) that explores some of these themes further (if not conclusively).
4. Some authors argue that trends in materalists' attitudes contradict at least some of Inglehart's analysis: see for example O. Hellevik (2002), 'Age differences in value orientation – life cycle or cohort effects?', in *International Journal of Public Opinion Research*, 14, pp. 286–302.
5. This could then be seen as the basis for a new ideological polarization, while the traditional left-right polarization has seemingly declined as forecast in the analyses of Bell and others concerning the 'end of ideology' – better the 'end of some ideologies'.

6. On the other hand, there are also opportunities for companies and sectoral groups arising out of new and more participatory forms of decision-taking. An interesting example of relevance to this study is that of the UK Government's recently-established 'Chemical Stakeholders Forum', which brings together a range of groups (including industry representatives, trade unionists and environmentalists) to provide advice to government on issues surrounding chemicals in the environment. Although the body is explicitly not an expert advisory committee, and members are present as representatives of a particular sectional interest rather than as experts in themselves, this committee engages which precisely the kind of discussion about possible environmental effects of particular chemicals that its parallel scientific advisory committee does. The body is meant to embody the range of viewpoints held by the stakeholder organizations represented, but is presumably still expected to come to a consensus in order to be able to offer advice to ministers. Whilst the model is a significant recognition of the fact that scientific and technological advice cannot be generated in a vacuum, the question remains of how policy-makers will deal with divergences between supposedly pure scientific advice from the long-standing expert committee on chemicals in the environment, on the one hand, and explicitly non-scientific advice from the forum on the other. Whatever the potential political problems, the stakeholder forum model seems to be in favour at least in the UK Department of the Environment, Food and the Regions (DEFRA). DEFRA is currently creating new stakeholder bodies to look at other contentious socio-technical topics such as hazardous waste.
7. Which may extend to user or consumer knowledge, increasingly recognized as important to innovation in knowledge-intensive business and service sectors as diverse as scientific or medical equipment and computer games.
8. See the Governance Web Site: http://europa.eu.int/comm/governance/index_en.htm for debates on governance initiated by institutional and non-governmental actors, debates on the Future of Europe and the Commission's portal on interactive policy-making.
9. For instance, the recent report of the UK Better Regulation Taskforce (BRTF 2003).
10. DG Enterprise (2003), *Innovation Tomorrow: Innovation Policy and the Regulatory Framework: Making Innovation an Integral Part of the Broader Structural Agenda*, report by Louis Lengrand & Assoc/PREST/ANRT, as European Commission innovation paper no 28, published Luxembourg: Office for Official Publications of the European Communities EUR 17052.

REFERENCES

Beck, U. (1992), *Risk Society, Towards a New Modernity*, originally published 1986, London: Sage Publications.

Better Regulation Taskforce (BRTF) (2003), *Scientific Research: Innovation with Controls*, accessed at www.brtf.gov.uk/.

Callon, M. (1986), 'The sociology of an actor net-network: the case of the electric vehicle', in M. Callon, J. Law and A. Rip (eds), *Mapping the Dynamics of Science and Technology*, London: Macmillan, pp. 19–34.

Castells, M. (1996), 'The rise of the network society', *The Information Age: Economy, Society and Culture*, Cambridge, MA and Oxford: Blackwell Publishers.

European Commission (CEC) (ed.) (2000), *On the Precautionary Principle*, COM(2000)1, Brussels: CEC.

CEC (ed.) (2001), *Sustainable Production. Challenges and Objectives for EU Research Policy*, expert group report EUR 19880, Brussels: CEC.

Coenen, R. *et al.* (2001), 'A methodology for appraising the sustainability implications of EC initiatives – the integration of economic, societal and environmental aspects', research report EUR 19xxx, Sevilla: ESTO/IPTS.

Fukuyama, F. and C.S. Wagner (2000), 'Information and biological revolutions: global governance challenge – summary of a study group', RAND Science and Technology Policy Institute report MR-1139-DARPA.
Funtowicz, S., I. Shepherd, D. Wilkinson and J. Ravetz (2000), 'Science and governance in the European Union: a contribution to the debate', *Science and Public Policy*, **27**(5), 327–36.
Gibbons, M., C. Limoges, H. Nowotny, S. Schwartznan, P. Scott and M. Trow (1994), *The New Production of Knowledge: The Dynamics of Science and Research in Contemporary Societies*, London: Sage Publications.
Glynn, S., K. Flanagan and M. Keenan (2001), Science and governance: describing and typifying scientific advice structures in the policy-making process – a multinational study.
Inglehart, R. (1971), 'The silent revolution in Europe', *American Political Science Review*, **4**, 991–1017.
Inglehart, R. (1977), *The Silent Revolution: Changing Values and Political Styles*, Princeton, NJ: Princeton University Press.
Inglehart, R. (1990), *Culture Shift in Advanced Industrial Society*, Princeton, NJ: Princeton University Press.
Inglehart, R. (1995), 'Public support for environmental protection: objective problems and subjective values in 43 societies', *PS: Political Science and Politics*, **15**, 57–71.
Inglehart, R. and P.R. Abramson (1995), 'Economic security and value change', *American Political Science Review*, 336–54.
Kanagy, C.L., C.R. Humphrey and G. Firebaugh (1994), 'Surging environmentalism: changing public opinion or changing publics', *Social Science Quarterly*, **75**, 804–19.
Kuhlmann, S. (2001), *Management of Innovation Systems. The Role of Distributed Intelligence*, Antwerp: Maklu-Uitgevers.
Lash, S., B. Szerszynski and B. Wynne (eds) (1996), *Risk, Environment and Modernity: Towards a New Ecology*, London: Sage Publications.
Louis Lengrand & Associes/PREST/ANRT (2003), *Innovation Tommorrow: Innovation Policy and the Regulatory Framework: Making Innovation an Integral Part of the Broader Structural Agenda*, European Commission Innovation Paper no. 28, Luxembourg: Office for Official Publications of the European Communities, EUR 17052.
Miles, I. (1975), *The Poverty of Prediction*, Farnborough: Saxon House/Lexington Books.
Molina, A. (1990), 'Transputers and transputer-based parallel computers: sociotechnical constituencies and the build-up of British-European capabilities in information technologies', *Research Policy*, **19**(4), 309–33.
Molina, A. (1993), 'Networks, Communities, Systems and Sociotechnical constituencies: analysing failure in a European technological initiative', Programme on Innovation and Communication Technologies working paper series, University of Edinburgh.
Morgan, K. and A. Morley (2003), *Relocalising the Food Chain: The Role of Creative Public Procurement*, Cardiff University: Regeneration Institute Report.
Pierre, J. and B.G. Peters (2000), *Governance, Politics and the State*, Basingstoke: Macmillian.
Renn, O. (2002), 'Foresight and multi-level governance', paper for the EU-Presidency Conference, The Role of Foresight in the Selection of Research Policy Priorities, Sevilla, 13-14 May.

Rhodes, R. (1997), *Understanding Governance: Policy Networks, Governance, Reflexivity and Accountability*, London: Open University Press.
Rotmans, J., R. Kemp, M.B.A. van Asselt, F. Geels, G. Verbong and K. Molendijk (2000), 'Transitions and transition management for a low emission energy supply', a study for the 4th National Environmental Policy Plan (NMP4) accessed at www.icis.unimaas.nl/projects/nmp 4/downs/00_35_ab.pdf.
Rotmans, J., R. Kemp and M.B.A. van Asselt (2001), 'More evolution then revolution: transition management in public policy', *Foresight: The Journal of Future Studies, Strategic Thinking and Policy*, **3**(1), 15–32.
Stoker, G. (1997), 'Local Governance', *Public Administration*, **75**.
Tübke, A., K. Ducatel, J. Gavigan and P. Moncada-釀Paternó-Castello (eds) (2001), 'Strategic policy intelligence: current trends, the state of play and perspectives', research report EUR 20137.
Tylecote, A. and E. Conesa (1999), 'Corporate Governance, Innovation System and Industrial Performance', *Industry and Innovation*, **6**(1), 25–50.

PART 6

Conclusion

14. Industrial ecology and spaces of innovation: emerging themes

Sally Randles and Frans Berkhout

INTRODUCTION

Building disciplinary and strategic research bridges across industrial ecology (IE) and innovation studies (IS) has much to commend it. Many theoretical and empirical avenues open up by allowing the two research fields to interact and there may also be opportunities for contributing to policy debates. Such cross-fertilization between disciplinary perspectives is already a feature of both industrial ecology and innovation studies. In this, our concluding chapter, we attempt to find common ground between these already hybridized fields of research.

However, this is not as simple as it might at first appear. As the chapters in our book show, there are a number of areas where each domain displays blind spots. Furthermore there are areas where fundamental conceptual incompatibilities produce non-trivial problems when the intellectual foundations of IE and IS are carefully compared.

In this short chapter we will reflect on four areas where difficulties, as well as opportunities, are revealed when potential bridges between IE and IS are explored. The four areas are: (a) the validity and compatibility of the underpinning conceptual metaphors; (b) the question of scale; (c) conceptualizing knowledge and information, and understanding 'information failure'; and (d) assumptions about agency and the role of the agent. We will respond to the question of bridge-building across industrial ecology and innovation studies by reflecting on how the two domains handle these themes. We conclude in a spirit of cautious optimism noting that, despite the problems encountered, there is more to be gained than lost from welcoming and furthering the objective of inter-disciplinarity which can only enrich both communities and facilitate the development of new insights, methods and contributions, each to the other.

VALIDITY OF THE METAPHOR

Industrial ecology springs from mixed intellectual roots (Erkman 2002; Erkman and Ramaswamy, this volume; Fischer-Kowalski 2002). This heritage presents some problems of consistency, giving rise as it does to a range of quite different, and arguably incompatible conceptualizations, even within industrial ecology. Accepting that the use of metaphors is a necessary device and strategy for simplifying and reducing the real world into 'models' and 'entities' which frame, assist and indeed construct our comprehension of real world objects and phenomena, we need to appreciate that the deployment of weak, inappropriate or mixed metaphors raises important questions of validity and coherence.

From physics, IE borrows from thermo-dynamics by stressing entropy and disorder over time. From engineering, by contrast it borrows from mechanical systems, where the mapping of engineering systems leads to a more deterministic approach to management. Conversely from ecology and biology, IE borrows the idea of order through reproduction and maintenance. The notion of an industrial eco-system borrows from natural eco-systems seen as efficiently utilizing the resources of natural life-cycles (Tibbs 1992). The natural eco-system as a model explicitly assumes transferability into social worlds of structures, relationships and flows as they are conceived in natural systems. Thus, we need to be aware of the dangers that metaphors bring and scrutinize far more rigorously the variety of metaphors that IE deploys and their incompatibilities.

Turning from conceptual representation to diagnosis of the system, mainstream IE depends heavily on quantitative diagnostic tools (Mass Balance Analysis (MBA), Resource Flow Analysis (RFA) and so on). This family of methods has the effect of analytically closing down the system, depicting it as momentarily (and diagnostically) impervious to external, unanticipated, shocks or intentionally excluded influences. This is an approach which borrows from the laboratory, or other closed-system contexts, and transfers them into the social realm and into social research. Closure, like its counterpart 'equilibrium', draws heavily on Newtonian physics and dominates traditional economics: a co-contributor in the development of industrial ecology. The drive to closure however lies at the heart of many of the conceptual problems encountered when we try to reconcile industrial ecology and innovation studies. For innovation studies, openness – across histories, technologies, industrial 'sectors', institutional multiplexes and locations – is central to an analytical concern with transition and transformation. Innovation scholars are largely in agreement on the centrality of variety and diversity (of agents, of interactive networks of agents, of institutional forms and patterns across time and place) as giving rise to the sorts of struggles over regulatory authority, governance, sovereignty, legitimacy, and political, economic and technological power which both drives and enables innovation and change.

When innovation scholars express affinity with a 'systems' approach, the metaphor is deployed quite differently to the systems depicted in industrial ecology, to capture the idea that innovation springs not from the actions of individual autonomous agents, but rather through the interactions of agents operating in networks of individuals and organizations within a variety of institutional settings. Attention has been paid to the emergence of new actors (entrepreneurs, small innovative firms and so on), the shifting places of different agents within systems, and to questions of path dependency, irreversibility, and in particular the social construction of systems of expert knowledge (Bijker *et al.* (1987) especially chapters by Hughes, Callon and Collins). Attention to a 'meso' level where different organizational forms are linked through systems provides further potential for common ground between industrial ecology and innovation studies, even if relations between system components are very differently handled: one focusing on physical and energy flows, the other on flows of knowledge and capital.

In addition to the question of closure versus openness, industrial ecology with its emphasis on transference of concepts from the natural and physical/mechanical sciences into social worlds has yet to fully incorporate notions of social structure, social order, stratification and reproduction, indeed social structuration processes. These kinds of problem are arguably exerting greater influence in innovation studies. Sociology is only just beginning to speak to industrial ecology (see contributions by Warde and Medd and Marvin to this volume), but others trace the origins within the social sciences of ideas which have come to underpin industrial ecology, for example the concept of metabolism (Fischer-Kowalski 2002).

QUESTIONS OF SCALE AND 'MULTI-LEVELNESS'

Like openness, the question of 'multi-levelness' presents fundamental conceptual problems for industrial ecology which typically represents its systems as existing and operating at a single 'bounded' scale. In the name of analytical simplicity, it also holds down the superimposed influences of changes occurring at all other scales on the scale in question (Randles and Dicken 2004). Industrial ecology systems (and indeed innovation systems) are typically depicted as uni-scalar and bounded rather than multi-scalar and intertwined. Whether the 'scale-of-choice' is international, national, or regional/sub-national, this scale is given primacy over others in analysis and in the prescriptions that may arise for policy and management.

Paradoxically, as industrial ecology drives towards scale simplicity, the significance of multiple and interacting scales becomes more apparent in social and economic research (Flanagan, Miles and Weber in this volume) and in particular is receiving a great deal of attention from human geographers (Brenner *et al.* 2003) and political economists (Jessop 2004, chapter 5). While innovation studies has begun to search for ways of talking about the interactions between change at multiple levels (Geels 2004; Elzen, Geels and Green 2004), industrial ecology has sought to deal with issues of scale with discussions about 'system boundaries'. What is allowed into, and what left out of the analysis will often have a critical influence on the results. Typically the appropriate boundary, and therefore the number of systemic elements considered, will be determined by which question is being asked and by whom. A challenge for industrial ecology is to recognize that the shape and dynamics of industrial systems is often determined by social relations that cut across the boundaries of the physical embodiments of these systems.

KNOWLEDGE, INFORMATION AND 'INFORMATION FAILURE'

Reflecting its heritage in engineering, industrial ecology places a great deal of faith in data – its collection and its analysis – and invests a great deal of effort in the development of sophisticated tools and methods for both the collection and analysis of data. The underlying logic is: (a) that the system in question can be completely 'known'; and (b) that access to this knowledge comes from the collection of quantitative information. Research 'failure' in this context equates to 'information failure' of one sort or another, to be remedied by improvements to the data gathering exercise, or to data gathering tools, or to the level of investment of time or money available to improve the tools or gather more data.

In sociology, it is by now commonplace to regard all information as partial, incomplete, contestable, and employed strategically by knowledgeable agents (Bijker *et al.* 1987). These objectives take a number of forms. First, the notion that technical advances will render the social system completely knowable as an objective phenomenon, independent of the researcher, is debated across the social sciences, yet industrial ecology retains a strongly positivist orientation. Although knowledge claims may in themselves be independently valid, their salience to specific social or economic choices – such as a transition to greater sustainability – is a question that remains open and contestable. A second argument relates to the question of systems in equilibrium. In IE traditionally, systems in equilibrium are depicted as both ontologically correct and normatively desirable. However, following Schumpeter, much IS is ontologically and normatively dedicated to just the opposite – depicting systems as undergoing continuous self-transformation (Metcalfe 2001). In very schematic terms, IS holds that even if we assess a socio-techno-economic system to be, at some point, in equilibrium, this does not substitute for understanding the underpinning nature and causes of change which may render a static but known phenomenon, changing and poorly-understood a moment later.

The third argument is entirely practical and indeed is well illustrated by the excellent chapter by Mirata and Pearce in this volume. In their case studies of industrial symbiosis in the UK, lifted from theory (and indeed from the exemplar case of Kalundborg) to field implementation within various regional, institutional and sectoral settings of the English regions, they describe an important lesson. They find that there are unanticipated benefits to be secured from abandoning the aim of maximizing the quantity and quality of information collected by potential partners in industrial symbiosis projects. Paradoxically, data collection acted as a deterrent to partners, whose first priority was to deliver against their business objectives – parameters which were frequently determined by decision-makers located outside the region and out of the control of local managers. Equally the researchers learned that factors other than information distinguished likely-successful from less successful industrial symbiosis projects. Among these were the more socially-based criteria of securing the commitment and active participation of a project champion. A blend of commitment and uncertainty typify these moments in the innovation process, when there is little added value to new information.

This simple description of the real, applied, experiences of implementing industrial symbiosis in England illustrates an important difference in the role of information (as a key or subordinate dimension of 'knowledge') across industrial ecology and innovation studies. In industrial ecology, we would argue, high quality information and assumptions about rationality in the design of technological systems play a fundamental role. In innovation studies by comparison, as several of the chapters in this volume have illustrated, adaptive and continuous *learning* is the key to knowledge. Hill describes the process as 'learning our way forward' (Hill, this volume). The emphasis on learning as a dynamic process reflects an assumption about innovation as a process of change. Change is seen as both an outcome and cause of further change. Innovation is therefore a restless phenomenon (Metcalfe 2001) and innovation studies is the study of restlessness.

A substantive, theoretical and normative interest, often indeterminate, which potentially connects IE and IS may be learning organizations, learning regions, indeed learning societies, including the differential capacities of societies to recognize and respond to new circumstances and opportunities.

AGENCY AND THE ROLE OF THE AGENT

It is perhaps not surprising but it is still a crucial omission that industrial ecology has no theory of agency (Andrews 2001). The primary objects of analysis being system structures and flows, this omission is understandable. Questions of agency are not thoroughly dealt with by innovation or technology studies either, suggesting both disciplines share to a greater or lesser degree a 'blind spot in this regard. Innovation studies has drawn on a range of contributory disciplines – economics, psychology and sociology – for ideas about the nature and behaviour of agents. Again unsurprisingly, this has sometimes led to confusions in explanations of behaviour.

The autonomous, rational, decision-making individual, *homo economicus*, remains of course the dominant figure of mainstream economics, and compares with the equally unfavourable unreflexive, socio-structurally determined individual, *homo sociologicus*. A third model of agency and economic action comes from 'new economic sociology' which attempts to situate economic action within a sociological perspective which is neither entirely economics-determined nor overly sociologized. (Granovetter and Swedberg (eds) 1992; Smelser and Swedberg (eds) 1994.) Likewise sociology and psychology have long disagreed on the nature of agents, with the latter oriented towards the cognitive individual of independent motivations. By contrast, sociology has long witnessed its own internal battles between advocates of socio-structured agents and advocates of reflexive agents capable of acting autonomously (compare Giddens 1979, 1984, 1991, 1992). Borrowed by innovation studies again is the culturally embedded agent with cultural differences emerging as historical phenomena which explain agent heterogeneity and therefore provide the national systems of innovation literature with one source and explanation of national difference (Lundvall 1988, 1992). These debates, which aim to tease out the differences and flaws within each conceptualization of human agency provide a central bone of contention between different disciplines which is unlikely ever to be resolved, but at least questions of agency, agency-structure relations, and explanations of human action are asked. Industrial ecology by contrast, arguably has no theory of agency at all (Andrews 2001).

As a result of its eclectic multi-disciplinary history, innovation studies is itself both confused and ambiguous about the nature of agency and its importance. Nonetheless, to be consistent with its commitment to dynamics and open transformative systems, at least some agents, at least some of the time, must have the capacity to learn, by monitoring and assessing their own situation and that of others, and by adjusting behaviours over time. Agents in innovation studies have to be heterogeneous; so a causal link between the nature of agent heterogeneity and innovative 'performance' can therefore be theorized, even where that heterogeneity is the outcome of slow historical processes. A potential connecting point between industrial ecology and innovation studies on the question of agency emerges. Industrial ecology already has a powerful normative base – it wishes to intervene in social systems to make them more ecologically and environmentally sustainable in terms of their usage of material (including economic) resources. But is has no theory of agency. Innovation studies has traditionally paid less attention to intervention, and more to understanding the innovation process. It employs a range of theories about agency. If some bridges can be found theoretically around heterogeneous learning agents, then common ground may also be possible around the way agents seek to influence innovation processes towards more desirable outcomes and away from less desirable ones.

CONCLUSION – A WAY FORWARD?

We began with metaphors. Metaphors are important because they bind together the mental maps and expectations of scholars and other agents. The metaphor frames problems and determines the range of possible solutions. One of the main strengths of the way industrial ecology traditionally framed its own problems in metabolic, ecological, and holistic terms, is that this combination of underpinning ideas posed new problems and highlighted old problems in a new way. It also offered a fresh look at possible solutions. The question now is whether this has been sufficient, since decisive choices within industrial systems are still likely to be dominated by social, economic and institutional conditions. This book has been concerned with trying to explore whether industrial ecology and innovation studies can be brought closer together.

Sometimes the best way to do this is by looking at a concrete example. One such empirical question might look at the way innovating agents behave, but set in the context of a specific resource flow system. Innovation studies does not have a binding metaphor, or such a clear normative basis. It does anticipate the use of metaphors by actors themselves. The generation and articulation of multiple and diverse metaphorical representations of systems could provide an interesting object for future research. In innovation studies, the process of problem identification and the search for solutions emerges from the system participants themselves. Visions and metaphors have a place in helping actors to learn their way forwards into more sustainable futures.

We have discussed the use of systems perspectives in both disciplines, and noted their incompatibilities as well as a degree of consistency in language and representation. Innovation studies concerns itself with the institutional context of change, be it incremental or disruptive, and explanations of why change does not occur. Industrial ecology on the other hand is primarily interested in system appraisal in energy and materials terms, and is less concerned with the historical processes that brought it about. Industrial ecology holds central the role of information and is dominated by technical practices of information collection and analysis. Innovation studies by contrast is more concerned with processes of change, how change is or could be incentivized, and the limits or barriers to shifts in desirable directions. In pursuit of this objective, a branch of innovation studies is concerned with the study of everyday practices (what people actually do, Randles and Warde in this volume), the existence and nature of incentives, the possibilities of split or incompatible incentives, and the phenomena of risk, trust, and risk aversion as barriers to change. Such questions are poorly understood in industrial ecology and arguably provide an opportunity for innovation studies to enrich industrial ecology. We have noted through the chapters of the book how different parts of the innovation system interact and are co-constructed (for example, how provision, regulation, and technological infrastructures are co-constructed with practices and therefore consumption).

We have noted that multi-scalar perspectives are blind spots in both fields, both in terms of temporal scale (where arguably innovation studies needs to pay more attention to time and dynamics) and spatial scale (where both require a greater attention to the interaction of different scales). There are still important methodological and theoretical challenges ahead in making sense of scale and dynamics.

Finally, we can add to the discussion on networks and system parts examples of phenomena and flows that are missing from systems as framed in industrial ecology. Institutional processes that organize and regulate resource and material flows, flows of knowledge, capital, risk and power need to be represented in some way. Their importance is that they have the effect of structuring dependencies, relations and the asymmetries in systems in ways that cannot be understood by focusing on the material and energy dimensions of the system alone.

REFERENCES

Andrews, C. (2001), 'Building a micro-foundation for industrial ecology', *Journal of Industrial Ecology*, **4**(3), 33–51.

Brenner, N., B. Jessop, M. Jones and G. Macleod (2003), *State/Space, a Reader*, Oxford: Blackwell.

Bijker, W., T. Hughes and T. Pinch (eds) (1987), *The Social Construction of Technological Systems*, Cambridge, MA: MIT Press.

Callon, M. (1987), 'Society in the making: the study of technology as a tool for sociological analysis', in W. Bijker, T. Hughes and T. Pinch (eds), *The Social Construction of Technological Systems*, Cambridge, MA: MIT Press.

Collins, H.M. (1987), 'Expert systems and scientific knowledge', in W. Bijker, T. Hughes and T. Pinch (eds), *The Social Construction of Technological Systems*, Cambridge, MA: MIT Press, pp. 329–49.

Elzen, B., F.W. Geels and K. Green (eds) (2004), *System Innovation and the Transition to Sustainability: Theory, Evidence and Policy*', Cheltenham, UK, and Northampton, MA, USA: Edward Elgar.

Erkman, S. (2002), 'The recent history of industrial ecology', in R. Ayres and L. Ayres (eds), *A Handbook of Industrial Ecology*, Cheltenham, UK and Northampton, MA, USA: Edward Elgar, Chapter 3.

Fischer-Kowalski, M. (2002), 'Exploring the history of industrial metabolism', in R. Ayres and L. Ayres (eds), *A Handbook of Industrial Ecology*, Cheltenham, UK and Northampton, MA, USA: Edward Elgar, Chapter 2.

Geels, F.W. (2004), *Technological Transitions and System Innovations: A Co-evolutionary and Socio-technical Analysis*, Cheltenham, UK and Northampton, MA, USA: Edward Elgar.

Granovetter, M. and R. Swedberg (eds) (1992), *The Sociology of Economic Life*, Boulder, CO: Westview Press.

Giddens, A. (1979), *Central Problems in Social Theory: Action, Structure and Contradiction in Social Analysis*, London and Basingstoke: The Macmillan Press.

Giddens, A. (1984), *The Constitution of Society: Outline of the Theory of Structuration*, Cambridge: Polity Press.

Giddens, A. (1991), *Modernity and Self-Identity*, Cambridge: Polity Press.

Giddens, A. (1992), *The Transformation of Intimacy*, Cambridge: Polity Press.

Hughes, T. (1987), 'The evolution of large technological systems', in W. Bijker, T. Hughes and T. Pinch (eds), *The Social Construction of Technological Systems*, Cambridge, MA: MIT Press.

Jessop, B. (2002), *The Future of the Capitalist State*, Cambridge: Polity Press.

Lundvall, B. (1988), 'Innovation as an interactive process – from user-producer interaction to national systems of innovation', in G. Dosi, C. Freeman, R. Nelson, G. Siverberg and L. Soete (eds), *Technical Change and Economic Theory*, London and New York: Pinter Publishers.

Lundvall, B. (1992), *National Systems of Innovation: Towards a Theory of Innovation and Interactive Learning*, London: Pinter.

Metcalfe, J.S. (2001), 'Restless capitalism: increasing returns and growth in enterprise economics', in A. Bartzokas (ed.), *Industrial Structure and Innovation Dynamics*, Cheltenham, UK, and Northampton, MA, USA: Edward Elgar.

Randles, S. and P. Dicken (2004), '"Scale" and the instituted construction of the urban: contrasting the cases of Manchester and Lyon', *Environment and Planning A*, **36**, 2011–32.

Smelser, N.J. and R. Swedberg (eds) (1994), *The Handbook of Economic Sociology*, Princeton, NJ: Princeton University Press.

Tibbs, H.B.C. (1992), 'Industrial ecology – an agenda for environmental management', *Pollution Prevention Review*, Spring, 167–80.

Index

PROLOGUE

After 24 years and a couple of months, I closed the front door to *Chez Knollys* for the last time. The removal van had departed and nothing was left but bare floorboards and the odd dog hair missed by my fervent cleaning. I walked from room to room, the sound of paws clattering behind me, surprising myself by my lack of emotional attachment to the house.

"It's just bricks and mortar, Mum." Callum had said when he left, patting the brickwork in a sort of salute. Yet, this was the only house Tony and I had ever bought together and upstairs was the room he had died in. Dogs had wandered in and out of the gardens, wagging tails and digging up rose bushes – the ashes of many of them lay scattered in the soil. We'd brought our new-born son home from the hospital through this same front door and watched him grow into an amazing young man.

I resolutely turned the key in the lock and walked down to the waiting cars resisting the urge to look back. But I did tap the brickwork.

It goes without saying that leaving any home for a new one requires a little bit of organisation; a skill I'd normally have left to Tony as I'm more of a 'fly by the seat of your pants' kind of girl. But when one's destination is 1,223 kilometres away in a different country and your travelling companions include nine woofers (as the dogs were collectively known) and a metre high Peace Lily, it helps to have a plan.

Step one: Buy a motorhome so you and your four-legged companions (and the Peace Lily) could get from London to the South West of France in comfort.

Step Two: Get a co-driver who hasn't got four paws. I'd roped in my best friend Rene [after a few glasses of wine] – after all she needed a holiday and the woofers adored her.

Step three: Pick up the motorhome from the dealer; after the sudden realisation that you need a lift to get there and a place to leave the woofers.

Step four: Get former client, now great bosom buddy Natalie (known as Nat) and her partner Iain, to pick you up from London and use their house in Surrey as a drop off base. Oh, and one of them can drive you over to the dealership to pick up 'The Mothership' as the motorhome had been christened..

Step five: Once said Mothership had been driven back to Nat and Iain's place, stay overnight and whizz off to France the next morning.

And so it was, on one of those perfect blue sky mornings – the kind where you just have to wear sunglasses and flip flops, Iain, Nat and Rene were waiting for me as I walked down the short garden path leading to our equally short but [as we were in London] highly sought after driveway. “How the hell are we going to fit everything in?” I mumbled to myself, glancing over at the two estate cars parked up on the verge. No-one else, including Rene, seemed to share my concern however.

“What on earth is that?” Nat said as I hoisted my treasured plant into the bigger of the two vehicles, one an estate car and the other, naturally Nat style, was a horse box..

“It’s a Peace Lily,” I replied stating the somewhat obvious. “Tony and

I were given her as a house-warming present and I'm not leaving her."

I couldn't explain what she represented to me but Lily was coming with us, pot and all. It took us a while to cram the confused woofers and the remaining suitcases into whatever space was available, either back seat or boot – taking care that Evee Chihuahua wouldn't be suddenly hurtled through a window by an over-excited Arry German Shepherd. My neighbours, Michelle and Trevor, came over to say their farewells – the sole representatives of the inhabitants of a road Tony and I had lived on for a quarter of a century. It was just as well, as the estate agent hadn't turned up to collect the keys. I rang the office and informed whoever was manning the desk that they'd have to get them from next door; some of us had a schedule to keep.

I squeezed myself into Iain's car, Arry's delighted slobber dripping down my back as Alice Border Terrier launched herself into my lap. As our chauffeurs turned the cars around to head off to Surrey, I noticed that the council hadn't collected the mattress from the driveway – the one Tony and I had shared for 25 years. I directed my gaze to the front; there was no looking back.

Goodbye *Chez Knollys* and farewell London.

CHAPTER 1

How we managed to get back to Natalie's Surrey farmhouse in two cars, filled to the brim with dog hair and mostly unnecessary items, Lord knows, but we did. As Nat put it, "They do have water bottles in France you know". Nat is a horse person so the property had a side access to the parking area big enough to get a wide trailer through. Like most equestrian establishments, you opened one gate, drove through and closed it behind you before entering the next. I'm not sure if this was for the horses or the number of rescue animals Nat took in. Once through the last gate, there were stables, several hitching posts and a tack area to one side and adjacent to them, the parking lot. Nat had fenced off the sizeable garden to the back of the house to allow dogs to run free without encountering horses or her much adored chickens. I knew her place well as I came down occasionally to let Arry loose in a paddock or two; we'd set up a little agility area and with both Nat and I having a passion for rescue dogs, there was always one or two cases for us to discuss. Arry knew her dogs so shrieked with excitement as we parked up, the other woofers following suit just because…

Having off-loaded the woofers and sent them to romp around with the resident dogs, we sat down over a cuppa to decide who was coming with me to get the Mothership. It was at this point I remembered a little flaw in my plan. I had to wait for the bank to receive and clear the payment for Knollys before I could go and pick up my new wheels. Time was ticking by and I was starting to panic (this is not an unusual state of mind for me; my mother calls it the worry gene). I had to get to Reading before the dealership closed, an hour away, and I'd booked us a Eurotunnel slot at some ridiculous pre-dawn time the next day. Stupid I know but I've always had the belief that an early whizz under the Channel means getting to Paris before the rush hour. It never works, trust me.

Anyway, here we were bursting to the brim with copious amounts of tea and Arry charging around the garden with a live hose in his mouth, soaking everyone within range, whilst I checked my phone every few seconds. Arry, or Aragorn to give his full name, is one of my failed fosters; a dog who'd had two homes before he came to me at six months old and if you met him, you'd know why. I once ended up in hospital after he got a ball stuck in his throat, having to have plastic surgery to repair the nerves on my hand he'd ripped to pieces whilst I successfully pulled the bloody thing out.

At the last available minute, my banking app pinged and the race was on. We had voted in favour of Iain driving me, mainly because Nat could handle my dogs better than anyone, so the two of us bombed down several narrow, winding Surrey back roads across country to get to the Mothership on time. We did.

I'd had the notion that buying a motorhome was a really great idea several months earlier. At the time I owned a much-loved Land Rover; The Black Rose as Tony had called her. She reminded him of a pirate ship but there was no way I would have been able to fit the woofers and belongings into her and be comfortable at the same time. On a whim and having nothing better to do, I Googled motorhomes and found the Mothership; a massive four berth Bessacar with kitchen and shower (and a rather comfortable double bed) in the back.

Chris' dealership was off the beaten track somewhere outside Reading; you couldn't miss it with an endless assortment of beds on wheels lined up like a vehicular housing estate. The Mothership was on the forecourt, my first thought being that she looked a lot bigger than the last time I'd seen her. More like a semi than the motorhome. When I went to view her, it was bang in the middle of the Covid lockdown and with social distancing, I had wandered through her repertoire alone. The miniscule shower was a plus but the semi-circular bed at the back made my potential woofer carrier a must. Unfortunately, I had had to leave Arry in the Land Rover that day as Chris owned a Boxer – utterly gorgeous but the one breed my idiot Shepherd detests above all others. The cab itself (lingo for the driving seat area) wasn't much bigger than the Black Rose and I did feel rather powerful sitting up high; visions of pootling down the autoroute with the woofers spread out in the back, Rene and I taking turns at the wheel singing along to our favourite tunes came to mind.

My second thought was 'how on earth am I going to get this around a corner?' but I donned an air of faux confidence and decided that the inside wasn't as scary as the outside and before I could talk myself out of it, I'd bought her. And now I had come to take this beast of a bus across France, noting that Rene, being quite a bit shorter than myself, would never reach the pedals. My little flutter of anxiety was caught by Iain.

"Take the turnings wide and don't worry about how slow you're going. Everyone gets stuck behind a motorhome."

I took Iain's sage advice and after having handed over a small fortune, followed him with trepidation down the country lanes back onto the motorway bound for our first stopover. Apart from the odd thumbs-up from lorry drivers – a blonde driving a home on wheels must have been a novelty — the return journey went so well, I was brimming with confidence as I negotiated the small gate into Natalie's horse yard. Waving nonchalantly to my waiting friends, I promptly drove into one of the hitching posts onto which the back left wheel arch stuck itself quite nicely. It always helps to have friends who know someone who specialises in unhitching hitchings and within the hour, an unamused mechanic named Andy had turned up to remove the arch from its place of refuge.

That evening and having sung the praises of Nat's bathroom and super power shower, Rene and I gratefully took up our host's offer of a few glasses of wine as we all sat round the outdoor table. The woofers fed and exhausted, plopped their tired little bodies under our feet and snored happily on sun-warmed paving slabs. None of us really felt like eating which was just as well as none of us had thought about food. I seem to remember cream crackers and Pringles being the order of the day, but for me it was the equivalent of the perfect beginning to an ending. Here I was, in the Surrey countryside, laughing about crashing my new wheels whilst discussing the benefits of a certain horse cream Nat had applied to the grass scratches Mo and Coco had covered themselves with. Both being disabled and needing their made-to-measure carts which I had decided not to unpack, they had managed to rub their bellies raw on the dry summer grass. To this day, I still use the same cream.

Eventually Rene and I headed off to our new home from home, both exhausted but wine-sated, for our first night's sleep sharing the Mothership's double bed with any woofer that could make the jump. As morning broke, and having drunk our coffee, we waved our goodbyes to our most gracious innkeepers. It was at that moment Nat shouted out, "Oh by the way, we're moving to France too next year! You gave us the idea, thank you!"

I nearly took the other wheel arch off.

Laughing, I gently inched the big white bus out of the gates once more and headed out of Surrey en route for the Eurotunnel.

It was August 2020. With the world in Covid crisis, we got to the terminal in record time with no traffic to delay us. Dutifully covered in masks and disinfectant, we manoeuvred ourselves past customs and passport control without any problems. No-one asked questions such as, “Sorry how many dogs are you carrying? And why have you got a massive plant hiding in the bedroom” kind of thing. Apart from a few swear words as I missed the turning to get into our loading lane, I managed to drive the Mothership safely onto the train without hitting any posts.

Now, as anyone who has ever travelled this way will know, you’re asked to place your vehicle in Park and open your sunroof. This I dutifully did, or at least I thought I did. The Channel tunnel is pretty level but as you enter Calais, it begins to angle downward slightly. It is at this point that I should mention that we were the first to be loaded onto the carriage so were at the front. As we approached the French terminal, the Mothership started to inch forward towards the huge exit doors. With seconds to spare, I rammed my foot on the brake hoping no-one had noticed my faux-pas – I had put the ship in Neutral instead of Park. If I hadn't run over the wheel stop the attendants had placed in front of the doors, they probably wouldn't have. "Could I reverse slightly?" said a perplexed young man. By sheer luck, none of the other cars had been damaged and with wheel stop removed, we headed up the ramp – Rene gently pointing out that we should remember to stay on the right.

CHAPTER 2

The journey to L'Horte was a familiar one, which was just as well as I hadn't noticed until we headed out of Calais that the dashboard didn't have a Satnav. My family's home, L'Horte, was located in the Aude region of South-West France, just outside a little village called St-Hilaire. My parents had lived there for almost the same number of years Tony and I had been married and we'd spent our 25th wedding anniversary at L'Horte; the last holiday we had together. L'Horte was my sanctuary, the one place I felt connected to – my little piece of heaven on Earth.

Normally, I would have driven straight down to St-Hilaire in around 12 hours, only stopping for petrol and pees but seeing as we had the time and the Mothership, Rene and I had thought that we would make it a more leisurely trip. We ear-marked a few campsites with scenic views and charging facilities – being technologically deficient I hadn't realised we needed to plug in to get the electrics to charge our phones amongst other things. Apparently campsites had such modern facilities available.

The route was pretty simple, once you got through the Paris périphérique at least. The plan was to get onto the A16 autoroute as quickly as possible and head for the capital and once out of there, head down the A71 for Orléans, A20 through Limoges, Brive and Cahors until we got to Toulouse. From there it was the A61 to Carcassonne and the all too familiar D118 home. Rene, being a lot more organised than myself, had suggested we stop somewhere in the Loire Valley past Orléans, an area filled with wineries and fancy chateaux.

Experience has taught me the importance of refuelling before you get anywhere near Paris, because if you get stuck in traffic on the périphérique [outer ring road], and it is very likely that you will, you're in trouble. It always amazes me that when you look on the map, the distance between London and Paris doesn't seem that far but it's a good 5-hour drive even on the autoroute. With this consideration, we pulled in at a service station soon after we had left the tunnel, to fill up and let the woofers have a toilet break. As I stepped down from the cab, I couldn't help but notice more than a few eyes on us as people peeked around their cars to get a better view. Well we did stand out a bit – an over-sized camper van with the woofers singing their own version of 'Band on the Run'. Checking first that I wasn't going to embarrass us all by putting the wrong type of petrol in, I slid my debit card into the payment slot only to find that it was declined. I tried again, same thing. Frantically I tried to remember the PIN for my credit card but by this time my brain had left France.

Obviously wondering about the delay, and embarrassed by the woofers singing on repeat, Rene appeared by my side [it took her somewhat longer than me to climb down from the cab; she'll probably take up mountain climbing later with her new-found skill] and held out her card. Thank the Gods for co-drivers, whose cards work. I carefully opened the petrol cap with the necessary key, filled up and went to close it again. Well, at least I tried to but I couldn't get the key out of the damn lock.

"Where's the manual?" said Rene, trying not to laugh,

"No idea – I didn't ask!" I replied, the language following my reply was not lady-like.

The two of us twisted, pressed, wiggled and, in Rene's case, had a fit of giggles, until suddenly the bloody thing decided to extricate itself from the cap.

Keeping my gaze averted from the onlookers who were no doubt commenting on the post-Brexit idiocy of Les Anglais, I jumped back into the driver's seat and drove off to find a suitably isolated green space where we could let the woofers stretch their legs; or at least those who could. There was no way I was going to search for Mo and Coco's carts buried in the boot locker, so they were to be left in their nappies. whilst Rene and I wrangled the other seven into their harnesses and I finally made my way out of the door down the steps that led from the 'living area'.

The steps were very narrow with a drop off the end so it was impossible for them all to come at once. Plan discussed, we attempted to organise them into two groups; I would take Arry (simply because like his mum, he has no patience for anything), Simi (the easiest of them all) and Alice, and then replace them with Neo, Evee, Gizmo (miniature Yorkie who Tony insisted we rescue because Evee needed a husband) and Yogi Bear. This was a smaller group and therefore less likely to get squashed by Arry.

All I can say is that 20 plus years of training dogs and 2 years at university studying the nature of such feral creatures, did not equip me with the skills needed to get desperate dogs out of a Mothership without entangling leads or breaking ankles. That being said, I did somehow manage to escape injury and get the first lot out and having let them sniff about and do what they needed to do, I switched to the next lot. By this time, we had gathered quite a crowd, probably wondering if the circus had come to town, so I tried to get the last pee done as quickly as a tree could provide. Neo refused to pee, anywhere. He was like a deer stuck in the headlights and my heart went out to him; my broken boy who I'd rescued after he'd been left for dead by the side of a road with a fractured pelvis and a rope embedded in his neck, was too scared to pee. But pee stops were of necessity quick, so I popped the second round back up the stairs and having made sure the door was securely locked, hoisted myself back behind the steering wheel. The Mothership was full of fuel and we were ready to hit the road, and the Paris périphérique.

The Boulevard Périphérique rings around central Paris with access roads running alongside – a lot of them. It's like a four lane bumper car ride with added motorcyclists hooting and weaving through the traffic. The only plus side to entering the chaos in the Mothership was that I could see the cars ahead, a blessing since I was pretty sure I stood out as an Englishwoman in a motorhome who'd crossed the Channel in the middle of a pandemic.

Driving cautiously in the slow lane, I glanced back and noticed that the woofers weren't taking advantage of the palatial rear seating area. They were starting to crawl into the footwells. Now I'd bought this very expensive motorhome so the woofers could be transported in comfort to their new home; well I might as well have bought a Mini. The only one of the pack who actually lay down on the window seats was Simi, a dog I had rescued from Romania several years before who, despite being a street dog, saw any form of bed as hers. The only downside to her personal lounging area was that every time we went around a roundabout, the cushions started to slip off so Rene's co-driving entailed shoving them back when we hit a straight road so as not to ruin Simi's slumber. After an hour or so of crawling with the traffic, we finally left Paris with Evee chihuahua safely tucked under Rene's feet, Yogi Bear (Norfolk terriers hate being ignored) on her lap and Arry brooding in the space between our seats – a few more hours and we could stop for the night and recharge.

Logically, one would have thought that a campsite or motorhome park-up in France would have been easy to find in mid-August. Rene had downloaded a 'caravan and camping' app on her phone and, after looking at the map (something I rarely do) I had found a few en route. What neither of us had considered however was that: a) it was peak holiday season so the first couple we found were full and b) the world was in the middle of a pandemic so those which weren't over-crowded with baguettes and barbeques were closed. After veering off course an hour or so to the west, I was getting tetchy, and so were the woofers. With the night drawing in and everyone in desperate need of respite, I aimed for the closest Aire.

Aires, or places to park up, are commonplace along the autoroutes. Amenities range from petrol stations, restaurant services and picnic tables to the very basic sites with 'hole in the floor' toilets. Naturally, any sensible traveller will grab the best spaces, knowing that they will get something to eat and a clean loo, fairly early on into the evening. This left us with little choice but to park up between two lorries in, let's just say, the least desirable lot to spend the night. Despite my love of vampire literature and all things, Dracula, I'm not a fan of horror movies. The place was devoid of any human life unless you counted the truck drivers, and they were tucked up in their cabs. What little lighting there was over the concrete building housing the toilets reminded one of a scene from the Victorian East End at midnight with no-one but thieves and body-snatchers about and the passing headlights from the cars along the autoroute nearby didn't make you feel any better.

While Rene hurried off to use the facilities in the concrete monstrosity (we might have had a toilet in the Mothership but there was nowhere to empty it), I allowed the woofers to hurl me down the stairs so they could do their thing. If any of the lorry drivers had passing ideas about two single women alone in an over-sized caravan, these would have been flattened by the appearance of Arry, Neo and the rest of the pack barking through the undergrowth and marking trees with their own version of social media. They had no qualms about bogeymen jumping out of the woods and if one had, they'd have peed on him anyway such was their desperation. Having stretched their legs and emptied bladders, I fed the dogs and waited for Rene to return so that we could break open a bottle and put our feet up,

"Oh my God! Oh my God, Sophi!" my co-driver and all round best friend shuddered as she climbed back into the Mothership. "Have you seen those toilets? They're beyond disgusting!" she cried as she poured half a bottle of anti-bacterial spray onto her hands.

"Glass of wine?" I responded.

My darling Rene, who was the first person other than family who I had called after Tony died (and she'd spent that night holding me), the only person I ever wanted to make this trip with and I'd subjected her to a trucker's toilet.

Between us we had packed enough cold cuts, cheese and crisps to keep us going for a month; a blessing as both of us were too tired to figure out how to turn on the cooker (I still couldn't find the manual). Having slipped into our summer PJ's, we curled up between the woofers on the window seats with our groaning picnic and a bottle of wine, when I suddenly realised we were eye to eye with the lorry drivers parked on either side. This sent both of us into hysterical laughter as we pulled the curtains to protect what modesty we had left.

CHAPTER 3

We both slept reasonably well despite having to share the same bed with dog hair and woke to an empty Aire. The lorry drivers had left before the crack of dawn so we had the place to ourselves. We must have been exhausted as neither of us, or 9 woofers, heard them leave. Wanting to try out all the amenities in my newly-purchased motorhome, I squeezed into the shower; a tight fit but worth the quick wash and Rene miraculously managed to get the hob working so we could have morning coffee. Those who know me, and she certainly does, are aware know that I cannot speak sentences until I've had a cup of coffee, two is preferable. Sipping my second cup, a thought crossed my mind. How had Rene managed to get the hob to work? As she pointed to a load of buttons near the stairs leading out from the living area with 'water on', 'gas on' clearly marked, it dawned on us that we had misled the transport police at Folkestone. We'd unwittingly stationed ourselves on the Eurotunnel train with our gas canister tap open.

Caffeined up and in good spirits, we headed out of the lorry park and hit the autoroute heading back towards Orléans and straight down to Toulouse. I was beginning to feel 'at one' with the Mothership as we covered the kilometres and was rather enjoying the power of monster vehicles. We sang along to Rene's vintage music collection (we grew up in the '80's and '90's) as I touted my favourite views.

We didn't have time to visit the grand chateaux or the vineyards that brought the tourists flocking to the Loire Valley but the forest through which the A20 takes you, past Limoges and on to Brive, is breath-taking to see. The dark green of the trees covers the hills for miles and if you ever pass through there at dawn, the rising sun appears to set fire to the forest below. The bridges that cross over the wide valleys and rivers, particularly the one that takes you above the La Roselle viaduct however, are a bit of a challenge for someone with vertigo like myself but the Mothership guided us safely across them. Finally, after a couple more pee and petrol stops, we arrived in Toulouse, with only a hundred or so kilometres to go to L'Horte and home.

In my defence, I had been driving that day for around ten hours and despite the smell of the horse manure along the vines banking the straight road being as familiar to me as the route itself, I missed the turning into South-West Carcassonne. The one I thought I knew blindfolded. Annoyed and muttering more than a few expletives, I had no choice but to head towards the centre of town, something I really didn't want to do in a big white bus. Taking on English country lanes was one thing but I really didn't relish the prospect of having to negotiate a French roundabout with my steering wheel on the wrong side and a wide behind. Temper flaring and several more missed turns later, I got back on the route thanks to Rene and her ever-present phone map and headed towards St-Hilaire. Reaching the final toll booth, my foot missed the brake pedal. Just as the Mothership was about to make mincemeat out of the gate, the toll tag beeped and the bent bar rose unsteadily allowing us through; I wasn't stopping to check for damage and floored it the last ten kilometres home.

Any first- time visitor could easily miss the entrance to L'Horte. On one side of the road there are rows of vines as far as the eye could take you and on the other, Le Laquet – a tributary of the River Aude. We'd always tell guests that once you passed the huge sign indicating Domaine Martinolles, a hundred metres or so around the bend, and if you kept your eyes peeled, you'd spot the tiny one that said "L'Horte" on your right. Oh and there was a massive telephone pole next to it.

Considering there are several Domaines or vineyards dotted along the road, it would be easy to forget which one you had to look out for. But if you were lucky, you'd slam on the brakes and take the sharp right into the driveway and weave your way down to the parking area.

Set on twelve acres of varying terrain, the property had two houses with a short walkway and parking area between them, plus a double garage which usually only had one car in it and the remnants of several bits of machinery, and a workshop. My father (always known as Pop) and my younger brother Timothy (known as Moth) had rebuilt the main house from the shell it had once been when my parents had first bought the place. The previous owners had used the smaller of the two houses as their domicile, keeping horses in the larger one along with a colony of bats. The horses went with them and Moth, being an ecologist and wildlife expert, constructed several bat condos on the outer walls. It took a couple of decades to finish but eventually we had the perfect family home.

Built over three floors, there was a kitchen and sunken dining room on the first, a bedroom, bathroom and an impressive lounge which accessed a wide terrace on the second and another two bedrooms, a bathroom, Pop's office and a library at the top. Needless to say, the office and the library doubled up as sleeping areas when needed – especially when all the kids were there.

The second house, the Stable as we called it, had only needed a few alterations and usually served as the overflow sleeping arrangements when the whole family descended for the summer. After all, it only had three bedrooms and one bathroom but it did have a rather cute little bar kitchen and lounge; both only ever used for late night drinking purposes I seem to remember because most of the appliances didn't work.

L'Horte was the place where the family reunited for the holidays; my family being my older brother Simon and his wife Alba, his three children Louis, Maxime and Kate. Then there was myself and Callum, my sister Bong and her husband Andrew along with their children Ellise and Jojo and last but not least, my younger brother Moth with his partner Hilde and their son Tiago (known as Tigger).

The land hosted an orchard, a beautiful English lawn overlooked by the big terrace and plenty of open meadow for the kids and dogs to run wild in. They used to run along the path under the fruit trees and gorge themselves on apples, plums and pears before legging it down across the meadow to the big house in the vain hope the bugs missed them. Oleanders, winter jasmine and irises of all colours framed the bank that enclosed the lawn and roses wove their way up onto the terrace balcony which was also home to numerous flowering pot plants. Behind the garage and next to the tractor shed, Pop had constructed an irrigation system that was fed by the river, all very 'nature' as he wanted but it did tend to attract moles. I distinctly remember wooden sticks with plastic bottles (recycled of course) poking up all over Mumo's prized green lawn but to no avail. I think it was the earth tremours caused by the kids bouncing on the trampoline that finally got rid of the little rodents.

In the evenings, us adults dressed up for 'apero's' (aperitifs) on the terrace whilst the younger generation whizzed about playing football or cricket (or both) on the lawn below, as the sun went down over the hills beyond – you never tired of looking at the view. L'Horte had everything you could dream of, but the best bit was the river.

Wrapping around two-thirds of the perimeter like a horseshoe flows Le Lauquet, a tributary of the River Aude and the place where Callum and his cousins had spent their childhood summers kayaking, swimming and zip-wiring above and through sparkling waters with dogs cavorting around them. Those were the halcyon days; lazy days spent under the grapevines on the Stable's outside deck until the heat got the better of you and cooling off in the river was the natural thing to do.

Although there was a dam part-way down stream, Pop built a wooden jetty in a shallower place along the reed-covered banks. It was also safer. The dam had a long thin stone wall with deep winding steps on one end leading to a small gate and vineyards on the other. It always terrified me and with good reason; as you negotiated the steps, on your left the river was usually calm and inviting but on the right there was a several metre drop into the dark waters below. That never scared me, but the stone buttresses that held the dam did. Water rushed over them, creating waterfalls and rapids below, crashing over huge rocks until it made its way down river. Pop fell off there once after trying to clear algae off the path and ended up in a coma. If it wasn't for my nephew, Louis, dragging him out of the water, he wouldn't have survived. But, like everything else the river had to offer, those waterfalls were heaven sent and if you were intrepid enough, or you were a dog, you slid down the rocks next to the steps and made your way down to the 'beach' below.

Oddly, Tony never really took to L'Horte – he wasn't a fan of the 40 degree heat for a start and the mosquitos always seemed to liken him to an 'all you can eat' buffet. He didn't much like putting any part of his body in the river either as in his words; "It's got a slimy, rocky bottom and I can't see what I'm about to put my foot on," But I loved it and so did whichever woofers I took with me including Macgyver, my soulmate and devoted German Shepherd's whose ashes I'd scattered under the big tree on the lawn. For him, his brother and sister (owned by my parents) and all the kids, it was the best of times.

When I turned the final corner and saw that faded white signpost with L'Horte carved into it, I felt a moment of utter calm wash over me. I'd done it! I'd survived all that the Gods had thrown at me over the last two years – now the woofers and I had another life to begin.

CHAPTER 4 - The beginning 2018

I wasn't always defined by widowhood. Back in 1990, I got a bit drunk at a work party and went home with a blue-eyed, dark-haired fitness trainer by the name of Tony Stewart. Our friends gave us six months – we married two years later, got a dog and moved into a nice four-bedroom house in Streatham South London complete with garage and off-street parking.

A few years later, in 1998, I gave birth to our son Callum and we settled into family life. Knollys was a friendly road in those days with the occasional chat over the garden fence; one of our neighbours even bought us a little Peace Lily plant as a house-warming present. We were the typical suburban couple with Callum, two dogs, two incomes and a small mortgage. Callum went to the local prep school in nearby Dulwich where I made friends with a group of mums in his year group. We took it in turns to pick up each other's kids and give them dinner every Friday evening, always ending with mums being retrieved by their long-suffering partners after a few glasses of Cava. It was during one of these raucous evenings that I met Irene, or Rene as I have always called her. A tiny, half-Columbian, dark haired beauty who can make any man swoon at twenty paces, she became the Yin to my Yang, or the Yang to my Yin – my best friend.

If the weekly get-togethers with the girls were my release, then Tony's was his Sunday morning football matches in Tooting. He'd cycle the short trip over to the local pitch every weekend without fail; he loved playing almost as much as he loved watching his beloved Liverpool FC.

Although we had talked about having another child, Callum being diagnosed with Type 1 diabetes at the tender age of 3, put Tony off. Despite the doctor assuring us that Callum's condition wasn't hereditary, Tony didn't want to take the chance. So, I focused on work and by the mid 2000's the Dog Hollow Behaviour Consultancy was a name to be reckoned with in the business and Tony decided to take a back seat to my career. He turned our garage into a fitness studio hoping to 'work from home' but Dog Hollow was the only one to draw in a decent income. As much as we were happy in our new roles, our marriage wasn't without arguments. I had to supplement our income by taking in boarding dogs, something he resented as it meant holidays were few and far between. I'd also started fostering and rehoming dogs for a charity which, of course, meant we ended up adopting quite a few of them. We had our ups and downs like any relationship but somehow life together just worked.

In 2018, my world fell apart. Tony and I had decided, since Callum had more or less moved out of the family home and was now working in Shetland (yes, right up there above Scotland), it was time for us to think about upping sticks ourselves. We both needed to start living a life together without the stress of London and my work. But Tony wasn't well. "Just a tummy bug I've picked up." he said.

I wasn't overly concerned as Tony did tend to worry about anything health-wise and had regular visits to the GP because of his high blood pressure (he married me after all). But the bug didn't go away. I persuaded him to put our house on the market, thinking he'd be so much happier and healthier living out in the country. But he got worse and started complaining about leg cramps.

I remember the two of us going out for a pub lunch after viewing a particularly nice property in East Sussex. It was way over budget and too big for the two of us but we liked the location; a quiet little village with all the basic amenities and this pub.

“Let’s order steak and chips,” I said. A rare treat, as I had banned it from the house since my father had read a book about the environmental impact of mass beef production.

“I love you,” Tony suddenly blurted out with a look of desperation on his face.

“Ditto,” I replied slightly confused. ‘Ditto’ always followed ‘I love you’ after we had been to see the film Ghost.

It was odd. Whilst neither of us was ever unaffectionate, we’d grown into the sort of couple who, after 28 years together, didn’t blurt out those sorts of sentiments.

He didn't eat much of his steak that day and over the next week he went rapidly downhill rarely leaving our bed. Finally, I had had enough. I came back from viewing another house and found him in our bedroom with the curtains closed; there was a yellowish tint to his skin.

"You have to go and see the doctor, you're turning yellow," I ranted at him. It wasn't the first time I'd badgered him about going to the medical centre but this time I was adamant.

"I'll call him in the morning," he replied and that was it.

He did call the doctor and was promptly told to go to the hospital and get checked out. I wanted to go with him but work got in the way and he dismissed my worries. I didn't hear from Tony again until that evening

"I've got cancer in my pancreas and it's spread to my liver. It's terminal."

Terminal. The sheer magnitude of that word shut off the emotional part of my brain, the ‘you aren’t going to die, there are always options’ lobe still working. Tony was a healthy 56 year old non-smoking, teetotal, vitamin-taking veritable he-man. Surely the doctors could just cut out the tumours and zap him with chemotherapy? But Tony didn’t just have cancer, he had pancreatic cancer – virtually impossible to diagnose in the early stage and even if it had been, the chance of surviving more than 5 years is around 3% (www.cancer.net 2022). By the time Tony had been admitted to Kings College Hospital, just down the road from us, the bitch had spread to his liver and surrounding organs. He had end stage cancer but I still didn’t believe there wasn’t hope, even when he was cramping up in pain, he’d get through this. He was my Tony.

I had to call Callum.

“Darling, Dad has cancer and it’s bad.”

No mother wants to hear their twenty year old son cry.

“I’m coming home as soon as I can, I’ll give my notice in now, a couple of weeks at most,” he sobbed down the phone.

He didn’t get home in time.

Tony was in hospital for a week. I went to see him whenever I could, although trying to find which ward he had been moved to was often a challenge. When I did find him, there was never a doctor available for me to talk to which left me frustrated and angry and Tony’s attitude didn’t help matters. Our conversations were invariably one-sided, I chattered away about life and football results but he was always distant. I felt like he didn’t want me there. I bought him in a radio that normally he would have been delighted with but instead he told me it made too much noise. I bought him puzzle books he never used and Liverpool FC magazines that he never read. I asked him several times why he hadn't been moved to a cancer ward and why hadn't a doctor asked to see me but he evaded my questions – I felt brushed off and helpless. He was in constant agony, not only because of the cancer but the leg cramps had turned out to be a thrombosis so he had to have medication for that by injection yet no morphine was given for his pain. Why?

Phil, one of Tony's closest friends, was a regular visitor and I got the feeling he knew more than I did. The two had known each other for more than 30 years, Tony having been one of Phil's tenants in his Chelsea flat back in the day. Both had a mutual love of football and cinema, often getting together for one or the other. Phil was, and still is, one of those people you naturally confide in so I understood even if at the same time I felt pushed away. I learnt from Phil that Tony was booked to have chemotherapy, something I was positive about but my husband wasn't exactly enthusiastic. Extending his life was a good thing wasn't it?

Finally, after all the frustration of 'guess which ward I'm in today' hospital visits, Tony called me to say they were sending him home. Phil was there when I arrived which gave me a bit of a mental boost; at least someone could help me find out what the next step was. The nurse had told us that we had to wait for the doctor to sign him off, another plus as I had a list of questions to ask. But he never appeared. Eventually another somewhat harassed nurse arrived with his exit papers and hurriedly showed me how to change the dressing on his stomach where the thrombosis injection went in. Everything was so rushed; I had no notes telling me what I was supposed to be doing or what was going to happen next. When was his chemo due to start? What happens if he gets worse? Who do I call in an emergency? Nothing!

I was screaming inside but not wanting my darling husband to see me lose it – I kept my mouth shut. Someone was bound to ring me when we got home.

The same over-worked nurse, really an angel in uniform, helped us get Tony in his wheelchair out to the car park, risking her job in the process. Apparently, nurses weren't allowed to accompany patients to their cars and neither were wheelchairs. Health and safety and all that bureaucracy! I asked Tony if he had a list of the medication he needed so we could stop by the chemist; nothing except paracetamol he replied. Where was the morphine?

Having got him home and comfortable in Callum's old bedroom, the next few days were spent trying to get Tony to eat, bathing him and waiting for the doctor to call. He didn't. Ever. There were no visits from cancer nurses or palliative care teams – I was on my own, scrambling through the days trying to find a balance between caring for my husband and working to pay the bills. Except I wasn't completely on my own, I had two of the best friends a girl could have in Rene and Phil who were always on stand-by to sit with Tony if I needed to go to work, and a doctor, one of Tony's old football buddies, a urologist, who was always referred to as Doctor Tom. It was he who pushed the GP for more medication (Tony had debilitating constipation to add to everything else) and was always there to listen whilst I ranted and raved during midnight calls. Between the three of them, I stayed just about upright. I was angry; angry with Tony for not wanting to fight the bitch, angry at being put out to dry and angrier with myself for not being able to fix him. His GP was beyond useless; any attempt I made to contact him was ignored.

One lunchtime, I asked Phil if he could come and sit with Tony while I headed into Kensington to drop the day-care dogs off. Having a few dogs in a couple of times a week meant I could be at home most of the day. Tony seemed brighter and even managed to eat as we giggled over something ridiculous.

"I love you" I said giving him a smacking kiss

"Ditto" he laughed back

In an upbeat mood, I left the two of them and drove off into town. Less than an hour later, Phil called. Tony had got up to go to the toilet, fainted and was now slurring unintelligibly,

"Call an ambulance, I'm on my way back," I shouted down the phone as I turned the car around still loaded with dogs

Phil called again as I was trying to weave my way through peak hour traffic in South Kensington,

"Soph, you'd better get back quick, the paramedics are here and it's not good."

What did he mean, 'it's not good'? Were they taking him back to hospital? The paramedics rang me, did I want a police escort home but I declined as I was just around the corner and anyway I could just follow them to the hospital if need be? I turned into our road only to be confronted by police cars and ambulances outside our gate; it looked like a crime scene. Hurling my Land Rover into the driveway, I somehow managed to get all the undelivered dogs into the kitchen and ran upstairs. Tony was lying on the floor, motionless with paddles banging down on his chest and a defibrillator with a flat green line on its monitor beside his feet. I planted myself behind his head and cradled it in my lap, those beautiful blue eyes looking vacantly up at me.

On the 17th May 2018, the man I had promised to love through hell and high water died in my arms. His heart stopped at the same time his watch did, twenty to five in the afternoon.

CHAPTER 5 - The Goodbye

There are no textbooks on 'What to do when you suddenly become a widow' and even if there was one, you wouldn't read it. In my blog that I started at the beginning of 2019, I wrote about 'Surviving the Splinters' (11th March 2019)

It was a feeling I shall never forget. As I sat on the floor cradling Tony's head, the paramedics calmly whispered to me "he's gone." I felt the shards of glass snap into place around my heart, numbness enveloped me like a steel cloak. As long as the armour was in place, I could function

And that's what I did, I functioned. Like one of those robotic vacuum cleaners, I went through the motions mindlessly just doing what I was told. Someone took my arm and gently removed me from the room and I asked where Yogi Bear was. He had taken up a 'Greyfriars Bobby' attitude to Tony's bed, rarely leaving his spot on the floor and now I couldn't find him. Phil opened my bedroom door and there he was, shivering – I lifted our little Bear up onto my lap and just sat on the bed hugging him.

Where was my phone? I had to call Tony's parents, Jenny and Jeremy and his sister, Frannie who was currently somewhere in Australia with her partner Steve. Frannie who I always refer to as my 'sis' as she is so much more than just a relative by marriage and someone who adored her only brother. Oh God, I had to call Callum! I called Jeremy first, I knew he would get hold of Frannie and then I called my older brother. Callum was working out his notice in the Shetland Isles at the time which was also where Simon lived. I didn't want to break the news over the phone and in the middle of Callum's shift so my big brother came to the rescue. I left my parents until last; if I was going to break down, it would be to my Pop on the other end of the phone. Phil must have called everyone else, I don't remember.

There is a sort of amnesia connected with the days after that, I just kept going until the batteries ran out and perhaps sleep would come. Callum arrived the day after his father died, wrapping me in his arms as he walked through the door.

Ask any English person what they remember about a particular year and they will inevitably give you a weather report. "It was the winter of '76," or "We got married in the autumn of '92." The recollection of events always preceded by, "It was a warm day," or "It was cooler than average for that time of year," "The forecast said snow" – that kind of thing. The May of 2018 was one of those months that at any other time of your life would have had you basking in the glorious sunshine but I was frozen inside.

People sent flowers; so many that my wonderful neighbours, Michelle and Trevor, had to keep bringing round various containers for me to put the foliage in, and countless sympathy cards poured through the letterbox. Poor Michelle had been the first to knock on the door after Tony died wanting to know why there were policemen on my front lawn. The phone rang constantly with friends and clients wanting to check up on me and offer condolences. I was grateful for the contact because it kept me busy. The palliative care team arrived the day after Tony died and I slammed the door in their face – they were a little late I spat at them. Jenny and Jeremy came up that first weekend along with Tony's best friend and executor Dickie. I had known him for as long as I had known Tony, nick-naming him 'sexy legs' for no other reason than he was always the life of the party. Having met back in the late 80's when Tony was a lodger with Dickie's neighbour, in Dickie's words, "They hit it off straight away." They were total opposites – Tony being the more serious of the two but Dickie never failed to make my husband laugh, usually by leaving long, rambling messages on the answering machine. I wish we'd kept that thing. Tony adored his friends but Dickie was the one he chose as Callum's godfather.

I had put the TV on as it happened to be Harry and Meghan's wedding that day, thinking it might provide a distraction from the awfulness of it all. It didn't of course. I felt awkward, here were Tony's parents who had just lost their only son and I couldn't do anything to make it any easier. Dickie took me out into the garden, pulled me into a hug and told me not to worry; Tony had made sure I was taken care of. But he hadn't had he? He wasn't there to take care of me.

Family and friends. They kept me from disappearing under the ever-threatening waves of self-pity and hopelessness. Rene and I sweltered under the midday sun along with the crowd at the Chelsea Flower Show and after an almost disastrous shopping trip in Bromley with another one of my 'mum friends', Lareine, I managed to find a black summer dress in the last store at the end of the shopping mall thanks to her determination to find me something, albeit a little on the miniscule side and probably not entirely appropriate. Funereal colours were not on trend that summer apparently. Lareine is the bubbliest blonde you could ever meet and always the last one left on the dance floor. No-one was more suited to dragging me around the centre and making me laugh at the less than widow-like outfits she made me try on.

Over a lunchtime bottle of wine (I was finding it difficult to eat anything but drinking was easy), Lareine looked at me quizzically. "You haven't …

cried yet, have you?"

It was a question I found and still find, difficult to answer. I couldn't, it was that simple. I would lie in bed every night waiting for the outbursts to come but they never did, my emotions had shut down. I didn't have any tears to cry.

What I did have was a veritable army of furry medics to keep me from going doolally. As I once quoted in the blog, "Keep your foot on the pedal." (13th January 2019), *Woof oil. For all the times they have hogged the bed, barked at next door's cat, pooped where you step and left dog hair on every piece of furniture, they are there when you need a paw. They make you get up every morning because you have to go to the park and act like you've never seen a park before. They kiss you at the moment you least expected it but needed it the most. They snore, fart and bark in their sleep but you know you aren't alone at night.*

Arry, Alice, Coco Loco, Danny Boy, Evie, Gizmo, Mo Cridhe, Neo, Simi and Yogi Bear; our dogs gave me some sort of normality to life after death. Whether I liked it or not, I was an integral part of their daily routine. I remember going to the park the day after Tony died, waving in the distance to the regular dog walkers and thinking – they don't know and why should they? My life has just been turned upside down and all I can do is smile and mouth hello. We got up every morning, went to the park, ate and crashed. Wake up and repeat. I owe my sanity to those 10 dogs.

There was another thing I had to think about, I had to plan a funeral. The last time I'd planned a major event was Tony's surprise 50th birthday party, the irony. The paramedics had sat down with me and explained what would happen to Tony immediately after they took him away and left me with several pamphlets listing everything that needed to be legally sorted, most of which I put away in a drawer for later. Tony had stated in his will that he wished to be cremated so the obvious choice was the West Norwood Crematorium, being only five minutes or so away from Knollys and having my favourite pub just outside its gates. Suddenly I had something to focus on, make lists, call people – a purpose. I needed a purpose.

Streatham and West Norwood might have two different postcodes but, like many suburbs in South London, they had moulded together. As Knollys Road bisected the two, I was familiar with both high streets, often passing by various undertakers on my way to meet friends but never taking much notice of them, after all they aren't the sort of place you just pop into. Mercifully, Cheryl, a recently widowed friend recommended one that just happened to be located on Norwood High Street opposite the crematorium so I rang and booked an appointment the next day. It wasn't until I walked through their doors and sat down that I realised how out of my depth I was or perhaps clueless might have been a better explanation. Over a table bestowed with catalogues of coffins and floral tributes, I was asked where Tony's body was. I hadn't a clue. I seemed to remember something about him needing to go the coroner as he had died suddenly at home but at the time, I wasn't exactly in listening mode. Angie, the funeral director and perfect example of why you need people like her, calmed me down.

One phone call and she had located his whereabouts and with another phone call, booked the crematorium. A word to note here, crematoriums get booked up so you have to plan ahead, ironically. They managed to squeeze us in for a slot in two weeks' time (pretty sure it wasn't a cancellation) and we moved on to choosing coffins and those floral tributes.

An hour later and having exhausted my debit card, I was back home feeling totally miserable. Everything was so expensive! What little savings I had would barely cover the costs and I still had bills unopened on the kitchen counter. Tony's parents had insisted on paying for the 'wake' which was a relief but until my husband's estate was sorted, all I had was what was in my account. But then Dickie had said Tony had made sure I was taken care of so why was I panicking? I had a little chat with my 'worrywart' inner self; we'd be okay in a couple of months so keep calm and spend. Someone up there must have felt sorry for me because a couple of days later, Dickie called me to say that the money in Tony's current account would be transferred immediately. Now, at least for a few months until probate was sorted, I could breathe. And spend. I tasked Lareine with booking The

Great North Wood pub (the one next to the crematorium) and Cheryl with organising the flowers. My mother-in-law, Jenny, wanted cornflowers in the wreath but they were out of season. I had a brain wave, I would put a packet of blue wildflowers in each order of service so everyone could plant a little memory garden.

I had already decided that there were to be three charities to donate to; Pancreatic Cancer, Tsavo Trust (he was passionate about elephants) and Surfers against Sewage. The next step was to get Tony's death certificate before I could secure our spot at the crematorium. Naturally this particular piece of necessary information had been given in the pamphlets I had shoved in the back of the drawer.

Having dug out the various bits of paper, I went online only to find out that I had to make an appointment at Lambeth Town Hall, a bus ride away in Brixton – usually a two week wait but someone up there was being nice again and I got one almost immediately. The offices were bang next door to where the once infamous nightclub known as the Fridge used to be, Tony loved house music. I arrived early which was just as well as it was pouring with rain and I had no idea which part of the monolithic building I was supposed to be in. Google maps only go so far. Eventually I found the reception desk only to discover that it was the same place that registered marriages and births. Here was I waiting for a document that not only would declare me a widow but also allow for an official burial (no back garden landscaping here), surrounded by babies and newly-weds. Hundreds of numbered cubicles lined the corridor with no indication on their doors as to whether they were going to congratulate or commiserate. I sat on a bench outside my numbered portal, next to a young couple arguing over how to fill in their baby's forms. If only they knew.

My number was called and I was ushered into what could only be described as a crypt. Nothing but white walls and a single blue flower on the officiate's desk. No pictures, no inspiring quotes for you to look at whilst your entire life and that of your husband's is verified on a computer. Was this what his life came down to? Just another statistic? I paid £8 for a few copies of paper proving that my husband was dead. There in black and white type were the simple words; cause of death – Metastatic Malignant Neoplasm of Pancreas with a Bilateral Pulmonary Embolism. It was all so impersonal, sterile even. As I left the cubicle and the door gently snicked closed behind me as if to say, 'you're done now — next' I noticed an elderly gentleman sitting on the same bench looking as vacant as I felt. I didn't say anything as I passed by, what could I? By the way, don't expect hearts and flowers when you go in there?

CHAPTER 6

Having sorted out the certificate and all the other stuff the council could help me with, I turned my attention to the guest list only to find we had a problem. Tony had had a little red address book that he carried everywhere with him, listing everyone's contact details but it was missing. I searched everywhere, even calling the hospital, but it had disappeared. His phone and computer were password locked and apparently I had no rights to access either – who knew? Mind you, the powers that be didn't know Callum's tech abilities so whilst he was busy hacking into Tony's files, I called Dickie and Tony's close friends, Stu and Strikey. Tony, Stu and Strikey had met at Exeter University and become instant friends. In the early days of my relationship with my would-be husband, it wasn't uncommon to spend a wild weekend with one or all of them. Despite being teetotal, Stu was almost always the one that got us into hysterics and Strikey's cocktail parties were legendary. Over the years, distanced by work and bringing up families, the three friends always kept in touch and I knew they could help me get the invite list out. Between us we sorted out who we would call and in a moment of genius (rare I know) I turned to his Facebook page – that I had access to. Within a few hours, my phone was beeping off the proverbial hook, everyone wanted to come. From Australia, New Zealand, The U.S and Ireland, they were coming and the numbers grew. Strikey had messaged all Tony's university compatriots and then there were others who had worked with him through his years in London who sent messages forward – Tony was loved.

The funeral was a resplendent affair if one could say that about the day you cremated your husband. The tiny crematorium in West Norwood was over-flowing with people – far more than anyone could have envisaged but then again, Tony was the best friend a person could have. Even the then Mayor of London, Sadiq Khan was there out in the foyer – he and Tony had been football buddies long before Sadiq took office and I'd trained the Khan family dog.

The day hadn't started well, I was late getting to the hearse as Michelle next door was doing my hair (I might have been wearing black but this girl wanted wild colours in her hair and had a professional hairdresser as her neighbour) and then there was the 'incident'.

Aside from being a well-known short cut between Norwood and Streatham High Road, our street like all the adjacent ones, had parked cars packed bumper to bumper on both sides of the pavement edge. It was also a fairly narrow road, meaning cars would often have to find a gap in order to pass one another. Now, any sane and caring driver (and there were many that day) would have stopped their vehicle and waited for the grieving widow and her entourage to be escorted into the cavalcade of black limousines blocking the road, but 'there's always one'. As I came down our front steps and began walking towards the sombre looking procession, a near punch up had started below me. Dickie, Strikey and Stu, were about to hang an idiot from the nearest lamp post who was raving about being late for an appointment. Angie the funeral director saved the day and the idiot when she pointed out to him, "Didn't I bury your father last year?"

Looking back, I probably should have been more distressed about the situation brewing ahead of me but my mother had given me a large glass of 'fortitude' in the shape of wine on my as usual empty stomach.

The idiot meekly got back into his vehicle and sat there stony -faced as we drove off to the service; Tony's family, Callum and myself in the first car and my parents with Tony's 'boys' in the other. I had been asked prior to the day if I wanted to take a particular route to the crematorium, were there places special to my husband? I declined, unless you were going via Tooting and the football field which wasn't exactly 'en route', there wasn't anywhere else. Now I wished I had, my mind started to panic. "I shouldn't be here," "I'll wake up in a minute," "Why is everyone looking at me?", "Wake up wake up!" We arrived just as the throng had started to gather at the entrance to the chapel and I grabbed Callum's hand, before he took his place along with Tony's friends to carry his father's coffin into the hall. He was wearing his father's mourning suit which was a bit big. "Try not to trip hey?" I whispered as he got out of the car. He smiled at me; he has his Dad's smile I thought. I watched him as he took his place as a pall-bearer, my brain screaming, "He's too young for this!" I knew everyone's eyes would be on me so I did an Audrey Hepburn, put my dark glasses on and took Frannie's reassuring hand. If I was going to do this, I wanted her calm, unruffled self to lean on. “Come on girl, give me some spine,” I whispered to myself as we walked through the doors together into the place where I would have to say goodbye.

About a week before the big send off, I’d had a sudden panic. Having chosen the coffin and what clothes Tony should wear (and please can he take his glasses), it occurred to me that he would be travelling alone so to speak. Tony hated being alone so I’d had a thought; he needed a dog with him. Not just any dog and certainly not one that was currently still on this Earth, it had to be Ash. Ash had died the year before from lung cancer and Tony had been there for him right to the end. Ash was an old soldier; a collie mix we had rescued from Romania who’d loved nothing better than lying at his master’s feet as Tony tapped away at his computer. With this in mind, I rang Angie, could this be done? Apparently yes, asking for your pet’s ashes to accompany those of your beloved is not an uncommon request. I just needed a certificate from the pet cemetery that had dealt with Ash and all would be fine. Now Tony was a hoarder, he kept receipts dating back to 1984 but do you think I could find the certificate? I found numerous others from dogs that had been and gone but not one for Ash.

But there are angels in this tale and Judy-Ann, a trainee vet at the time and a friend for almost two decades, came to my rescue. One phone call and I got a replacement certificate an hour before the deadline. Ash's ashes could go with Tony.

Having just got over the stress of sorting out Tony's travelling companion, I was called one morning by Angie, "Would we (Callum and I) like to visit Tony 'lying in state' before the final farewell?" I wasn't sure I wanted to see Tony in a coffin but my younger, and much wiser sister Kendra (fondly known as Bong) who had dumped everything on her saint of a husband and flown in from New York a few days before, was adamant.

"You should go Sophi, it's your last chance to say goodbye," she said as she hugged me. "Callum needs it too even if he doesn't tell you."

So, we did. The three of us walked arm in arm down to the funeral parlour to say goodbye. Why on Earth do they call them parlours? It isn't the most genteel of environments, let's face it. There we were, waiting to go and see my guy lying in rest behind a vivid purple velvet curtain; the room reminded me of the ladies' toilets in Harrods. Each of us took our turn to say our bit, having been instructed not to touch him (ice cold we were told) and not to add anything to the coffin. I remember seeing the photos I had asked to be placed around him; his great Uncle Bill whom he adored, one of Callum and a picture of the two of us. His glasses case was in his hand but empty because they wouldn't allow the metal frames (paradoxically, they cause sparks) and of course, Ash. My husband looked so peaceful in his suit and the tie I had picked out because it matched his outfit, a slight smile on his face as though he had been listening to one of my awful jokes. “Sleep well,” I whispered. It wasn’t until many months later that Callum told me he had touched his dad – tell a child not to touch an electric socket and they’ll stick their fingers in. That day was the last time I saw him cry over the death of his father.

Tony would have loved his send-off; I had requested that the last song would be ELO’s ‘Blue Sky’ as he knew every word, and as his coffin slowly disappeared behind the curtain, the skies shone clear above him. The service was brief but poignant, and as the congregation made their way out into the Spring sunshine, I hugged, consoled and chatted to one after the other before they headed down the crematorium’s long drive to the pub. Finally, after the last few had wandered down to join the others, I was left to make my own way there. I recall taking my shoes off, needing to feel the security of the ground beneath me. “Deep breath, Soph, you can do it,” came from behind me as Sophy and Sean wrapped their arms around me. Friends are always there when you least expect them and most need them. Sean cracked jokes as we walked down the winding hill from the chapel leading out of the crematorium and Sophy carried my shoes, both of us commenting on the gravestones nestled in the gardens. It’s funny how you remember some things but not others.

At my request, photos of Tony and personal memorabilia were scattered around the pub’s private room. The giant television screen above the bar was supposed to stream endless pictures but had managed to get stuck on only a few. I drank enough not to care. I commiserated, played the host and talked to strangers. In one corner of the room Stu, Dickie and Strikey had been joined by their wives and several of Tony's old University friends. I made my way through the throng to greet them when I noticed Stu, Strikey and most of the other men wearing matching ties. There was something very familiar about their neck wear;

"Stu, what is that tie you're wearing?" I asked

"It's the old St. Luke's P.E. tie," he answered. "Why?"

"Because it's the same tie I dressed Tony in," I replied.

I didn't know it was his old college tie, I just liked it – spooky.

Funerals are a bit like weddings in an obscure kind of way. You spend so much energy preparing for the day and when it comes, like a blink of an eye, it's over. And you're left thinking what do I do now?

I hadn't got a clue.

CHAPTER 7

"Tony would have been so proud of me last week. I took the punches but kept the temper in check. He used to say I fought first and thought later. I would rant and rave whilst he calmly stood there (probably watching the footie over my shoulder) until I stopped. Then he would pull me in for a hug against that gorgeous warm chest of his. God, I miss that chest" (Blog 4th February 2019)

Like most organised people, Tony had written his will early on in our marriage and long before Callum came along. Mind you, if he'd written it later, he would have added the clause, 'Don't let Sophi go and blow the lot on a new kitchen.' And although I had no idea what was left monetary-wise, I did know that he wanted his ashes placed next to his Great Uncle Bill; a man he adored but who had died long before I came into the picture. Jenny and Jeremy spent weeks looking for the plot only to find that it no longer existed – I think houses or something similar had been built over it. Without a headstone or marker denoting where we were supposed to carry out Tony's final wishes, we were stuck for ideas. Here was I with a tall box in the bedroom carrying his ashes (it reminded me of one of those expensive whisky cases, which amused me since Tony was teetotal) and nowhere to put the contents.

"I saw online that some widows have the ashes made into a necklace," I mused over the phone to Jenny.

"Ooh, that's a bit weird don't you think?" she replied. "Wearing a dead person around your neck."

She had a point, it did sound a little vampire-ish but I didn't like the idea of living with the box in the bedroom either.

Between us we found a solution that suited everyone. Tony loved the sea so Callum and I would scatter some of his ashes off a beach near his family home. Prinsted Beach, a few minutes' drive from Emsworth Harbour and Tony's parents' home in Westbourne, was a place with special memories. Every visit we made to his parents, Tony and I always found time to walk along the pathway above the rocky sea wall as it looped around the shingled bay formed by the ebb and flow of the South Coast's deep blue waters. Naturally we took a few dogs with us and whilst they were getting wet, we looked for 'treasures' amongst the rock pools. It didn't matter what season it was or how bad the weather (which it invariably was), we loaded the car up with wellies and towels (for the dogs) and drove through the back lanes of Westbourne to Prinsted. Although the beach was used regularly by walkers and fishermen, it never seemed crowded and once we had parked in the miniscule designated area, we either walked down onto the beach or took the path to the sea wall – depending on the tide. Followed by over-excited wet bits of fur, Tony would throw sticks or seaweed into the sea – it was one of the few places I ever saw him totally relax and be in the moment. I knew he'd be happy with my decision.

The rest of his ashes, thanks to the local vicar, could be buried in the gardens of The Church of St John the Baptist, the same one that Tony and I got married in. The church, only a short walk from his parents' house had a beautiful and somewhat famous Yew tree avenue leading up to the main doors. It was planted in 1545, making it the oldest one in the United Kingdom. I remember it vividly, me swanning up in my Thai silk wedding dress to meet my future husband whilst my younger brother, Moth, being late as always, was changing behind a gravestone. Unfortunately, the churchyard being as old as it was, it no longer had space for us to have one for Tony but we would know where he was. Right next to that walkway.

And so it was that on a frosty morning in September 2018, Callum and I went off to the beach with a few of our dogs and something that looked suspiciously like a wine bag, to say our last farewell to the man we loved. I asked Callum if he would like to help with the deed but he declined saying it was a little gruesome for his tastes so I gamely opened the box and grabbed a handful of its contents whilst standing knee deep in the freezing salt water. The scattering of my beloved's ashes would have been more traumatic if it hadn't been for the wind; it was blowing in the wrong direction and as hard as I tried to sprinkle Tony out to sea, most of him ended up on the over-excited woofers who we then had to persuade to go for a swim in the chilly waters. "I think I've just inhaled half my Dad," Callum choked out which sent us both into a fit of the giggles, collapsing onto the pebble beach. We sat there together for a few minutes, collecting our breath and looking out onto the deep blue water as the dogs shook themselves dry. We might have been a tad late getting to the church to meet Jenny and Jeremy but the emotional outlet had been just what was needed.

The final ceremony was a whispered affair (we weren't really allowed to dig another hole in the church's grounds but the vicar was very understanding) and after a prayer and a mental note as to where he was, we returned to the house. I may not have been able to carry out his final wishes to the letter but I think Tony would have liked the locations. And he would have laughed his head off watching us on the beach.

CHAPTER 8

A few months later, I was loading the car for the usual morning dog walk when my phone rang. The late January weather had put me in a less than genial mood and not recognising the number on the phone screen, I was about to answer it with a, “Don’t you think it’s a bit early for PPI calls?” snap until I heard my mother in law Jenny’s voice. My father-in-law, Jeremy had had a heart attack (of all places to have one, he’d had his in the medical school of Oxford University) and died. I looked up to the skies and whispered, “Why me?” and immediately felt guilty. Jenny had not only lost her son but her husband too, in the space of eight months. And I was going to have to tell Callum.

We drove down to Sussex and Tony’s family home as soon as we could. The last time Callum and I had been there it was Christmas, my car loaded with presents and wine – now all I had was an orchid for sympathy and several boxes of cheesy biscuits because Jenny loved them.

That weekend we talked, she cried and we reminisced. A thought crossed my mind as to how differently the two of us would approach widowhood. Jenny had had almost 60 years with Jeremy, I had had 28 with Tony. Unlike me, my mother-in-law is never one to stick her head in the sand, she's a 'get on and do it' sort of person but it was heart-breaking to hold her whilst the tears fell. I found myself being the expert widow and giving advice even though it felt the wrong way around. Jenny was and still is, more than just Tony's mother. I adore her. She can turn any negative into a positive and despite her diminutive stature, has an inner strength and ability to just get on with things. She's also an incredible cook; Tony refused to eat any mince pies except his mothers and I won't mention our first post-wedding argument about the way I cooked roast potatoes.

The weather on the day of Jeremy's funeral, unlike Tony's, was overcast and bitterly cold. I had managed to buy a black coat and trousers online just before the day, which was just as well even if the coat was a little tight over layers of thermal underwear. The church service was an emotional one; Jeremy had not only served as a Captain on the Ton Class Minesweepers but had been an active member of the association after he retired. The procession of banners held steadfastly by his old naval friends and then the mournful 'signing off' whistle brought the tissues out; the Naval cap atop his coffin and Frannie breaking down as she read her piece finishing everyone off.

As we congregated outside afterwards, waiting for the cars to take us to the wake, it hit me that the last time I had seen most of the extended Stewart family was at my husband's cremation. That day I'd felt part of them but now I felt like an alien intruder sneaking in on a private affair. The only link I had was with Jenny, Frannie and her partner, Steve. I'd always loved Jeremy, apart from anything else he had a wickedly dry wit and taught me about wine. He'd adored his only grandson; the one I was watching helplessly as he tried to get through another wake. He'd have rather been on Mars. I got drunk at Jeremy's funeral as well.

But the Gods hadn't finished with me yet. Two months later, my mentor and close friend, Keith died. The cancer bitch had got him too. Keith had been my champion, my shoulder and my vet, for over 20 years. Tall and debonair, he ran his Kensington practice for over 50 years and knew everyone, from royalty to the guy who owned the sandwich shop around the corner. There were very few behaviour consultants in London when I started back in the late 1990's and even fewer women in the field. If there was someone I needed to know, he introduced me to them, praising my work and not once seeing me as a 'newbie' to the profession. And he never talked down to me; he'd discuss the latest veterinary findings with me and if I made a mistake, he'd always say that's how you learn. Dog Hollow would never have been as successful as it was without Keith.

I had spent every afternoon I could spare by his bedside at The Royal Trinity Hospice, just a few minutes from Streatham, listening to his advice about my future and making plans for doggy world domination. Keith was always upbeat and nagged me constantly into believing I could do anything. His room was always full of old friends – most of whom had four paws. I'm still not sure if I would have been where I am now if it wasn't for Keith and his wife, Serena.

Back at *Chez Knollys* (as I had christened our house) and having to deal with the reality of life without

Tony, there wasn't much to do except work and wait for all the financial fiddly bits to be sorted out. I'd spend the daylight hours, when I wasn't walking the dogs, sitting in front of the television or moving furniture about. I needed background noise all the time – voices talking. Everything felt so lifeless otherwise. I rarely saw Callum, he'd taken to spending his days and nights holed up in his bedroom playing video games online and it was starting to worry me. He might as well have moved to Australia for all I knew. The man/boy who had been my rock since his father died had become morose and withdrawn, snapping at me if I dared ask if he was okay. The only time I would catch a glimpse of him was when he came down for a very late breakfast and when I'd made dinner – the latter being fleeting and peppered with complaints about my cooking (he still complains about my cooking). He never went out, slept until midday and stayed up until the early hours chatting to strangers online but never talking to me. The bond we'd had since the day he was born was getting thinner by the day and, I know it sounds daft, I was terrified he'd do something stupid – I'd read stuff. From the first night I'd bought my baby home to the time he'd left for Shetland, I'd always checked on Callum every night to make sure his chest was going up and down – the worry gene again but I hadn't recently as he was awake long after I'd gone to bed; I could hear him. Not wanting to intrude into his private space, I tried to talk to him in those brief moments we actually sat down together at the table but all I got was, "I'm fine", "Stop fussing".

I got so used to hearing the clicking of computer keys and Callum's swearing at something or other with someone or other every time I went upstairs, the silence from his room one particular afternoon scared me. Callum wasn't gaming. I poked my head around his door. I did knock first, I am a mother after all, but when I didn't hear him grunt his usual, “What do you want?!” I stepped into his bedroom. He was lying on his bed, headphones on but all I saw was a face distorted with pain. I sat by his feet and just looked at him.

“What’s going on Cal?” I whispered

“Nothing,” he whispered back before he looked up at me with those beautiful blue eyes filling with tears. And then broke down.

I’d forgotten I was a parent somewhere down the line of events. Whilst I had been cursing the Gods and cursing Tony, my son had been left to sort things out for himself. I crawled up beside him, why hadn’t I considered his feelings? Callum who’d held onto me when I was about to crumble and all I thought about was the woofers? I reached for him and we just curled into each other as he cried – not about his father or Grandfather but about his own life going forward. He hadn’t a clue what to do, felt even more alone than myself and even worse, his self-confidence was non-existent. In his words, "I'm just a useless piece of s**t." I wanted to deck him but I tickled him instead and he laughed, I needed to see that smile. And then we talked and talked. I suggested that he might want to think about college? He could make friends there? But Callum hadn't had a great history with education, bullied by a teacher when he was young (Tony and I didn't find out until after he left that school) and was left with little or no self-esteem in his abilities.

I got out my computer and started looking at courses nearby.

"What about carpentry?" I suggested tentatively (you must always tread carefully when giving advice to twenty-somethings). The course just popped out to me for no other reason than I thought it was something he could change if he didn't like it.

"Carpentry? But you'd need an A' level don't you?" he replied, grabbing my laptop. Pff A' levels, I made my way through university and had a successful business with only one decent grade and that was in English Lit.

We found a course in nearby Bromley. Actually, even though it was nearby, the train schedule was up the swanny and meant an hour's journey but at least it was something. He applied and got a place starting that Autumn and suddenly my boy/man/child was back. He stopped the games, slept, helped me sort out the garden and started to prepare for his future. And we kept the lines of communication open. Things were on the up and having just celebrated Pop's 80th the year before, we now had Mumo's to look forward to.

Pop's had been a big family get-together just before the scattering of Tony's ashes. My sister Bong, always the planner, had chosen a venue around the corner from where he'd grown up in Islington, North London and had booked a rather swish house for the weekend so that all the immediate family could be together.

Knowing that it would be the first time I had been with my family since Tony's funeral and having work commitments, I hadn't stayed there – I had the woofers to consider was my excuse but to be honest, I hadn't wanted the 'widow' to take away the focus from his big day. So, I'd dressed up in the funkiest catsuit and avoided any conversations about anything Tony. Once again, Bong had come to my rescue, placing a glass of champagne in my hands and moving me away from any awkward hugs. Just as well, as lack of appetite and bubbles in a glass make for a very merry widow.

Mumo's big do was back in France so I whizzed over for 24 hours, flying in for the afternoon and leaving the following morning. As much as I loved being with the family, I couldn't ignore the growing pile of bills. And I had to go to Keith's funeral.

I had started writing my blog in January 2019; why, I don't remember. I called it 'The widow plus woofers' for no reason other than it sounded catchy. Every Sunday, I would curl up with my furry family and pen a weekly diary filled with inner turmoil and whatever amusements or disasters that had happened. I made a decision after Tony died to dedicate Sundays to laze in bed days, the woofers needed the rest as much as I did and they never seemed to mind. They would sprawl across the duvet as I drank my coffee and penned my pages. To this day, I still write it regularly although reading through those first few years now is like reading someone else's biography.

A week or so before the first anniversary of Tony's death and knowing my girlfriends were taking me away for the weekend, I met up with my mother-in-law Jenny at the Victoria and Albert Museum in London. After we had mused over decades of Dior dresses specially put together for an exhibition, we did what all posh ladies do in Kensington and went off to have a glass of wine and a bite to eat in the restaurant upstairs. It was over a decadent glass of Pinot that Jenny posed a question,

"Why do you call the blog, 'The Widow plus woofers'?"

Now I have to admit, the answer seemed an obvious one but I replied saying I thought it was a rather catchy title.

Jenny hadn't finished, "I meant, why do you call yourself a widow? I don't label myself with such a word;

I'm just me," she countered, "and it reminds me of a spider."

I nearly snorted my wine over the pristine white tablecloth.

Later, I pondered over her statement. Was it this word that was stopping me from sorting out my life? Let's face it, there's no way of injecting happiness into the word 'widow'. You can have a happy marriage or an amicable divorce but losing a life partner is not a chosen path. Unless you dispose of your other half in the back garden that is, then you're a murderer instead of a widow. Even in other languages, the word sounds depressing; '*veuve*' in French, '*vedova*' in Italian, '*viuda*' in Spanish. Was Jenny right? Shouldn't we just be defined as who we are and not as the product of a life-changing event? Would Jenny still be my mother-in-law or would that change now that I’m not married to her son anymore?

Letter to my guy 17th May 2019

My dearest Wormy,

It's been a year since I held you in my arms. I still can't get my head around the fact that you are no longer here, I feel like I'm living in some sort of parallel universe and at some point everything will return to what it was. I can still hear you, feel you lying beside me in bed watching your DVD's and smell your 'life's okay now you're here' smell. I can still see your incredible blue eyes light up when you laugh and your lip tic because you can never lie. I miss you. I would like to point out that the football hasn't helped, I'm sorry Liverpool didn't quite get there but you still have the

Champions League . I hope you haven't broken any chandeliers up there with your leaping about.

Life would be a lot less stressful if you were here by the way, just so you know. Your will and estate alone has caused me no end of anxiety and sleepless nights, I never knew you were so complicated darling. Mind you, you'd probably tell me to be patient and it'll all get sorted eventually. I'm trying. Dickie as ever has been your best friend and a calming influence for you know who. I'm sorry for swearing at you so much but I don't have your chest to lean on.

The Coven girls are taking me away this weekend. They have been the best friends a girl could ever have. I wish you could have met Alex, you would have loved her and Tom. They have been a tower of strength for both me and Cal. And Cal has been accepted on his college course, I'm a beyond proud mum. You have amazing friends too but you know that. The footie gang, Phil who refuses to let me wallow in grief, the best neighbours in Michelle and Trevor, Cheryl, Simon and Oliver. They've all badgered and prodded me to make sure I'm okay. And our families of course.

I'm sure you've noticed the renovations on Chez Knollys, you would have liked them but not the money I'm spending. Oops. I'm sorry about the Camellia too but hopefully it will get over the trauma and blossom again. I'm sure you are pleased with my decisions on the work front, very overdue but I know you would be proud of me for taking that step. It's a step without you though. I was a bit nervous about your reaction to my tattoo but I like it and you know I never do what others expect. The dogs have been my backbone over the last 12 months; I would never have kept my sanity without them. Mind you, they've driven me to insanity on a regular basis too. Yogi is better now that Callum is home but I think if he could talk, he'd need a therapist. I will never know a dog's thoughts after they watch their human pass away in front of them. I hope you don't mind the lilac colour in the bedroom, it calms me.

I miss you babe. It's a weird existence without you but I thank you for Callum. He's my rock just like his father. I suppose I'll come to terms with the word 'widow' eventually, your Mum is right, it's a horrible name tag. At least I have the woofers to add though. Say hi to your Dad and all my furries up there; Scout, Hero, Rupert, Jordi, Amor, Ash and my beloved MacGyver. Always know that I love you babe and you are remembered forever by all of us down here.

I love you always and forever.... (ditto)

Sonic xx

CHAPTER 9

"I don't think anyone is prepared for widowhood, it's not something that's taught in school or practiced like a driving test. You learn on the job, rushing around like a sugar-shot six year old for the first month or so sorting out funeral arrangements, direct debit changes, account closures and mountains of paperwork. And then suddenly, everything is done, your job is finished." (Blog 21st January 2019)

I called it the 'grey' stage of widowhood. You can't move forwards until probate is finalised and your life as part of a couple has ended. Here was I, in a big house, alone except for the woofers. Callum had left his car up in Shetland so he and two of his cousins decided to take an extended road trip after his dad's funeral to bring it back to London. Although Callum never said it, he needed the space to sort out his head. Unfortunately, I didn't have that option.

A good friend of mine and fellow widow, Laurie, told me not to make any changes in the first year. Don't sell your house, don't change your job, don't move country. Considering I was clueless about anything financial (I barely scraped through GCE Maths for those old enough to remember such exams), let alone knew what day it was, such advice suited me just fine. The only alterations I made were rearranging our bedroom and painting its walls. I had no idea where I was going or how to get through the next day but wherever the destination, everything looked better in lilac. There had been plenty to do before the funeral, just not a lot afterwards. If Tony had battled a little longer, perhaps I might have been better prepared but hey, wishes and dreams and all that.

I had to keep my foot on the proverbial pedal because if I stalled the engine, I'd break down. Probate was out of my hands, my future controlled by faceless legals – I'd used what money Tony had left in his bank account to pay for the funeral and go on an interior design binge. With all the thunderbolts the Gods were throwing at me, it was time to confront the demons with every bit of arsenal I had left.

Having had an argument with EDF, the electric company, about my over-priced light switches, I managed to find a very understanding lady who promised to sort out my bill so I moved onto 'cleansing' the house of all its negativity (I read it somewhere) and found a book on Feng Shui. This ancient Chinese art of placing objects around the home to bring luck and give the Gods something to think about was going to turn my life around. I painted the bedroom, added wistfully floating curtains and found one of the many unused birthday presents I had given Tony in the form of a water fountain which I put in the living room. I think it was facing the right direction but I failed Geography A-level.

While I was absorbed in Feng Shui, Callum returned and went on a wrecking spree, ripping everything out of his bedroom, the one his father had died in, until nothing was left but the bare bones. Out went the radiator, the plasterboard that covered the walls and most of the furniture. Admittedly, Tony and I had never done much in terms of updating *Chez Knollys* but I didn't think it was that bad. Perhaps Callum was trying to purge a ghost or just wanted something to do; I was just grateful he left the windows intact. Thankfully a friend of mine, Alex, sent her professional painter husband, Tom, to help him and he became my son's first boss. Poor Tom also got bitten by Neo for all his kindness. On the plus side, Callum was about to start his new career in carpentry so he took on his role as apprentice very seriously and earned much praise from Tom.

I was obsessed with the idea that I could swing the universe in my favour but instead we came to blows. Danny Boy (our rescue Husky cross) and Arry, having always been the closest of friends, had a huge fight resulting in a large veterinary bill – something I really didn't need.

As Callum laid waste to walls, I needed to lay something else to rest too. Anger over the way I was left out in the cold by the medical profession was threatening to bubble over. I needed to talk to someone who had been present when Tony was in hospital. Against the advice of most of my friends and family, including Phil, I managed to get an appointment with the Patient Liaison team at Kings College Hospital through Rene. List in hand, I arrived at the meeting room to be greeted by a group of doctors and nurses I didn't recognize; except one, the nurse who had kindly helped Tony to the car. I was glad she hadn't got fired. My questions might as well have been written in Chinese for all the feedback I got. Oh, they said all the right things like, 'Sorry for your loss,' 'We didn't know he would die so quickly,' 'Yes, we will take care in future to remember the spouse.' All whilst smiling in faux benevolence. I felt nothing as I walked out half an hour later, no peace, no closure, nothing. Perhaps I should have done a Callum and ripped out a few of those walls – at least then I couldn't bang my head against them.

A month or so after the bedroom blitz and the Gods' total disregard for positivity, Callum and I took a collective deep breath and decided to tackle the huge attic and the twenty-plus years of debris that Tony had hoarded up there. If we were going to do anything with the house, we needed to clear up and throw out. Box after box of receipts (circa 1984), tax spreadsheets and Liverpool FC newspaper cuttings were rammed through the shredder, actually two as the first one gave up hope.

We unearthed six computers (I wanted to list the old Viglen monstrosity on Ebay but Callum nixed that idea) along with four printers and two video recorders. The Antiques Roadshow would have had a field day up there but everything had to go. I found a box of memorabilia from our holidays which also housed wedding and christening invitations. Tony kept everything; Valentine cards, birthday cards and even the cards sent to us on our wedding day. As much as I would have liked to hold on to them, Callum was the sensible one. "Mum, it's just bits of paper from the past, what would you gain from keeping them stuffed away in a cupboard somewhere?"

Hard as it was to shred so much history, he was right. If I was to keep going forwards, I had to get rid of the cardboard. And we didn't stop there. Carpets were ripped up and floorboards bought back to life with varnish, new curtains now adorned all the windows and the bathroom had had a makeover thanks to Tom the painter. Now the question was, did I really want to stay here? Tony and I had always dreamed of living out in the country, perhaps that would be a better place to start again? I could use what money Tony had left to find a little cottage somewhere with a bit of land for the woofers?

One evening, Callum called me into his sacred space, his newly-refurbished bedroom. He had a surprise for me. Finding our old camcorder up in the attic, he had (I have no idea how) managed to transfer all the old footage onto his computer. Video footage from times spent with friends, our wedding, honeymoon and beyond. I thought I would find it too difficult to watch; I still couldn't watch any films involving Macgyver, my late German Shepherd but I had a weird need to have a look.

A couple of hours later and Callum and I were laughing at Tony's antics, the 90's fashion faux pas and his then very youthful parents. There were so many happy memories on those tapes and not once did I break down into tears. Sadly, the camcorder had been replaced by our mobile phones by the time Callum was born and a lot of the content lost. Callum loaded everything onto discs for me and they are still packed away in a box marked ‘TONY’.

There were days after however when the glass shell inside me started to splinter, as though shards were piercing my chest – the pain was indescribable. I missed Tony so badly. My world was crumbling around me and I didn’t care. But I never cried. One night, I noticed Yogi Bear (Norfolk terrier extraordinaire) lying on the floor next to Callum’s bed. Callum had gone back to L’Horte to help my parents clear the damage a flood had done to the property a few months before so there was no reason for a bear to be in there. Yogi looked utterly dejected and as I picked him up, he curled into my arms. I knew why. Yogi had been with Tony in that room when he had the cardiac arrest although the paramedics had put him in my bedroom whilst they were trying to resuscitate my husband. It was the first time I had really thought about how much the woofers were affected by everything that had happened, especially a little Bear.

And he wasn't the only one. Arry and Danny Boy were now fighting on a regular basis. With so many trips to the vet, Danny was beginning to look like a patchwork quilt. Everyone was on eggshells, me included. Something had to give.

CHAPTER 10

As I mentioned, back in the autumn of 2018, our particular area of France had been devastated by flooding, L'Horte included. The River Aude had burst its banks in several areas and Le Lauquet followed suit. It ripped through everything in its path, taking cars and even someone's swimming pool with it, we were hit hard. The water had reached the second floor of the main house, luckily my parents had been on the third at the time but the pictures they sent of the devastation shocked the family badly and Callum had flown out immediately to help with the clear-up and rebuild.

In times of major crisis, the French government mobilises a special team known as the Sécurité Civile who specialise in clearing up after natural catastrophies. The men, led by a tiny female, were amazing, moving fallen trees and diverting the remaining water but Callum knew this land better and knew what needed to be done.

I drove down a month later with Arry, Alice and Yogi Bear – I needed a break and no doubt Danny Boy did too. Steven, my dog-sitter, was on hand as always to look after Danny and those I'd left at home. Steven had been my anchor since Tony died, looking after the woofers when I needed him and never complaining about the journey between his and mine. Steven and I had met several years before in Hyde Park and became instant friends. With his rugged, tall Yorkshire gruff and hob-nailed boots, you'd think the dogs would be scared of him but it was the opposite. The woofers adored him on sight, even Neo, and there was no-one else I would have trusted to take care of them -or me.

I arrived to see the debris the river had left in its wake – tree trunks split across the land and river bank but the scene was strangely therapeutic, as if both

L'Horte and I were still standing after what the Gods had thrown at us. Oaks upturned with their roots like fingers grabbing at the sky, decorated with the remains of plastic bags and car parts. Mounds of redistributed earth now lay along the once shaded path down to where we used to tip our toes into the waters before a swim (it was never warm). But there was still life amongst the wreckage, tiny wild orchids and cowslips popped their heads up as the autumnal sun warmed the soggy undergrowth.

Arry, Alice and Gunner (Pop's German Shepherd) leapt over the now-piled up remnants of those oaks and fruit trees as I basked in the sunshine and the serenity that warmed my soul. We spent the days scouring the river for family heirlooms – Callum found one of our tractors downstream and managed to haul out the mangled machine. I located the wooden jetty that Pop had built so we could get into the river with some sense of decorum, stuck in a bush half an acre away from its mooring. Ironically, considering how much furniture, crockery and machinery we lost down the river, the kayak didn't move from its storage underneath the terrace.

Once the waters had receded, all that was left was thick mud covering floors, walls and paintings, even the piano. As tough as it was to see so much damage done to the home Pop had built, I threw myself into helping clean and restore whatever I could. Here at L'Horte and with carnage everywhere, instead of my mind being full of financial worries and mental stagnation, I simply lived in the moment and relished every second of it.

My sister, Bong, was to tell me later that the pictures taken of me that October were the happiest she'd seen me since Tony died. A chance opportunity was about to fall right into my future.

I'd always loved the smaller of the two houses at L'Horte, the one that hadn't needed much renovation when Mumo and Pop had bought the property back in 1995. The name, L'Horte, means 'garden' and originally ours was the one that made flour for the Abbaye in St-Hilaire. The original chapel at the Abbaye was built in the 6th century, before some Benedictine monks built a church there a couple of hundred years later. But it was the wine they produced that made the Abbaye famous. Long before the method for producing fizzy stuff was known, the monks accidentally created Blanquette, a sparkling wine that was the precursor of champagne.

The Stable, as the smaller house was known by, had been the main residence of the previous owners as mentioned before, and horses were kept in the bigger house. And the bats. The best thing about the stable was the grapevines hanging over the pergola outside the back doors, which we feasted on every summer (before the flood waters had washed them away) and the concrete deck outside was a star-gazers heaven.

An idea had begun to root itself in my brain. What if I moved here, with the woofers and made it a home? I could look after Mumo and Pop and perhaps have some sort of business. My French was bordering on the pathetic but I could learn? I could rent out *Chez Knollys* and have an income, maybe I could write a book? About dogs of course, it was my specialist subject. We had several acres so I could fence off part of the land; my parents' two dogs, Chrissie and Gunner had never needed containment devices but mine would, and I could start a new life.

I talked to the rest of the family and everyone thought it was the perfect solution although Pop did suggest, more than once, that it might be more financially savvy to sell my London pad. I disagreed. One needs a place in London because if you sell, you could never afford to buy there again. This was it, the panacea to my pain. I was going to move.

Little did I know that at the time I was discussing kitchen design and colour palettes with Mumo, six months later the French government would decide that the houses damaged by the river through St-Hilaire, including L'Horte, were deemed to be too dangerous for habitation. The Fonds Barnier (a government subsidiary that dealt with all things land-related) offered to buy up any homes that were at risk of being flooded again. I say offered but it was a bit like 'The Godfather', you couldn't refuse. We could keep most of the land but they were going to raze the houses; everything Pop had built. The family began legal proceedings; after all, L'Horte had been standing for hundreds of years through many floods and the Stable hadn't been affected but to no avail. Mumo and Pop were going to have to move.

But the Gods still hadn't finished messing up my life.

Having returned to *Chez Knollys* in a more positive frame of mind, the local Animal Welfare Officer called me to say that the law for boarding dogs was about to change and I needed to be ready for an inspector who would be coming to check I had all the required 'ticks' in my boxes. Naturally, I farmed all the woofers out to various friends for the day and read in depth, sort of, all the necessary bits of paper needed to pass 'the test'. If the truth were to be known, skimmed through them would have been more apt, after all I had been boarding dogs for over a decade and had never had a complaint. The inspector in question arrived, Sharon her name was, and decided this widow was going to be her victim. She went at me like a Pitbull on acid, relishing in her ability to reduce me to near breaking point, demanding to see my degree and accusing me of not providing enough stimulation for the dogs.

Considering I lived 5 minutes from the park and had both a front and back garden, this was a little extreme but then again there were no educational toys lying around to help them relieve their boredom (what boredom?) – I daren't point out that any toys in the house would have been destroyed within a minute; I knew these dogs.

I had also failed to put up signage referring to emergency fire exits, as far as I'm aware dogs can't read, and neglected to write individual diaries for each boarder just in case my clients wanted to know what their little darling was up to. I found out later from my other 'boarding friends' they had never been asked any of these questions, had to produce a chewed piece of plastic toy or asked about their qualifications. Sharon thought she had a victim – she didn't. I just carried on regardless and my clients were happy.

There were those days that I felt like I was in a really bad action film – you know the ones where the heroine keeps getting punched and shot but gets back up even when half her limbs are missing? I had to evaluate my life. It wasn't so much the desire to go off and travel the world because Tony's life was too short and we hadn't had the opportunity to do it together, I wanted to be doing something I wanted to do and not because I had to do it. The soupy fog that had clouded my brain was starting to lift, that I owe Keith for.

After Keith's funeral, another beautiful remembrance this time in the heart of London at which I got drunk again – it was becoming a habit at such events, I decided I needed a mental health kick up my derriere. This self-pity closet was becoming suffocating, I had to open a new door and jump through it. I once described the scenario in a blog as leaping through a one -way portal, you just hope the ground on the other side is solid. I was ready for L'Horte and its new challenges, so buffered by Keith's wisdom I decided to hand over control of Dog Hollow to my business partners. Dog Hollow had been my baby for almost 22 years but had also been what Tony had called, 'the third person in our marriage'. I wished I hadn't had to jump solo though the portal – it was like sitting on a plane that had been grounded forever and suddenly takes off. I wasn't sure if I was ready for the journey ahead and I hate flying.

Making big plans without someone to bring you back down to Earth is, let's say, not advisable especially when I was the one flying the plane. Hoping that probate was on its final approach, I had, for some ridiculous reason, an idea that *Chez Knollys* needed a face-lift. If I was going to rent out the house, a ground floor extension would add to the appeal with the addition of bi-folding doors into the back garden. Tony and I had often talked about removing the dated archway and wall that divided the kitchen and dining area. It always made the downstairs look dark and the utility room's only asset was the pokey loo in the back. I had had a dog grooming salon added on which you could access via the utility room's back door but to be honest, it was a bit more of a lean-to. My idea was to knock all the internal structures out and create one big space with an abundance of kitchen units and a huge island for entertainment. I'd change my canine coiffure area to a more bespoke space adjacent to the new living area and pinch a bit of the back garden to extend the whole thing. Possibly convert the attic space into another bedroom whilst we were at it?

I brought in an architect and plans drawn, contracted a building company to do the work. The woofers and I were relocated upstairs whilst the work was being done, including fitting a new heating system right in the middle of an unseasonably cold Spring. With no kitchen, I had taken the microwave with me and a shelf full of easy-cook meals plus all the dog food. Not only were we freezing up there, the woofers and I were living on top of each other. To add to all the stress, I had managed to find the only builder whose inability to budget was worse than mine.

Tommy had a habit of disappearing 'on other jobs' for weeks on end, naturally in areas with no phone signal, leaving me in half a house and a very cold one at that. What was supposed to have been finished in a couple of months ended up taking five months and I was flat broke by the end of it. And Tommy wasn't the only one draining what little pounds and pennies I had. This fast-track course; 'Widow 101, DipEd, BSc, PHD had missed a couple of essential subjects off the syllabus – 1) Probate is designed to wait until you have nothing in the bank and 2) you really should pay attention to household bills before your other half passes away. And there was one particular bill that was going to change everything.

CHAPTER 11

"Making a decision about sharing any part of my future with someone else other than Tony makes for mixed emotions, after all we were a partnership that had lasted 28 years and frankly, I had assumed there would be 28 more. I trusted him completely and the idea of ever loving anyone else is the furthest thing from my mind. But here's the thing. At 25, you are attracted to that certain someone that makes your heart go boom and your nether regions go bing. At 52, you want a companion to share common interests with but not necessarily, your bed" (Blog 22nd June 2019)

As I mentioned, three of my girlfriends had decided to take me away for the weekend of Tony's 'first anniversary'. Rene had booked an apartment in Ramsgate, a seaside town in Kent. We had a little balcony overlooking the beach, a couple of twin bedrooms and a huge lounge with a kitchen to one side. Obviously the owners had had the same idea as me about knocking down unnecessary dividing walls. As both Rene and Georgie wouldn't be able to get there until after work hours, Lareine and I travelled down on the train together having bought a little 'picnic' mostly containing wine and a few nibbles. It was just as well because Dickie called me in tears at one point in the journey and then, having arrived at our designated station, Lareine and I discovered we'd have to walk a mile or so with our suitcases in tow.

Having wheedled our way into the apartment (neither of us could work the key safe but some nice neighbour let us in) we dumped our luggage and headed off for the nearest supermarket. Actually, there was only one and I think we bought enough food and alcohol to last us into the next millennium. Rene arrived not long after, naturally she knew how to work the key thing because she's always organised. At precisely twenty to five that evening, we raised champagne glasses in a toast to my Tony as we sat on the little terrace. My phone buzzed with messages from friends and family and Phil called to make sure I was holding up okay. By the time I'd put it down, Lareine was a sobbing wreck (possibly because she'd exhausted her debit card in the supermarket) and Rene just gave me a smile. She knew I couldn't cry. Thankfully, Georgie arrived a few moments after the first glass livening up the sombre party – Georgie is always late and always apologetic but this time her timing was perfect. I was looking forward to leaving my troubles at home and just spending a weekend amongst friends.

Let's face it, Ramsgate doesn't have a lot to offer 'off season' aside from the UK's biggest Wetherspoons (glorified pub with restaurant) and far too many

Brexiteers but somehow we managed to let our hair down. Walking along the long (very long) path that ran adjacent to the sea, we laughed about everything and nothing. We mused about where we'd do the next one, Bahamas maybe? And danced the nights away dressed up in ridiculous outfits to anything we could sing along to. Yet, as much as I had looked forward to that weekend, looking back – I don't think I was ready for it. I missed the dogs and home and Tony. On our last evening in Ramsgate and at the poshest restaurant in the town, discussion turned to debate about my future and possibly sharing it with another.

It wasn't the first time that the subject had entered my head. Not long after Tony died, I joined an online support group for widows and widowers on which this was a fairly common 'should I, shouldn't I' thread. I was beginning to hate the lonely evenings, Arry and Alice may have shared my duvet but debating politics over a pizza is not a dog thing. I needed to get out more. Apart from those people I met over birthdays and deaths, it was dawning on me that the only other beings I was having vaguely normal conversations with were the woofers. I was turning into Sophi Valentine.

A month or so before the Ramsgate outing Georgie had invited me to join her and the family for Sunday dinner. Of course, I accepted – an evening with good friends was the perfect medicine and I could actually talk about something other than, "What have you got in your mouth?" or "Leave that alone" – dog owners understand such rhetorical questions. Later that night, I wondered if my sub-conscious knew something because, unusually for a night out with friends, I decided to drive over and forgo the booze. Halfway through the meal and with everyone chatting around me, I had what could be only described as a minor panic attack. I don't know why but I had to leave (the woofers are always a good excuse) so with apologies to everyone at the table, I did. That night, once home and suitably dowsed in dog drool, I curled up on my bed. I felt like I felt the day after Tony died, disengaged from life and numb all over again. What the hell had happened?

A few years later I was having a conversation with Mumo and I knew the answer. No-one knows how you feel when your other half dies. The conflicting emotions about loneliness, guilt, coping with family gatherings when everyone else has their partner to banter with are unique. No two widow or widower situations are the same. I had the most wonderful group of friends but, let's face it, they had their other halves to go back to, I didn't. I missed being a part of someone even if, at the same time, I was starting to like being just me.

One evening I met up with Alex, Tom the painter's wife in The Great North Wood Pub, the same one that had held Tony's wake. It had become a regular haunt for me and my girlfriends because not only did it serve great food and wine but they allowed dogs in too. The bar and eating area had large wooden tables and mis-matched comfy chairs, along with New York style bar stools and high tables looking out onto the High Street. Having drunk one too many glasses of Pinot Grigio on one such stool, I blurted out my concerns about entering the 'man' pool to Alex. Now Alex is a born again romantic, having met Tom through the Lonely Hearts pages after a not so great marriage and divorce. The arrows flew, she said. Determined to take control of my destiny and boost my flagging ego, she suggested online dating. I balked at the idea. Dating? I hadn't been on a date since 1990, what did I know about 21st century etiquette? I'm not exactly a head turner; over 50, no boobs, big nose, oh and more dogs that most people think is normal on their sofas. But Alex was having none of my self-pity and the wine was skewing my protests. Before I could think, she whipped out her phone and took some pictures and then between us, we wrote a halfway decent profile. I did try to delete everything the next day but I had somehow managed to sign myself up for a 3 month membership in Pinot-flavoured ink. Mind you, Alice was in most of the photos and she is rather a catch for a Border terrier.

Switching on my phone the following morning, I couldn't help but admit I had a little smile – I had messages! Okay, so half of them were from twenty-somethings that either couldn't read or thought an aged widow might have money but there were one or two that didn't make me want to break into hysterical laughter. I called Alex;

"Send me their profiles before you do anything," my matchmaker declared.

Feeling a little guilty for even thinking about someone else in my life, I called Phil. Perhaps he would be the voice of reason and put me off;

"Don't worry Soph, they take all sorts," was his answer to it all

Oh well, might as well jump in the deep end then. Alex came around that afternoon for a more in-depth discussion and by the time she left, I had somehow managed to book an evening at the same pub the following week for my first date. I was still a little worried as, despite my protests, Alex felt it wasn't necessary to mention the number of woofers I had – only that I had a 'few'. And that wasn't the only thing making me nervous, should I be venturing into the 'man' territory since Tony had only been gone a year? I asked Callum what he thought;

"Mum, why should you spend the rest of your life alone? Go out there and have some fun, it's not like you have to marry them." Or words to that effect. He might have also added a few requests; must have hair, a bank account and absolutely no dogs of their own. The last was definitely a deal-breaker.

CHAPTER 12

My first date in the new millennium was with Jim. I took Yogi Bear with me since I felt a big German Shepherd like Arry might be off-putting. He insisted on picking us up in his car which I thought unnecessary considering the pub was a 10 minute walk but it was gentlemanly all the same. My girlfriend trio had jokingly said they would sneak in and spy from the other side of the bar, at least I think they were joking – I did do a quick scan of the area when we arrived.

Jim was easy to talk to; an ex-policemen turned builder with a heart of gold and loved dogs. Unfortunately, he had three of his own so I knew this wasn't going to tick Callum's boxes and not a whole lot of hair either. I had had a few conversations with him over the phone before our meeting; he'd asked me if I'd been put off by his photo. Obviously, I hadn't been – the fact that he was taller than me and around the same age had gone a long way in securing the date but it did make me think how times had changed since I met Tony. Back then, there was no swipe to right (or is it left?) or judgement by photograph. Tony was a gorgeous specimen of a human being but it was his laugh (and his blue eyes) that tipped me over, not just the fact that he was a tall, dark-haired hunk.

Jim and I had a few dates and I enjoyed our constant chatter and his easy-going nature, especially where the woofers were concerned. Neo, always the sceptic, had welcomed him with open paws and I always felt very relaxed in Jim's company. I had surprised myself on that first date with my self-confidence, my growing independence was giving me a sort of 'take me as I am' attitude even with the wrinkles and wobbly bits. But as much as I liked Jim, I knew that we would be better friends than anything else. Aside from the dogs, we had little in common; I was and still am, a runner, he was definitely not and I couldn't see him dropping his dogs and kids to move to France. Jim also loved to compliment me constantly on how I looked which made me uncomfortable. I wanted the banter, not the romantic mush.

Another possible candidate was a rather handsome (okay so I did look at his photo) Frenchman by the name of Pierre. He lived in Northern France but often travelled to the UK so we arranged to meet, this time in a coffee shop in Kensington because he could only do an afternoon and I never drink alcohol during the day. This time I did take Arry, mainly because Pierre hadn't mentioned anything about dogs so I felt the biggest would mean either a quick exit or a 'wow, what a fabulous beast you have' kind of reaction. I arrived at the designated café and nonchalantly walked back and forth by the windows for a few minutes just so I could see what Pierre really looked like, not easy when Arry decided that the white fluffy dog on the other side of the street needed talking to.

Let's face it, Pierre did not look like his photo. Yes, he was tall and dark-haired but the days of wearing a sweater tied around your neck over a monogrammed golf shirt went in the eighties. Nevertheless, I had been brought up well so I put my game face on and went in to meet my date. As I opened the door, I realised my first mistake in choosing this venue, the internet search had said it was dog friendly but I don't think that applied to those over two kilos. The chrome, circular tables were barely a few inches apart and the place was packed – how on Earth was Arry supposed to lie down and not become a trip hazard in the process? Pierre had chosen a corner table although, by the look on his face, he'd assumed I was bringing Evee chihuahua and not a rangy, over-sized hulk of a dog. I could see him shuddering about hairs on his pristine white sweater already.

And it went downhill from there. Over a previous conversation or two on the phone, I'd asked Pierre if he minded speaking French to me as I was learning the language but he insisted on speaking English during the whole hour or so I had to suffer him. I couldn't get a word in as he droned on and on about his life and how wonderful travelling around Europe was even though I had blatantly put on the profile that I hated flying. I was starting to think about making an excuse about needing to be somewhere when the ground shifted beneath me – Arry had decided he'd had enough boredom for one day and was heading for the door, his lead still attached to my chair leg. I managed to grab him just before he'd reached full exit height and making my excuses (he needs to wee), ran out. Arry may be crazy but, as we charged off towards the park, I gave him the biggest hug. Pierre did try to rearrange other dates but I ignored him, Arry had put his paw down.

Over the next three months, I had a few more man rendezvous' but I'd yet to find someone who'd fit the brief so to speak. Mick was a great laugh but unfortunately shorter than me and loved his job and life in Essex too much to ever consider anything else. And he had a dog. But he has remained a good friend.

Shaun was a chauffeur for a wealthy family up in North London. As with all the others, I'd talked to him on the phone beforehand so felt comfortable about inviting him for coffee in my newly extended kitchen; with bi-folding doors of course. Tommy and company hadn't quite finished so I knew I wouldn't be alone and Neo has a habit of sorting out the men from the boys. Shaun was one of those with boyish charm and the chat to back it up but he did tend to go on and on about how dissatisfied he was with his life and how his employers treated him. He did try to kiss me once but I ducked at the last minute, his touch gave me the icks but that didn't stop him from asking to spend the weekend with me. I didn't see him again although he did ring me one more time to say that he hadn't been well, a nasty flu he said. Covid pretty much put an end to any possible romances after that.

I did have one more try just after the pandemic started which should never have happened except that Robert was very persuasive about wanting to meet me. About a week before we got chatting over the phone, I'd pulled out my front tooth whilst trying to open a bottle of dog pills – you know the ones with a plastic tab on them. I'd rung the dentist immediately but of course, they were closed just like the rest of the world because of the Covid outbreak. In desperation, I bought some dental glue online and after several attempts of trying not to attach my fingers to my mouth, managed to get the tooth in place. The problem was that I couldn't drink anything, let alone eat, without it popping back out again which left me looking like a haggard old witch. Obviously, I explained the situation but Robert wouldn't give up so I agreed to going for a walk in the park with Arry. I donned my mask and strode out to meet my date at the gate and off we went, literally. With the knowledge that he was getting an extra walk up to the park, Arry took off like a demon possessed which wouldn't have been so bad but Robert wasn't exactly dressed for the occasion. White linen trousers and loafers do not suit Arry's idea of a belting good time through the woods. Normally, my oaf of a German Shepherd would trot along by my side whilst we climbed the short hill to where he could run off his lead but not this time, he was going to drag me. By the time we got home, Robert was a frazzled wreck complete with shredded shoes and tan-coloured trousers. Poor man deserves a beer I thought so after he accepted, I left him on the front lawn with a droolingly happy Arry and returned with refreshment. Stupidly, I hadn't thought about two things; one that I would have to take my mask off, and two; that drinking anything would be complete dental disaster. I never saw him again; his quick exit would have matched Arry's for speed. I love Arry.

I never dated again after Robert and deleted my account with the app. Although the whole experience gave me the confidence to just be myself which I didn't know existed, I realised I just wasn't ready to have anyone in my life.

Rene's thoughts were that you had to kiss a lot of frogs before you either got sick of the taste or found the proverbial dish. Well, I hadn't actually kissed any of them but I found myself becoming politically incorrect about virtually everyone. Too old, too young, too fat, too short, too hairy or not hairy enough, too clingy, too much ego or just weird. I blamed Tony. I blamed him for a lot of things but mostly because he'd just set the bar too high.

CHAPTER 13

"Birthday on Halloween. Star sign Scorpio. Chinese Year of the Fire Horse. Basically, born in disaster mode. Yup, that just about sums up my mood this week although the inhaling of furniture spray may have contributed to my mental state. Just as I was beginning to feel like I was winning the war of independence, someone up there fires a couple of missiles. I sometimes wonder did I do something unforgivable in a previous life or is it just the luck of being a short-tempered, catastrophe prone witch" (18th August 2019).

I still had no idea when probate was ever likely to be finalised. The solicitor dealing with it had, Dickie told me, decided to take a holiday on a cruise somewhere. I was getting increasingly frustrated sitting in my proverbial boat chained to the dock with no key in sight; no doubt the solicitor had taken it with her.

I needed to do something, move somewhere, get out of this leaking vessel. The idea of moving whether it be L'Horte or a quaint little village in an affordable part of the U.K was becoming ever more tempting. I talked to Mark about it. I'd known Mark since he was a pimply twenty-something year old back in the early 2000's, fresh from Somerset and keen to start a dog related career in London. Keith introduced us; he'd known this gangly young man since he was a kid and wanted me to take him under my wing. I didn't need to, Mark and I hit it off the first time we met – he had the same fire under his derriere that I'd had when I started out. Not only could I confide in Mark about anything, he was and is still one of the best dog breeders in the country (he had bred Evie chihuahua after all) so it just so happened that our conversation about uprooting to greener pastures coincided with Alice Border terrier's season in June 2019. Tony had always wanted Alice to have a litter, I'd given her to him as a 25th wedding anniversary present the year before. Mark arranged the stud, a client of his had a gorgeous male, and 'borrowed' Alice for a week or so. Sadly, nothing came of it but at least Mark and I could share thoughts about prices in and around the South West of England versus moving to L'Horte;

"Worcestershire has some nice properties," I mused to my dear friend,

"Floods and rain, darling," he smirked, "better off in France I think."

"But it's a long way from friends," I countered,

"Hmph, country folks aren't any friendlier" Mark laughed, "and us friends would rather see you in sunshine than wellies."

On the plus side, my nightly 'why me' rages at Tony had been heard and the kitchen extension floor had finally been laid. Unfortunately, a leak in the new skylight meant we could only look at the space; a bit like seeing a new swimming pool with just a puddle of water in the middle of it. Summer had arrived and not being able to get out the back, the woofers and I were reduced to the front garden which wasn't ideal as soon as Mo (the more outspoken of my two disabled dogs) and Neo saw the front gate the same way as Trump saw the Mexican border every time someone passing dared to stop and throw a glance. I think the Gods for once felt sorry for me (and the neighbours) as a few weeks later, Tommy and the boys declared the extension safe to enter and the woofers charged out of the very chic bi-folding doors like caged animals released back into the wild.

Those who know me will say that when I decide to do something, I do it – the exception being paperwork;

that was still sitting in the same drawer I'd shoved it into. Ignoring the lack of pennies in my wallet, I ordered a couple of glass cabinets and a matching chest of drawers from Ikea to jazz up the new space. What once was a pokey little kitchen with a utility room attached and a dark green wall-papered dining room was now one large room with the grooming come laundry area off to one side. There were floor to ceiling storage units along two walls and a central island with large drawers on one side. The whole room was painted white apart from the units which were more of a pale caramel colour – everything chosen by me and me alone. Before assembling my matt black Ikea units, I unpacked all the kitchen utensils that Tony and I had amassed over the years, most of them useless, broken or duplicated that I had put upstairs whilst all the work was being done downstairs. I challenge anyone who has been married for more than a decade to say they don't have at least one plastic what's-it buried under several mis-matched napkin rings which they have no memory of ever buying or wedding presents that had never reached a full set. Our dinner service was incomplete but Tony loved the colour so much it stayed. The Ikea stuff was another matter entirely, especially the almost identical but used in completely different places nuts and bolts that you don't know about until you have deep welts in your hands and boiling blood pressure. It didn't help that Arry had eaten the English translation parts of the instructions, I could feel Tony tutting at my scan through at rapid speed approach to reading such things but Portuguese isn't my second language. A few hours later however, I'd accomplished something on my own – even if the stick men drawings in their booklets clearly showed such tasks were to be completed by two people and I appeared to have a few extra bolts left over.

Despite my spectacular efforts at interior design, something was messing with the lay lines at *Chez*

Knollys. There was, I felt, an invisible being in the house causing havoc. I kept misplacing items only to find them an hour later in plain sight. I could blame the disappearance of socks and shoes on Alice's odd maternal instincts but I'd drop things, break things or lose things without reason. Despite my 'lush' extension and newly furbished other rooms, the house just didn't have that happy feeling. My previous attempt at Feng Shui obviously hadn't amounted to much so being a naturally superstitious person, I concluded that we needed expert help in getting rid of the pesky poltergeist or whatever it was the 'Who' was.

My first thought, having spent half my childhood in Thailand, was to find a Buddhist monk who could come over and bless the place. Back then, an orange-robed man chanting blessings on your new home was a common sight but apparently the monks of South London don't go in for that sort of thing. So, I called Tess. I've known Tess for over 20 years and apart from being one of my closest friends, she's one of those people who knows how to sort the spirits out for want of a better phrase. She also wouldn't think I was losing the plot. Tess advised me to start cleansing the whole house by sprinkling sea salt in each corner of the rooms.

Me: "Does it have to be sea salt?"

Tess: "Yes, not the cheap table stuff."

Me: "Will Cornish sea salt do?"

Tess: "Oooh yes, I do love Cornwall."

Me: "How much do I sprinkle?"

Tess: "However much your instincts tell you"

Me thinking: "Just a sprinkle because this stuff is expensive."

Under Tess's instruction, I made my way around *Chez Knollys* fairy-dusting each corner with minute amounts of Cornish salt in the vain hope of banishing the Who to wherever it came from. Callum found the whole thing incredibly amusing, watching his Mum lose her marbles as she counted the grains and we had to put the woofers away for fear of them hoovering up any of my attempts at spirit-squashing. Their water bowls were suspiciously low for a few days afterwards though.

It wasn't just the house that needed cleansing, I did too. Without noticing it, my alcohol consumption had started to creep up to a bottle of wine a day and I'd started smoking again. Having been somewhat religious about exercising daily, I'd taken a sort of 'fingers up' attitude towards fitness of late, after all, it hadn't done Tony any good. But if there was ever a time to sober up, this was it.

I had a call from Mumo, Pop had been put in an induced coma in Carcassonne hospital. The doctors had found that his intestine had moved up through his diaphragm, strangling itself and pushing on his lungs. Immediate surgery was needed and he was in a coma to give him the best chance of recovering. The following weeks were spent on the family Whatsapp group so that we could all keep tabs on Pop. My father was a fit man despite that time he fell off our dam and broke 13 bones some eight years before, so none of us were overly concerned but I heeded the wake- up call, threw out the bottles and took up yoga instead.

Starting exercise again is never easy especially when you own too many dogs. I had sold off all the gym equipment not long after Tony died and rented out the garage/fitness studio to a nice young man who had asked me more than once if he could buy it. He wanted to turn the miniscule building into a home but the answer was always no. He stayed anyway and I got a bit of rent every month. So here I was, desperately trying to put one leg in the air and balance on the other whilst the woofers used me as an agility course. They would race in and out of the back doors, tearing up my newly-laid lawn and dropping various items under my nose that according to them would serve a much better purpose if thrown. The activity allowed my thoughts to wander at will, “Was France too big a step? Maybe I should find that little cottage in the English countryside instead?” I couldn’t tell the woofers about my ponderings but talk of it often came up with friends.

CHAPTER 14

Moving out of the city or thoughts about were frequent conversations at the local pub or at dinner parties. I suppose it was because we'd all reached that age where the kids were growing up and work wasn't as exciting as it was a decade ago. Talk peppered with complaints about noise, pollution, crime and traffic but we stay because we are comfortable in our world. At least until something tips the cart. A friend once asked me if I was running away from my problems by moving. "Nope," I'm going to find new ones," I replied. Half-jokingly, I told her I was considering moving to Costa Rica to save the rainforest to which she retorted, "They have flesh-eating bugs there!" Actually, the flight would put me off.

I was like a tortoise sticking her head out of her shell and then retreating; I wanted to move but was too scared to at the same time. A couple of encounters one week gave me the boost I needed. When people find out you are a widow, they invariably tell you of their own tragedies. We are like walking confessionals. I was standing by the driveway one day, Steven was painstakingly putting red bricks down in some vague pattern to revive it, when I got chatting to a lady who was going through a divorce that her husband wanted but she didn't. I'd had the love of my life for 28 years so I couldn't imagine how she felt but she admired my courage to move on. The second was a man delivering some hire equipment Stephen needed to get some sort of level in the afore-mentioned driveway. As we waited for the thingy to be unloaded, the man with a van told me that he had passed my house for many years and often saw Tony pootling around the front garden. I threw in a comment about wanting to sell it with a brief bit of the widow story and a side of "I don't like London anymore." Once again, he gave his admiration as he hadn't been able to take that step. His son had been murdered two years before, a knife victim, and he couldn't find the strength to move away. He had a dog though, he said, someone to talk to at the worst of moments. I had 10.

And I had friends. Friends who called me and cajoled me into going out even when I wanted to curl up under the covers. Phil who had been with Tony when he died, took me to Leicester Square and the cinema. Afterwards, we took a stroll through the crowds (tourists love Leicester Square) and ended up at Green Park. Tony and I had spent many of our early dating days eating wontons in nearby Chinatown and window shopping along Regent Street which takes you from the Square to the Park.

Rene wrangled me into seeing all sorts of exhibitions at various galleries dotted around the capital and I had more dinner invitations than I needed dinners. And then there was a visit to see my old friend, Sophy (yes we both like to challenge autocorrect). Sophy and her husband Sean had been the ones who had guided this bare-footed widow from the crematorium to the wake. This time Sophy was in Moorfields Eye Hospital located close to the financial heart of London, as she had a tumour in her left eye. This wasn't her first rodeo with eyes and tumours but the radiotherapy and its corresponding drugs was. Although hospitals weren't my favourite places for obvious reasons and I do have an aversion to seeing eyes poked about, Sophy's zest for life was catching. Even if, by her own admission, the Tramadol was working and frankly, she loved the taste of morphine. Here was Sophy, about to lose the sight in one eye and loaded up with painkillers, yet her outlook on the future was full of humour and of overcoming anything that got in her way. Her infectious spirit transferred itself onto me and along with Rene, I broke into fits of laughter as we unsuccessfully tried to find a sushi restaurant in

Farringdon after the visit. Passing the same pub three times and the ‘Guided walks start here’ sign stuck to a tree outside one of them, we ended up back in the familiar chairs of our favourite eatery, Café Provencal close to home in Herne Hill. There’s nothing like the sound of car stereos at full volume and ambulance sirens shrieking through the night on a five-way traffic junction to comfort the mind. Sadly this local landmark closed down less than a year later, another victim of the Coronavirus economic catastrophe.

Lareine and Rene. Two of the best friends you could hope for and if there was something worth seeing, we’d see it together. A school teacher, an NHS services manager and a canine behaviour consultant – nothing in common except our ability to find humour in almost any situation, and music. Back in 2016, we’d partied the night away in Hyde Park to Carole King (I got so drunk, I hit my head going down the subway stairs and nearly ended up in A&E) and in the summer of 2019, we reconvened to see Barbara Streisand in the same place. And what a night it was. Mindful of the fact that I didn't have anyone at home to look after me if I did a repeat of the subway stairs, I stayed sober although you wouldn't know it considering the amount of singing that went on. That night just so happened to coincide with the end of Gay Pride weekend so the crowd was right up Bab's street. Or Park. I'm pretty sure a very well-lubricated Lareine told everyone within shouting distance that I was a widow so I was hugged, waltzed around and sung to – if there was ever away to kick the blues into pink, that was the way to do it. In a few months, I was going to turn 35 backwards so the presence of several gorgeous, hunky men wearing little to the imagination and singing "Happy days are here again" was ten times better than any alcoholic pick me up.

Surprisingly I wasn't the least bit worried about turning 53 at the end of October. Menopause had just about left the building and I had some sort of vague plan about the future. Mind you, my annual visit to the optician did miff me just a tad. When asked if I had been using the prescription driving glasses, I had to be honest – I didn't feel comfortable in them. I mean, I hadn't needed glasses until I went over the 50 year mark so donning specs to drive wasn't a natural routine. “Wear them around the house, you'll get used to them,” she said, then added “You're just about within the legal limit to be driving without them." Not exactly a promotional campaign then. Rene took me out for a birthday treat; first to the Vogue Exhibition at the National Portrait Gallery and then onto a very classy Japanese restaurant called Roka in Aldwych near St James Park. She'd ordered my favourite wine, Pinot Gris – a girl remembers these little things.

With Pop in hospital recovering and probate finally sorted – just enough to cover Tommy's costs and nothing else, I made a decision. Despite Dickie's soothing words of "Tony made sure you would be okay," I was definitely going to sell *Chez Knollys*. I had umm'd and aah'd about it for too long and I knew I would have to start afresh if I was ever going to feel settled again. Over dinner one night, I shared my thoughts about the house with the girls;

“Soph, it’s like having kids. At some point they are going to grow up, fly the nest and make their own way in life,” Rene sagely told me

She was right. My home needed a new family now and Dog Hollow needed to grow with the times under a new team. As much as I was worried about leaving friends behind, that was all put to rest by Lareine who concluded that we could put the world to rights over a bottle of wine anywhere. Or lemonade.

Following Keith’s advice, I had a meeting with my two business partners at Dog Hollow. Having always been the senior person in our relationship, after all I had bought both Alison and Nicky in as fledgling trainers, I was shocked at how quickly the tables had turned. In a somewhat heated discussion, I suggested that it might be better if they removed my details from the website after I had written my goodbyes to our clients and I would hand everything over at the end of 2019. Well they did. The next day. After 22 years, I'd vanished. No chance to write a farewell blog, no leaving party or thank you for what you have done for us gifts, I was wiped off the face of Dog Hollow as if I'd never existed. It hurt, as much as I'd needed the push, I would have liked some recognition for all I had done for both Nicky and Alison. I did have one precious reminder of all those years; I owned the Dog Hollow name, that was too special for me to lose.

The next day, I took the woofers out to the park and watched them tumble and race each other across the open grass. As I threw a ball, Neo chased after it even though he couldn't pick it up because he was muzzled – the reason? Neo had worn one ever since I rescued him 6 years before as he had an aversion to people, especially joggers. I looked around the big field, people I didn't recognise and dogs I didn't trust with my lot of heathens. If there was ever a point to moving out of London, that was it.

A few weeks after my dismissal from dog training, I had an invitation to discuss my account at my local bank (one of those arm-twisting ones that you can't avoid or stick in a drawer). For some reason best known to such financial institutions and my loyalty to this particular one for over 30 years, I had been classified as a V.I.P customer – hence the private tête-à-tête. Bless the young chap, he was incredibly enthusiastic about what the bank could offer me as such an important customer but I couldn't help but wonder if he was looking at the same computer screen as I was. I was most definitely in the overdraft zone, something he appeared to overlook. After an hour or so of me agreeing to absolutely nothing but somehow signing an awful lot of paperwork, I left with a lump in my stomach and a case of what could only be described as financial phobia. I ambled back along the High Street having left the Black Rose at the local garage hoping I wouldn't get a hefty service bill and noticed an estate agent a few doors down from the bank. "Oh well, in for a penny and all that," I thought wryly as I walked into their reception.

Of course, like any estate agent, they were thrilled to have such a fine residence on their books and we made an appointment for the next day; which I almost cancelled. What did I know about selling a house? Is there another Widows 101, Dip Ed etc. course about this? What happens if *Chez Knollys* sells really quickly and I haven't had time to sort out me and the woofers? What if no-one liked the house? Ha, it's a beautiful house and me and the Who would be very affronted if someone didn't want it. The Who had been rather active of late, even after I'd poured expensive rock salt everywhere; I lost my car keys in the car, I got a stye in my eye and Arry head-butted my right hand so hard I thought it was broken. My hand that is, nothing is capable of breaking that bone head's skull. But I didn't have much of a choice about selling *Chez Knollys* because the drawer in which I'd stuffed all the bills and what's-its after Tony died had a nasty little surprise waiting for me.

One morning, I received a letter from our mortgage company. Apart from the usual direct debit, I hadn't paid much attention to the mortgage and just buried the paperwork in the drawer along with several other certificates I'd found in Tony's office for who knows what. We'd taken out a small mortgage back in 2001 to pay for a new roof and like most of the monetary stuff, I'd let Tony sort it out – I just signed on the dotted line and paid my bit every month. Hindsight would be a gift because what I didn't know or perhaps more truthfully, didn't read, was that we had taken out a 20 year interest-only mortgage and the company were about to ask for their money back. Money that I didn't have thanks to a brand new kitchen and dining area with chic bi-folding doors and a skylight. And I didn't have a full-time job anymore.

Just as I thought it couldn't get any worse, it did.

CHAPTER 15

A little after 1 p.m. on the 30th November 2019, I got a phone call from Mumo. Pop had passed away at L'Horte. The man who had cradled me, held my hand as I walked down the aisle to Tony and rocked my new-born son in his arms, went peacefully in the place he loved most with the woman he had shared more than half a century with. His eldest grandson, Louis, was close by and of course, his adored German Shepherds; Chrissie and Gunner. He was a beloved husband, father, Grandpop, step-Grandpop, animal lover, passionate environmentalist and engineering genius. We had a mutual love of dogs, especially German Shepherds; Chrissie and Gunner were related to my soulmate Macgyver and we'd spent many evenings walking and talking around L'Horte with our dogs chasing each other through the long grass and along the river bed. L'Horte was Pop and Pop was

L'Horte. My father had always been there, along with Mumo, forgiving my mistakes (of which there were many) and cheering my accomplishments (of which there were few). I've always had the habit of turning right when everyone else turned left but they never stopped cheering. And now I had yet another funeral to go to.

Only my father would orchestrate his during a French air traffic control strike. It became a sort of endurance test for his children and grandchildren. The aim was to get to L'Horte on the one day Toulouse airport had decided to be picky about which planes were allowed to land. The family Whatsapp group resembled an unauthorised radar centre directing my siblings from New York, London (one of whom ended up going via Frankfurt), Bristol and the Shetland Islands. Moth, my younger brother got as far as Orly airport in Paris only to have his flight cancelled but thanks to the kindness of one airline official, he made it. Bong became addicted to watching the flight radar apps, thankfully no games consoles were involved and even Callum and I didn't escape the hurdles. On my way home to pack and catch our train to Stansted Airport, I got a cancellation text from the airline. I was stuck in peak hour traffic in Chelsea (a bit of déjà vu) and unable to call the airline, I sent the text onto Callum at college. I don't think he broke a sweat and despite being in class, he managed to book us on the next available flight leaving 5 hours later. He raced home, we threw a suitcase together, got in the most over-priced taxi in London and sped to the airport. Even the Gods couldn't catch us this time. Family fortitude and adventurous spirit got us all there and we said goodbye to the man who had guided us through the rocky roads of childhood. Earlier in the day, I'd picked handfuls of wildflowers from the river bank to lay on his coffin and penned a poem to read at the service. It was a quiet one but emotional nonetheless and this time I stayed sober.

That Christmas would end up being the last one at L'Horte so it was fitting that the whole family gathered together. Callum and I drove down with Arry, Alice and Yogi Bear, me singing along to kitsch seasonal songs and him with his earplugs. Even though we made the journey through torrential rain and my back seizing up, we got to L'Horte complete with fruit cake, pudding and armfuls of presents. The following morning, I opened the shutters to cerulean blue skies welcoming me home. On Christmas day we all took it in turns to scatter Pop's ashes under a baby oak tree that Moth had planted just above the river.

"In the dusky early morning light,
Dogs weave through the trees just in sight,
On ground so green through years of care,
A man in thought, I see you there.
Diamond sparkled water in the evening sun,

A man and river unite as one.

No sound but that of a heron's cry,

But I hear you through L'Horte's lazy sigh.

Thus, as the days and evenings fade,

On land so loved, your ashes laid.

Your presence in our hearts to keep,

So, rest now our Pop in peaceful sleep"

(written by yours truly 08/12/2019)

"I suppose it's inevitable that as the decade draws to a close, we look to the future with a mixture of excitement and trepidation. Whether it be a small promise to oneself to eat cleaner, give up the booze or in my case, sell up and move to a different country, a New Year gives us the impetus to leap into the proverbial fire." (29th Dec 2019)

It was no secret that the last half of the 20 teens had been frankly horrendous for me so spending the week at L'Horte with the family was the best way to end the decade. I drove back to London straight after New Year feeling ready to take on my next venture. A few weeks before I motored down to the family homestead, I had been driving along the A3 on my way to visit Jenny when I heard a popping noise coming from the side of my beloved Black Rose (the Land Rover if you remember). As I looked out of the passenger side mirror, I saw the rear wheel trim (this would become a habit of mine) fly off into the traffic behind me. Before I carry on with this moral tale, I will admit to the fact that I knew the trim wasn't secure as some kid on the block had damaged it a few days before and I hadn't got around to getting it fixed. Anyway, as I saw the bit of black plastic tumble into the distance I had a fleeting thought to pull onto the hard shoulder and retrieve it. Idiocy, I know, but nevertheless the notion temporarily whizzed through my head. I didn't but did send up a prayer that no headlines would appear in the next day's news about pile-ups caused by irresponsible Land Rover drivers. It wasn't until a few days later that I realised such a scenario could so easily have been an analogy for my life post-Tony. I'd spent so long rushing forward, ignoring little things that needed fixing (especially the little thing that became a big thing called a mortgage) then wishing I could go back and rectify the problem so it had never happened. It was a "coulda, woulda, shoulda' kind of attitude that I'm sure many widows and widowers can understand. We are either looking in the rear view mirror at the parts we've left behind or constantly trying to whizz forward to a settling point. And I still needed to clear the debris.

Knowing that we were going to have to give up L'Horte, my siblings and I decided over the holiday season that we should spend the money on another property, as close to L'Horte as we could find. Somewhere the family would own but where Mumo and I could take up residency. Unlike in the UK, under French law all property passes directly to the children when a parent dies, although the surviving parent has the right to stay on in any owned property for the remainder of their life. Wills are mostly irrelevant when it comes to property, you can't disinherit a child and you can't kick your mother out – not that any of us would. The sale of *Chez Knollys* would give us the bridging loan needed to buy somewhere else until L'Horte was sold. The land was ours but the houses were not.

Then there was Danny Boy. He and Arry were at a fragile peace but with Mumo having the other two Shepherds, I was under no illusions as to what would happen if there was a fight there. I began to put out the word. There were also dog passports to sort out and vaccinations to check or update. I was still desperately short of money but for once, the financial fairy was on my side. A couple of months before, I had been idly watching the television when a segment came on about PPI, better known as Personal Protection Insurance, you know the thing you know nothing about but is added on to your store and credit cards? The expert being interviewed recommended that even if you didn't think you could claim, you should try anyway as there was a deadline looming for pay outs. I had applied for a PPI claim a few years before and got a tidy sum but the only other store card I had was well out of date. I filled in the online form anyway, thinking that who knows it might work even if I didn't remember my password or which email account I had used and pressed send. And forgot about it. That was until a nice envelope (no mortgage stamp on this one) popped through the letter box, cheque enclosed. At least the woofers could get to France.

In the meantime, the hunt was on to find that property that would suit both Mumo and I. And a lot of woofers. Two houses on the same land, like L'Horte would be ideal but diamonds were hard to find. There were a couple of properties Mumo had looked at so we had a chance over Christmas to see them. Let's face it, they were not what I had envisioned. The problem was, as I saw it, my siblings didn't realise how much space we needed for so many dogs; 13 at the time which, looking back, should have warned one of us surely. One house did have a stunning infinity pool with a wrap-around terrace; I pictured the scene: The writer sips her morning coffee whilst looking out on the view provided by the sun-swept deck, a calm breeze ruffling her hair as she listens to the gentle sound of water lapping at the edge of the pool. Except that the view was a road, the garden too small and I would have to squish into a tiny extension at the side of the house. The other potential property was worse although it did have 'separate accommodation', that being a tiny hovel with a bathroom in the kitchen and stairs so narrow you'd never be able to eat anything without getting stuck. And it had no land. I drove back to London feeling somewhat disappointed and wondering whether wellies in Worcestershire or Wales was a better option.

Having met with the estate agent in question (I interviewed a few but liked this one best), I'd put *Chez Knollys* on the market. The price was far more than I had anticipated, the new kitchen come dining area with fabulous bi-folding doors was a hit even if the rest was for someone else to 'see the hidden wonders' so the New Year held promise. A couple of mornings later, I noticed someone had attached a 'For Sale' sign to the front gate. Reality started to kick in and with it, my worry gene. The "what the hell am I thinking" kind. For the first time in my 53 years, I was going to be solely responsible for the decisions I had to make. It was a daunting thought – I couldn't bury my head in the sand anymore and expect someone else to wave a magic wand. Yet, I felt strangely liberated too. Forget a weight being lifted from my shoulders, it was more like a colony of albatrosses taking flight. And if I had had any lingering desires about staying in London, sitting in traffic one evening and looking at the stream of red brake lights in front of me, threw them out of the window. For almost a quarter of a century, I had worn this town over me like a suffocating blanket. I wouldn't miss the endless potholes and lack of parking spaces, the ticking of the client clock or the 10 second traffic lights. Kensington and Chelsea had been my 'manor' though; on virtually every street in this bit of the city lived, or had lived, a dog I'd trained. Memories filled my head of running 14 or so dogs over Hyde Park, Dog Hollow's legendary lunchtime 'meetings', watching puppies grow into well-mannered members of society and holding an owner's hand when that same puppy grew old and it was time to say goodbye. I might not miss the traffic but I would miss the people.

Telling clients and friends that you are about to shut up shop and move abroad, I discovered, went one of two ways. There were those who were genuinely thrilled that I had finally decided to make the leap but others were less than pleased. On the one hand it was nice to think I was that important to them but the moans about who was going to look after their pooch whilst they went on holiday or who was going to drop everything because their furball needed a last minute haircut took me aback. Friends were the same. Most were already planning visits to France and the sunshine but there were also a few who thought I was making a mistake jacking in my career and moving to a foreign land. But I had made my mind up, as that sign on my gatepost stated, all I had to do now was wait for someone to add ‘Sold’ to it.

Of course, it goes without saying, that any ideas I had about driving off into the sunset in the near future were bound to hit another traffic light. 2020 was about to drop the world into a pothole.

CHAPTER 16

"I know I have to be patient and keep my resolution to take life as it comes but it's not that easy. And I know I'm the impulsive sort but flying solo for the last 18 months has made me yearn for balance and calm in my life. Not just my life, the woofers and Callum need it too. Rollercoasters make you queasy after a while." (26th Jan 2020)

James, the estate agent, came back from his festive break full of enthusiasm and before I knew it, we were to have visitors. I went through the house like a demon possessed; scrubbing floors, opening windows and polishing everything and anything that caught my eye. Having 10 woofers underfoot whilst dusting light fittings was no mean feat but at least there were no paw marks on the walls or hair in the fridge (how does dog hair get in there?). I even made a sort of Avant Garde rockery in the front garden where Steven had piled up all the concrete from the now bricked driveway. I'm not sure Monty Don was swooning but the amount of compost I threw in would have given any plant a chance. I'd resuscitated an ancient Geranium that I'd found on top of someone's rubbish bin and since the roses in the back garden were in abundance, I relocated a couple of them. It still looked a little bare but at least it had colour. My stress levels were hitting the red zone and it didn't help that Callum was virtually bed-bound with some sort of flu.

Although I reasoned with myself that whoever wanted *Chez Knollys* would expect to put their own stamp on it, I began to see flaws everywhere. Did I put too much furniture in the extension? Would they see those scrapes on the bathroom counter? Why did I replace the artificial grass in the back with real stuff that the woofers had now turned into mud? Why didn't I do up the attic space and put a new bathroom in? The last one was quickly dismissed, I'd run out of money.

With bookings to view the house in the diary, I planned ahead. I put the woofers and a feverish Callum in the garage, now called the annexe in the brochure. I'd moved out the tenant a few weeks before much to his dismay. We had a side gate leading to the back garden which had a rather cute chalet at the end – something Tony hated but I loved mainly because my girlfriends and I could giggle our way through far too much wine in there. When a visitor wanted to see the annexe, I'd shift its contents up to the chalet via said back gate.

The first prospective purchaser was charming, at least he didn't comment about the décor or my odd landscaping. It turned out he even owned a dog. And he said "I like it" several times. Whether or not he did, I wouldn't know but my woofer plan went almost perfectly. Moving 10 dogs up the side of the house and through a back gate whilst the client and agent are walking out of the front door is like herding schoolchildren onto a bus, at least one is always unaccounted for. In this test case, it was Neo who let's face it, is the only one that needs to be avoided and he made a dash for the front door just as it opened. In a former life, I would have been a rugby legend.

The viewings started gathering pace and with the woofers getting used to being cattle-marched through the back gate, I started to have moments of panic. The dream about moving to France (even if I didn't know where) was becoming a reality and as I said before, for the first time in my 53 years, I was going to go it alone. Sort of.

The woofers still needed their travel arrangements and I had to start thinking about packing up. I booked several appointments with the vet, after all there is a limit to how many dogs they can handle at any one time and ordered several journey's worth of cardboard boxes. Knowing I wasn't going to continue with the grooming business, I called my old friend Nat to see if she wanted to buy some of the equipment. Nat (short for Natalie obviously) and I have known each other for over 20 years, she being a former client of Dog Hollow. The first time I met Nat, she terrified me. Not only was she one of those people that could spot an amateur a centimetre away, she never said a sentence when two words would suffice. Yet for some unknown reason, Nat trusted me and we became lifelong friends, she even adopted one of my Romanian foster dogs. Since Nat had got her boarding licence (without Sharon's involvement), she was going to take the dogs that I had been looking after, so we both thought the grooming stuff would be useful and she'd pick it up once I had an 'exit' date. During one conversation, she mentioned that I might want to consider getting a camper van or such-like to drive the woofers over to France, I nixed the idea saying it wouldn't be big enough. But maybe something bigger?

One evening a couple of families arrived to view the house. It was oddly emotional seeing a new generation move from room to room. Flashbacks of bringing a tiny baby home from the hospital to a freshly-painted nursery, the various dogs we had owned rushing through the gardens, the crazy summer parties we threw and the 'how's your day been' hugs. They were all here within these walls and now for another family, it would be their turn to make memories for the next twenty years or so.

As much as I wanted to go into selling with an open mind, my mood was beginning to wane. Having had a steady stream of viewings, no-one had made an offer and everything started to peter off. I'd become quite the expert at reading body language although dogs are my expertise and they rarely comment about wallpaper. The

‘I can’t really afford this but it’s nice to compare’ to the ‘perhaps she’ll take a really stupid offer because she’s a widow’ facial expressions. The latter was more common. Family and friends kept telling me that the house would be snapped up but I was deflated. I was reminded that if the house did sell, I needed to prepare myself for the big move.

Although, Dog Hollow was no more, I was still doing the occasional dog groom (I specialised in hand-stripping so was never short of clients) and the odd behavioural consultation. I would pick up a couple of furry clients in the morning, take them to my grooming ‘salon’ just off the kitchen and then drop them back in the afternoon. These ‘pay days’ I could juggle between viewings but I had stopped boarding dogs. I had had to tell my two trusted dog walkers, Vanya and Susan, that I no longer needed or could afford them. Vanya took it hard, I’d been pretty much her main income for two decades but I had barely enough in the bank to get through the next few weeks let alone possibly months. However, such inconveniences didn't stop me from thinking about securing my financial future once I did sell up and move out.

I had toyed with the idea of buying a flat in London ever since I decided to leave *Chez Knollys*. Not only did I need some sort of income, I wanted to keep a foot in the London property door and have some sort of inheritance for Callum. As I had no control over how long it would take to sell my house and with time on my hands, I made a few appointments with letting agents and went to look at some potentials. At the beginning of 2020, the property market was on the up so whilst it was great if you were selling, buying wasn't so good for my meagre budget. Apart from the shock over the price of a miniscule 'can't swing a cat but the rent is fabulous' studio flat, most were short-term leasehold so not worth the investment.

My initial optimism was near bottom until I saw an advert on one of the online searches. I knew the area well; Tony and I had lived in Battersea when we were first married and I'd often passed this particular estate. At that time, it was owned by the local council but now it had been turned into private flats, two of which were well within my budget. Feeling quite the investor, I arrived a few days later to meet the agent. I had the feeling he had been pulled out of retirement to meet the growing market, he was more than a little out of breath climbing the stairs and I wasn't totally comfortable either. I've never been good with heights, specifically those that have narrow walkways and railings overlooking concrete and both flats were on the third floor. On the upside, the first was a charming one bedroom affair with a newly renovated kitchen and bathroom, the second was slightly cheaper and a little faded. Having looked at both, I decided that the more updated one would be perfect and would have been except for what happened next. As we walked out of the security gate, an anti-terrorist van blew into the estate, dropping its commandos as it whizzed past. I think I read the poor letting agent's body language very well.

Feeling like one of those cartoon characters with a black cloud directly over her head, I decided to finish stripping the paint off an old fireplace in the smallest bedroom. It was a beautiful Georgian one with original tiles so I stuck at it for hours until later that evening I got sick. One minute I was making dinner and the next I was paying homage to the white bowl. Suddenly I went from someone who is rarely ill to a complete hypochondriac, my mind battered with Doomsday predictions. There was no way I was going to see the GP, he was never there for Tony although that health check they had offered seemed tempting now. It was out of hours anyway so I did the next best thing and went to bed with a cup of ginger tea (Tony always recommended it for a dippy tum) and curled up with the woofers. Sleep was the best medicine and I woke up the following morning feeling well enough to eat an entire bag of M&M's chocolate. But being ill scared me. I'd lost Macgyver, Tony, Jeremy, Keith and Pop in so short a time – would anybody be there for me? I couldn't expect Callum to stay by my side forever, in fact he'd found somewhere to live close to college thanks to Rene's parents, so would anyone be there if I really got sick? I Googled, 'what to do if I collapsed into unconsciousness?' 'Can one give oneself CPR?' 'Can one revive oneself with little chocolate buttons?' Learning to cope on your own is part of a widow's journey but knowing that doesn't make it any easier. I missed Tony more than ever.

Life was unfair and circumstances were about to prove just that. One morning I let the woofers out into the front garden as usual and heard a scream. I ran out the door and saw Arry and Danny locking heads, Danny wasn't going to win this one. Arry's jaws were wrapped around Danny's throat and as I frantically tried to disengage them I slipped in the muddy February slush. If it wasn't for Callum hearing me and grabbing Arry, I'm not sure Danny Boy would have survived. I had no choice but to get him out of the house and Nat yet again, came to my rescue. She drove up the same day and took him back to her place; I didn't even have a chance to blink. I'd had Danny since he was five months old and now 6 years later, he was gone in a blink. But where there's a hurt, there's a joy and within a few days, a wonderful lady called Diane offered him a home. I am still in touch with her to this day and Danny Boy has never been happier.

The Who must have gone to sleep because suddenly there were rainbows and silver-dusted clouds over my head. Mumo had called me to say she and my nephew, Louis, had found a possible property that would suit us both, and the woofers, in a little village a 5 minute drive from L'Horte. Aside from once popping into the service station below Rouffiac d'Aude, only to discover they didn't like my credit card, I knew nothing about the place. The photographs were promising, the house was on the main road and had a decent sized apartment attached, albeit its terrace was covered in vines and it had a bathroom which was a fetching shade of sixties blue. It might have been a bit of a doer-upper but the two-ish plus acre gardens were a woofers heaven, that you couldn't ignore. Before I knew it, I was about to buy a quarter of a house I wouldn't see until I'd actually moved in. And I could because someone made an offer on *Chez Knollys* and I accepted it.

I spent the first day after the 'someone wants to buy your house' phone call in a total state of panic. Suddenly, there were rows of boxes lining the hallway as

I laid into every room. When I wasn't filling out various legal forms, I was at warp speed shoving everything we owned into containers ready to be moved to our new home. As I was clearing, the new owners were popping by with builders to discuss the changes they wanted to make, my home was about to take on a new life as I was about to start mine.

CHAPTER 17

I'm generally a positive person and in spite of the need to relieve boredom, I'm going to stay that way. I don't read the papers and definitely not the doom and gloom on social media. And I'll get to France at some point in the not too distant future. I did consider buying a Winnebago the other evening if I can't get there at the end of April, now that would make a great book, "The Widow plus Woofers and a Winnebago." (22nd March 2020)

If there was one luxury I couldn't do without, it was having my nails done every month. False nails. I'd go down to the local salon and listen to the chatter whilst my hands and sometimes, my feet, were cossetted and painted. The shop was run by a group of young Filipinos who were not only incredibly fast but artistic too making the place very popular with the locals, mostly mothers having a bit of time out before they picked up the kids. One day, I was idly ear-wigging on a conversation when one of the clients started talking about a Chinese virus she'd heard about in the news. Perhaps she shouldn't be allowing these 'people' to be doing her nails (beautifully I might add)? The talk between the mothers started getting nasty and I knew I should say something but I didn't, I just smiled at the girl painting mine and pretended I hadn't heard. I should have stood up for them. Later that evening, I turned on the television and for once, watched the headlines. There wasn't much except a short piece about said virus infecting a number of people in a small part of China. A week later and the world's media came up with a thousand different ways in how to include 'unprecedented' into a sentence.

Since I'd pretty much packed up the whole house and boredom was creeping in, I decided to make the most of the time I had left in London. I met friends for dinner or drinks; Serena, Keith's wife, and I had a regular catch-up in a bar close to where they had lived and I made the best use of my Oyster card visiting as many museums and galleries as I could. I had memberships for the Tate, the Victoria and Albert Museum and Kew Gardens but had rarely used them when I was working. Rene and I would often meet up to go and see a new art installation or photographic exhibition and Jenny (my mother-in-law) would catch up with me at Kew Gardens. I became a sort of social butterfly fluttering between exhibitions, restaurants and pubs enjoying the company of friends and the city that I was about to leave, until one day everything stopped. On the 23rd March, we went into lockdown.

For someone who had spent so many years chasing the clock and appearing to thrive on being a control addict, the enforced stay-at-home time had an unexpected consequence. I started to relax. I still kept to a routine; the woofers were still walked every morning although it did mean battling a new set of pests otherwise known as 'Coronajoggers' – social distancing didn't seem to apply to the fleet of foot. I did my yoga and started French lessons online. I found a box of 10 jigsaw puzzles and waged war with The New York Times Spelling Bee which became an addiction. There were new ways to connect with friends via video apps so if I wasn't on the phone, I was on my computer. It was just as well that I was confined to home as in the first week of isolation if you recall, I had lost my front tooth. Like most of the self-employed workforce, my income had been reduced to zilch as being anywhere near humans was a big fat no-no but I was feeling surprisingly Zen and sleeping better for the first time since Tony died.

A few weeks in and my 'Dunkirk spirit' had been reduced to 'Dontknowwhatdayitisanymore spirit'. I was having recurring nightmares involving me, my nightwear and a desperate attempt to get the bins out before collection day. My brain no longer recognised the difference between Monday and Thursday and my less-stressed attitude waned with each passing day. With the solicitors dealing with the house sale disappearing into a vortex otherwise known as 'furlough', I didn't have much to do except make to-do lists. Lots of them. I split the woofers into two groups and started walking them to the local park instead of driving up. The car parks were now full to bursting point due to the sudden need for everyone to take up their allotted outdoor exercise time. Where once upon a time I would have met a familiar face, there were now inexperienced owners letting their dogs cause chaos as their 'walkers' couldn't take them out so unleashing all my lot at once was akin to putting an accelerant on a forest fire. I would time the walks so that one was done before most park enthusiasts got out of bed and the next whilst those who had been out, were having their elevenses. Not only was it a really great plan, it also used up the whole morning and the woofers were exhausted. Then I would spend the afternoons cleaning and packing away furniture, Googling Winnebago sellers and trying various cements that would keep my tooth in place.

London was at a standstill, the eerie silence becoming more noticeable as we headed into Spring. There were no cars hooting or stereos blasting loud music, no screaming kids or smells of burnt BBQ's – things you kind of got used to when a bit of sun appeared in Streatham. Even the skies above were quiet, no planes flying today thank you. I don't know if it was the angst of not being able to do much or the warmer days but I was suffering from a case of IMT. I miss Tony. I missed the way he always knew how to talk me out of a funk or the way his hugs always made me feel better. As I lay stretched out on a sun lounger one afternoon (one piece of furniture I thankfully hadn't packed) book in hand and headphones on, I could almost hear him say, "Soph, you're one of the lucky ones. You have your health, a house about to sell, an amazing son, wine in the fridge, toilet rolls in the cupboard and a collection of canines that make you shift your butt every morning." And he'd be right, I did. When life throws you lemons, a dog will always bring them back to you and make it a game.

May 2020 and the second anniversary of Tony's death. Knowing it was impossible to have a weekend away, as we did the first time around, my same girlfriends; Lareine, Georgie and Rene and I decided to have a celebration on the front lawn – that way we could do all the social distancing necessary and enjoy the sunshine. Phil surprised us all with a visit, bearing flowers and champagne – if I could have, I would have hugged him. We lay out on the grass; luckily this greenery was now out of bounds for the woofers, each with our own blanket, bottle and glass – like mannequins on public display. We may not have been in some exotic location as we'd decided in Ramsgate but with the late Spring temperature already in the 20's, we held up a toast to my guy and lazily emptied the contents of our purchases. And the champagne. I'd got through another year and for a moment, I felt a glimmer of hope.

Determined to remain positive, I started putting together ideas for the house I was about to move into, the one I'd only seen in photographs. Actually, apartment would be more appropriate as it was one, attached to the main house. I'd never designed my own place before; Tony and I always had very different ideas about interiors – he liked stuff and I liked minimalism, so I was looking forward to bringing out my inner interior designer. I knew that the place would need a new kitchen and my younger brother, Moth, felt the wall separating the kitchen and lounge should come down but I was far more excited about doing up the ridiculously blue-tiled bathroom. I'd always hankered after a rain shower and a powerful one at that, the one at *Chez Knollys* was a pitiful drip mechanism that constantly changed temperature whilst you were trying to rinse. At least once my London home was sold, I'd be able to renovate without pinching pockets but even that was taking far longer than it should have.

As the days passed, the silence from the solicitors started to wind up my worry gene. I started worrying about everything, forget making mountains out of molehills, I made an entire range. My superstitious Scorpio went into overdrive. Was it because I had packed the lion statue? Was Wookie unhappy about all the boxes in front room? Wookie, just so you know, is a carved wooden sculpture of an elderly man who gives me the heebie-jeebies but Tony loved him. Was the universe out to get me all over again? I over-thought everything. I was terrified that something would stop the house sale although the prospective new owner had told me that he was having the same problem locating his solicitor as I was. Any sane person would consider the Corona cause and effect and all those government guidelines about not moving but not me. I just stepped on the panic alarm. To top it all off, I had found the perfect motorhome online in which I could transport the woofers et al but I needed to put a deposit down which I didn't want to do if everything was about to go pear-shaped. I couldn't help being relieved when the said prospective owner asked to come by and measure up, Tony whispering "I told you so's" in my head.

The lockdown was turning me into a grumpy middle-aged woman however. I could easily have been mistaken for someone with Tourette's with a mood to match The Grinch during the morning dog walks.

Between having to move dogs around joggers and pavement cyclists who thought warp speed would beat the virus and oblivious dog owners who couldn't understand why they ought to keep their poodledoodle away from a crazy lady trying to contain more than her bodyweight on a lead, I was literally at the end of my tether. Neo almost had one jogger's ankle along with my "his lead is only half a metre long mate so one of us isn't observing social distancing" retort to his scream. To add to my dour disposition, the initial peace of lockdown had been replaced with one particular neighbour's need to blast Brazilian house music across the divide until the early hours. I was bored and frustrated by the lack of any movement with anything. The light at the end of the tunnel seemed to be getting further and further away and so was Tony. Two years and still never a day where I didn't think of him or wish he could just give me a hug. Or maybe he should have sent down a red card.

One minute I was ready to pack up and go and then the phone rang. My buyer was pulling out, Coronavirus had cost him his job. My positive bubble burst. “What did I do to deserve this?” I blubbed into Callum’s lap. “Why is everything so unfair?” But I pulled up my big girl pants and went into practical mode. Ring family, check. Ring estate agent, check. Ring removal company, check. Ring motorhome depot, check. I wasn’t going to go down with the ship, this girl had plans abroad. Within a few days, James the estate agent had the viewings rushing in. Hand sanitiser placed in every room and masked up, Callum and I seamlessly moved the dogs around whilst people poked and prodded. James informed me that, due to government guidelines, only serious buyers were allowed to view houses which helped the optimism. I ran around with disinfectant and gloves, wiping down door handles, answering questions and giving thoughts on improvements. One couple had even cancelled their wedding in Italy to see the house; it must have been a cheap one because their offer was embarrassing. With the Chancellor now dolling out a bit of money to those of us who were self-employed and the neighbour's boombox silent, something had to go my way. Nat had picked up the grooming equipment, adding to my purse and my old friend, Mark, had offered to buy The Black Rose. It was only fitting seeing as Mark had a small holding that my Land Rover should return to doing what she was born to do; carry sheep. I was determined, organised and moving forward and then, another phone call. Chrissie, my mother's beloved German Shepherd, was missing.

My nephews, Louis and Max, had been at their mother's house near the French Alps when Mumo rang and despite the travel restrictions (no-one was allowed to drive more than 8 kilometres from their home), drove 5 hours unchallenged through the night to search for her.

The village of St-Hilaire had turned out in droves to help but sadly, the boys found her drowned in the river; the same river that had taken L'Horte and in a way, my Pop, from us. Chrissie was Macgyver (my Shepherd)'s younger sister, a fiercely protective lady who loved family above all else. I'd bought her as a present for Mumo some 13 years before and they were devoted to each other. Louis and Max buried her next to Pop and Macgyver, the only difference being that they buried her whole body rather than her ashes.

In June 2020 we came out of lockdown and as if the clouds had lifted just for me, I had another (immediately accepted) offer on the house. It wasn't as high as the previous one but I wasn't going to ponder about it, I didn't trust the housing market and I needed out.

Letter to Tony 17th May 2020

I miss you and I always will. Even after two years, I can still remember your smell, your smile (and that tic) and the comfort of your arms around me. These are things that will never leave me no matter what the future brings. Thank you for all those one-way conversations, it does help you know even if I sometimes blame you for things that go wrong. I wonder what you would have thought about all this lockdown stuff? I think you would have taken it in your stride except for the football situation especially for Liverpool. Thank you for keeping an eye on Callum and all the family. Oh and thank you for looking out for Dickie. Say hello to Pop and your Dad and to all the woofers up there. I know Pop will be happy to have Chrissie by his side. I'll probably chat to you later as usual but for now, thank you for being you even if you aren't with us anymore.

CHAPTER 18

"I came very close to giving up this week. It seemed every time I took a step forward, another rug was pulled out from underneath me and frankly, I'd had enough. Just as I thought we were about to exchange on the house, problem after problem got thrown in my face to the point where I just wanted to go to bed and give up trying. And yet, no matter how many times that rug was pulled away, someone managed to shove it back under my feet again." (12th July 2020)

It is said that moving house is one of the most stressful experiences anyone can go through but I would argue, it's the bit in-between. The constant flurry of questionnaires from the solicitor asking me to tick a hundred and one yes or no boxes, did I want to leave the lightbulbs or take them with me sort of thing. I know it was necessary but did they have to duplicate everything? Callum had insisted on us taking a couple of the internal wooden doors from upstairs which the new owners hadn't minded about and in return, I gave them the made-to-measure curtains that covered the bi-folding doors but I wasn't going to worry about lightbulbs. Everything bar the kitchen and a couple of mattresses was now labeled and packed in the annexe in the vague hope that we might get out of London by mid-August. It was June and I was bored rigid. The only plus point was that lockdown was lifted so I could see friends in the flesh, albeit no hugs and socially distanced chairs but there was nothing like a face to face catch up over a glass or two. I had, of course, planned a big leaving party but that was out of the question unless the police wanted to wish me a swift goodbye.

As the old saying goes, 'out with the old and in with the new', my siblings and I signed the contract for the new house in Rouffiac just after I had accepted the offer on *Chez Knollys*. For once, timing was on my side. My older brother, Simon, amusingly pointed out that we signed said document on the 18th June, the anniversary of The Battle of Waterloo. And we all know how that turned out. It was hard not to get excited about moving but I had to keep grounded. Although I'd sorted out the logistics of getting from London to Surrey, thanks to Nat and her partner, Iain, and Rene was 'on board', no sale was complete without money in the bank.

If only sorting out all the administration at home was as easy as organising one's friends. I couldn't help but think how smoothly Tony would have handled everything; he would have had spreadsheets and white boards at the ready. All the documents needed for the house would have been filed in order and of course, he'd know where he had put them. Unlike myself. He wouldn't need to dig through wads of nonsensical paperwork trying to find a twenty-year old document about drainage, he'd have put it in a file starting with 'D'. This June should have been his, not only would it have been his 59th birthday but also because his beloved Liverpool FC had won the Champions League for the first time in thirty years. Nothing made him happier than a good football result, as Dickie and I both mused – he'd have been punching the ceiling.

And then just like that, it was done. Just a few words on an email and a signature. Contracts were exchanged and the home I'd shared with my husband for over two decades was about to become someone else's future and mine, well that was in a different country. Of course, the Gods and the Who couldn't help but stick their noses in – my printer decided not to work so I had to ask Michelle next-door to print off the documents and would she mind being a witness too. In return, Michelle asked me to be a trustee for her will – something I really should have sorted myself but had never got around to doing. By the time we'd finished exchanging signed bits of life's essentials over the garden fence, I had managed to acquire a will as well – even if I didn't need one in France, I wanted to make sure the woofers would be okay and Callum wouldn't have to hunt for my final resting place. Or throw me in the sea. I get seasick.

July arrived and knowing that we would most likely have another lockdown after the summer, I booked the removal company for the 9th of August – we'd leave the next day. They had already sent around one of their representatives to calculate what was going in where and how many breakable items they were going to have to repackage, apparently an art in itself that should only ever be attempted by a removal specialist. Since I had already filled the annexe to the rafters with boxes and bits of furniture – it wasn't a difficult task. The woofers' passports were stamped and signed; I'd even bought them new tags with their French address on, just in case they got lost somewhere between here and the Mediterranean. I posted lists everywhere; insurances that had to be cancelled, the Eurotunnel booking to be made and the Sky TV subscription to stop. The latter being the biggest headache. Apparently there would be a penalty fine if I cancelled before the end of the contract so I asked the poor minion on the other end of the phone if I could take it to France. No, they don't do France. I suggested they might want to explain to the new owners of my house that I was no longer able to move as Sky disallowed such heinous acts of illegality but I think the conversation was lost in translation. Needless to say, I just stopped the direct debit and let them try and find me somewhere in a country they 'don't do'. Mercifully, closing all the other accounts was easier if one didn't mind listening to endless repetitive music. Sorting out the motorhome insurance did raise my blood pressure a tad;

every self-respecting (and legal) motorhome must have a specific type of steering wheel lock which looked like it would have been better suited to an army tank. I had to lie repeatedly about how long I was going to be 'on holiday' in France. I didn't mention that we were being accompanied by 9 woofers either who had been known to bark when a bee knocked on the door. They might have added extra.

There was, of course, the small but important matter of Callum moving to his new digs. It wasn't the first time Callum had left home but this time he wouldn't be coming back. As Rene arrived to take him over to her parents' place along with his mountain of computer equipment and the several boxes of coins and stamps he had inherited from my late father-in-law, we said goodbye on the doorstep. I asked him if he would miss the house. "It's just bricks and mortar, Mum," he said and tapped the brickwork as he went. I watched him; this grown-up, the way he walked with a long loose stride down the garden path, thinking briefly back to the time Tony and I had gingerly carried our new-born baby up it. The years seem to rush through my head like a speeded up piece of film. The next time I was to close that door I knew I wouldn't look back. I was about to take a giant leap through the portal and leave the past where it belonged. Between these walls.

The woofers were beginning to notice something was afoot. Sets of eyes were watching me constantly as I moved from room to room, putting the final little bits in boxes. It struck me that some of them had never known anywhere but Chez Knollys, the exception being my rescued lot so I was grateful for their ignorance at what was about occur. It didn't take much to over-excite Arry and Alice so the less they knew the better and Evee and Yogi were too busy mourning the absence of Callum to notice the suitcases. Thankfully none of them were going to sneak in a few small get-togethers for their canine friends over the park. Unlike their mother.

Any lingering self-doubt about moving was put to rest after a couple of 'intimate' farewell gatherings. The first was spent with my old Dog Hollow crew; I'd extended the olive branch to both Nicky and Alison but only Alison showed up along with the rest of the gang. We spent the evening reminiscing about the old days and all the dogs that had come in and out of our lives. Alison was noticeably on edge as her and Nicky's new training company had dumped our former team mates and I was the only one drinking – it made for an awkward night. The second time I tried to say goodbye was with my 'girls'; the mums I had known since our kids went to school together. Lareine decided to bring one our kids' old school teachers with her which would have been fine except the focus seemed to be on her and not me. I felt excluded as everyone went on about future nights' out –

ones I wouldn't be a part of. The evening ended with Lareine getting totally plastered and falling asleep on a dining room chair after everyone left but not before she had berated me for leaving. Bless Lareine, she always wears her heart on her sleeve and tells you exactly how she feels. I had to call her husband in the early hours as I didn't trust her not to fall out of a taxi. I know her so well. Both experiences, however pear-shaped they had gone, only strengthened my resolve. I had made the right decision to leave London. I knew some of my friendships would last a lifetime (Alison and Lareine included) no matter which side of Channel we were on but others were bound to fade. If I felt a little sad at leaving the old neighbourhood after 24 years, someone in one of the adjacent terraced houses decided that Sunday mornings and Balalaika music played at full volume was what everyone needed as a goodbye gift. I took my phone up to the back of his fence and played "Let it Go" from the film 'Frozen' on repeat until he gave up. My parting gift.

Having spent the week saying goodbye to one and all, the last full day at *Chez Knollys* arrived. The removal van had come and gone, not without a hiccup or two of course.

"I'm sorry Mrs Stewart, we can't take this. It's illegal."

"Oh, don't worry, it's not loaded. The ammo is already in France."

"Well, we shouldn't but as long as you are certain, we'll take the risk."

I'm not sure when I should have pointed out that the gun case he was so worried about, didn't actually have a gun in it. That is to say, no ammunition enclosed in that very light canvas bag. One of the other removal men asked me if everything was to go and naturally I said yes. Why he took the empty holdalls that I was going to use for my clothes was slightly beyond me.

It seemed only fitting that my last night in the house should be spent with Phil; 2 years before he'd been the person by my side when Tony died. I will always be forever grateful to him for holding me upright and being there for Tony. We talked about their friendship and events leading up to that fateful day, raising glasses and mulling over memories. But Phil had something he needed to tell me. Tony hadn't wanted me to see his doctor or know anything about his prognosis. He knew I'd panic so had asked Phil to keep them away from me. The doctors had called to see how he was doing but had called Tony's mobile and not mine – Tony had ignored the calls. Something Callum had been confused about when he had broken into his father's phone but hadn't shared with me. He thought I knew. I would have expected to feel some sort of anger towards

CHAPTER 19

"I am often asked if I am happy here and do I think I made the right decision in leaving London. It's a difficult thing to answer because, as I've said before, there was my life then and there is my life now. There are times I have twinges of guilt; is it right to be happy when you have been widowed? For the first year after Tony died, I used to wake up most mornings hoping the nightmare was over and he would be there snoring by my side. Leaving London and that part of me behind was the best thing I ever did and yes, I think I do deserve to be happy. France and all its quirks is where I want to be. (Blog 18th July 2021)

The first time I laid eyes on my future home, I wasn't exactly bowled over by its décor. Stunned into silence would be more appropriate and not in a good way. From the overgrown vines creeping across the verandah resembling a spider condominium to the musty kitchen with chicken motif tiles, it was less of a do-upper and more of a demolition job. The main house didn't fare much better; there was mould seeping into the walls and the electrics were nothing more than an incendiary device without any earth wires. But then there was the land. Yes, it was a complete mess with half-dead trees, an old chicken run (I had the matching kitchen tiles)and a decrepit vegetable garden stuck in the middle of it all but I was sold. The apartment and the house would have the professional touch but this land was mine, I had plans for it.

The thing about living in a village whose residents number less than the whole of Knollys Road is that everyone knows you before you've even met them. The local corner shop, or épicerie is the central hub for any local news but our neighbour, Louise is the primary source. Louise is a born and bred local with looks that belie her 80 plus years and an encyclopaedic brain for everything Rouffiac. We might have a mayor but the Contessa, as I nick-named Louise is in charge. It was she who introduced me to Denis – a sparkly eyed widower who didn't and still doesn't, speak English. Denis became my confidante and so much else.

Being in the middle of nowhere France, I was determined to integrate myself into the community. It wasn't just because you couldn't pass by a single person in the village without acknowledging their "Bonjour", I'd spent over a quarter of a decade in a city where people would turn their heads or look down at their mobile phones to avoid any greeting. I wanted to be part of it all. There is a misconception amongst those living in the U.K that the French are an unfriendly, sour lot but nothing could be further from the truth, although I'm told Parisians are quite aloof. Of course, there were those who saw two English widows moving into a huge mansion (plus apartment) as easy pickings but Denis sent them packing with tails between their legs. And as Mumo and Pop had lived in St-Hilaire, a short drive away, for over two decades, we had their chauffagiste Max (heating and plumbing) and others that had been part of L'Horte life. But Louise and Denis know everyone and if you want something, they know who can do it. Or get it.

I didn't want to be one of those 'foreigners' who didn't speak the lingo so I had asked Cyrelle, who is part owner along with Sara of the afore-mentioned épicerie, to only speak French to me on our first meeting. This got around the rest of the populous immediately and before I knew it, I was in full on education – everyone wanted to be my teacher. Although I did do a couple of courses in nearby Carcassonne, the speed at which I have absorbed the language is thanks to those who said and still say “Bonjour” every day. Even if I once ordered half a kilo of oysters (huîtres) instead of trout (truite) for Christmas and I hate oysters.

Learning a new language and the dialect (a bit like being from London and having to understand a Liverpudlian accent) was one thing but learning the art of patience was another. Despite my relaxed ‘ex-city’ girl state of mind, thanks partly to taking up running twice a week over the hills and vines that bordered Rouffiac, there were and occasionally still are a few things that raise my blood pressure especially if they occur before my morning coffee. It has become a bit of an ongoing joke between the builders and Denis not to speak to me until I’ve finished my first cup. Although, after two years of living here in Rouffiac I have learnt that everything happens at its own pace, at the start something always seemed to test my patience before I could pour a second cup. Having spent enough time in Tommy the extension builder's world, you'd think I would be used to people not turning up on time or building materials being delayed but no, builders in France weren't any different, oh except for the lunch break. Bang on midday, they all disappear then reappear two hours later (in Max the chauffagiste's case, it's always three). I would be constantly messaging someone as to their whereabouts only to be told, "I'll be there tomorrow don't worry," which meant I might turn up next week.

On top of all that, I had to sort out my health insurance and residency permit over the phone. If you think you have a grasp of fluency, try talking to a French person over the phone – for some reason it's mandatory for them to talk to you as if they had just inhaled a semi-automatic weapon. But I did learn the art of letting it all go because I had and still have the perfect teacher; our home and the land that comes with it. Whenever I get the chance, I put my boots on and get lost in all things green, bright and beautiful. When we first moved in, there was so much physical work; pulling out metres of brambles, cutting down trees and ripping down crumbling garden buildings, it was a release from all the frustrations of muddled phone conversations and paperwork. The French businesses love paperwork, they'll send you an email with all your documentation and hand you exactly the same stuff before you leave the shop. And Denis still provides the humour, correcting my terrible local accent (I have improved) and cracking up when I say something that means something else entirely. The woofers took to him right away, even Neo, although Yogi Bear has always been my 'estate manager's' favourite. And it wasn't just the outside work in those early days that released my happy hormones, ripping plaster off walls and sanding beams had the same effect.

About six months after we had moved in, I was standing on the top of a ladder (who knew? I hate heights) and a thought struck me. A year before, I would never have seen myself, a girl who adored false nails and high heels, wearing dust covered boots and torn jeans laying waste to walls once again. That Sophi has been left behind in Knollys Road. As though it is healing itself, my mind rarely wanders to ‘those days’. That isn’t to say I don’t talk to a certain star every night and occasionally swear at him, especially when things don’t go according to plan, but the anger and pain has gone. And whilst life moves on, it hasn’t been without more heartbreak but there have been new beginnings as well. If the first two years were a battle against the Gods, these past two have been more about mediation. Yes, there have been times when I wanted to jump ship but I’ve learnt to navigate the storms instead. We may not always know where the tide will take us but the stars sometimes guide us to safe shores. And Serendipity…

……………..

Letter to Tony 16th May 2021 (the last letter I wrote)

My darling Tony,

It'll be three years tomorrow. Three years seems like such a long time but sometimes feels like only a few months ago. One minute we were laughing about growing old and chasing each other in Zimmer frames and the next minute, I'm sitting here without you. And whilst the tears dry with time, memories forever ebb and flow through my mind. The little things; your crooked smile, the way your lips used to twitch when you tried to lie and the hugs. I miss being able to mush my face in your chest whilst you wrapped yourself around me. I'm slowly letting go, every now and again I can feel you right there but it doesn't hurt anymore.

I don't think much about Knollys these days, it's funny how different life has become, almost like two separate people but the same me. The girl stressed out chasing her proverbial tail around London is now happy with a spade or hammer in her hand or running up hills with her Arry. Considering neither of us was any good at DIY, you probably laughed your head off at my bar table attempt. I know I should have been more accurate with the tape measure but then, you know I never read instructions and patience is not my strong point. Occasionally I still wonder how you would have coped with all this Covid stuff, not well without your beloved football I should imagine even if you have a front row seat up there. In my mind, you're most likely discussing Liverpool's ups and downs with someone like Bill Shankly or Gerard Houllier.

It would have been your 60th birthday next month. As much as you said you wanted Alice to have a litter, I'm not sure newborn pups are the top of your list – watch over her won't you? Ten years ago, I threw that surprise birthday party for you with all our friends and family and then whisked you off to Oman. Thank God, we didn't know what life was to throw at us.

There will be so many family and friends thinking of you tomorrow. I was just talking to Maggie the other day and she was saying how much Stu misses you. It'll be a tough day for all especially for your Mum, Frannie and Callum, Dickie and Phil too. I shall have my usual chat with you as the stars come out, I hope it isn't cloudy because I hate not being able to say goodnight. Say hello to Pop and Jeremy and Keith and all the dogs up there. Especially Evee. Thank you for getting me through the last three years even if you aren't physically by my side but tucked away in my heart. As always I love you or should I say 'ditto'...

The Woofer Files

Evee 

- Born 2nd March 2009. Tooting, South London (via Mexico). Bred by Mark Sturmey. Bought because Callum twisted our arms and Mark's car broke down so we had to look after his litter pups until he could pick them.

- Chihuahua female

- Personality: Bossy. All the sass of a Latino temptress with a bit of South London diva thrown in. Curvy (not fat), opinionated (especially if you are late at feeding time) and fiercely protective of her own (size doesn't matter). Small, feisty and definitely mighty. Known as 'Nurse Evee' as she has the ability to understand when other woofers are sick (or need their teeth cleaned). Born to rule men.

- Likes: Scaring Rottweilers and German Shepherds. Hogging pillows and snoring in your face. Callum.

- Dislikes: Being picked up by strangers, diets, being ignored.

Gizmo. 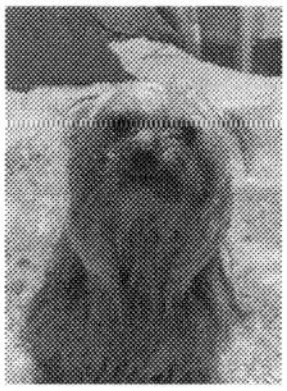

- Born October 2008. No idea where but bought at Harrods (not by us). Came to live with us August 2009 because his owner was moving abroad. Or perhaps because his owner lived in a 7th floor flat in the heart of Chelsea and couldn't understand why the pup she left for 8 hours a day peed on her cream carpets.

- Miniature Yorkshire terrier male.

- Personality: Hen pecked husband of Evee (think 1970's sitcoms). Despite his early childhood trauma, Gizmo sees the world as a full glass of bubbly. He is forever young, incredibly intelligent and loves to please everyone. He loves to greet you with a faded tissue or bit of plant – his idea of romance. And he loves to chat; he can be quite opinionated when the wife isn't listening.

- Likes: chasing and retrieving anything a human can throw. Talking. Curling up with Evee. Having long hair (makes him feel like Samson)

- Dislikes: having short hair. Baths. The robot vacuum cleaner. Being accidentally pushed off the bed by Arry. Vegetables.

Yogi Bear

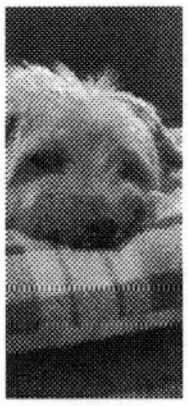

- Born 5th May 2010 in the Northamptonshire wilderness. Bred by Bear expert Linda Philip. One of only three of the pack that I bought as a puppy after I told Tony I was just going to have 'a look'.

- Norfolk Terrier (although according to Yogi, he is a bear not a dog). Male

- Personality: Bears are known for their deep thought and empathy. Yogi loves humans, especially his humans, and wears his heart on his paw. He gives the best bear hugs, wrapping his whole body around whoever needs one and prefers to lie on top of a pillow above your head. Basically, a love bug.

- Likes: People. And gardening. Especially people who like gardening. Tennis balls and not letting go of them even if you swing him in the air. Walking to the shops so he can meet more people. Food, food is definitely his love. Bears hibernate you know. Swimming in the river at L'Horte with the kids. L'Horte in general.

- Dislikes: Being left alone although technically he has the pack but needs a human. Injections. The time between meals. Other people's suitcases especially Callum's.

Simi 

- Date of birth on passport December 2010. Born somewhere in Romania. A Valentine's day plea and me being phoned on a freezing February morning resulted in two dogs coming to live with us in 2013. Simi was found with an older 10 year dog we named Ash (I am a huge fan of the Sherrilyn Kenyon books) in the middle of a particularly harsh winter. Ash had been shot, burnt and hit by a car but it was his devotion to a pretty young female than got Facebook talking. Ash became Tony's shadow. Simi found a bed, several of them. Ash died of lung cancer in 2017.

- Crossbreed (wolf or sheep in there somewhere). Female

- Personality: The ultimate hippy dog. The most laid-back rescue dog you could meet. Has the ability to blend into her surroundings to the point where you panic only to find her half a metre away. Has a laissez-faire attitude to life as long as there's a bed. Anyone's bed.

- Likes: Sun, food, belly rubs and a bit of a run occasionally. Soft, comfy duvets especially if a human has been in them first. Shedding hair.

- Dislikes: Cats (or should that be a like?), Once jumped out of the 2nd floor window at Chez Knollys in an attempt to get one. The clippers. Baths. Collar being grabbed, collars in general. Food being served late. Anything wet; rain, puddles, rivers, swimming pools.

<u>Mo Cridhe</u>

- Born 24th July 2013. Born in Romania and found on a busy roundabout. Suffered neurological damage so unable to use her back legs. I brought her in a foster dog in January 2014 but couldn't find a home for her. She stayed. Drives custom-made wheels. Named Mo Cridhe by a friend, meaning 'my heart' in Gaelic. Totally misappropriated.

- Crossbreed probably with a multitude of undesirable breeds. Female

- Personality: 'Disability, what disability?' Mo doesn't see a challenge anywhere and will take affront to any dog or human that tries to see different. She uses her 'wheels' like an extreme off-roader and can out-whizz many a dog to get a ball. Very verbal, loves to shout expletives at anyone passing the gate and loves to play rough with any dog that likes a throw down. Extremely disobedient and spends her spare time figuring out how to get her nappy off.

- Likes: Barking. Tennis balls. Donut beds (all of them). Her custom-made wheels and seeing just how well-built they are. Being outside. Especially butt-naked.

- Dislikes: People thinking she's a sweet little handicapped dog. Puppies. Being told to do something. Wearing nappies. Baths. Anyone else's dog. Vets.

Danny Boy

- Born November 2013. Found in a cardboard box by the side of a Romanian road. The first pictures sent to me were of a Malamute cross puppy with blue eyes, not the puppy I got but Danny Boy stayed. Under the late MacGyver's wing, he learnt obedience. Came to me in March 2014.

- Crossbreed probably husky and hound mix. Male

- Personality: Let's do something and do it now. Loves other dogs, using his brain. Called Danny Boy because he howls mournfully at fire engines' sirens.

Highly intelligent, he learnt from the best (Macgyver) but very cheeky when he wants to be.

- Likes: Goosing entire male dogs and then running away. Rolling in anything disgusting. Being chased (especially when it's an entire male Pit Bull), dogs in general. Being outdoors.

- Dislikes: being left alone. Children. Being cuddled. Sharing with other dogs.

Coco 'George Clooney' Loco

- Born 21st August 2013. Found in an outhouse in Romania with a broken back. Most likely caused by a broom or spade. Totally paralysed from lower back, x-rays revealed major trauma. Brought into foster with me in March 2014 but unable to find a home. Named Coco by his rescuers, and frankly a bit girly, the rest was added when his true sensitive but tough guy charisma shone through.

- Crossbreed. Male. Probably Dachshund mix. Same size as Mo.

- Personality: A sensitive soul, a homebody who likes nothing better than to curl up on his bed. Being an abuse survivor, things like mops and vacuum cleaners can bring up bad memories but with a bit of reassurance, he has learnt to avoid them. Loves attention from anyone and will claw gouges in legs to get it. Unlike Mo, he isn't that keen on his wheels ever since he got rolled over (accidentally) by two of the family Shepherds. The heart of a lion, the bark of a seal and the tenacity of a badger is our Coco.

- Likes: Mo Cridhe (although this is generally one-sided), women, cuddles and tummy scratches, dancing with Mum. Green beans.

- Dislikes: Vacuum cleaners, mops, brooms. Car rides. Being ignored when there's a crowd. Currently being in his wheels when the dogs are playing.

Neo

- Born 6th March 2012. His rescuer in Romania thought he was dead when she found him, broken pelvis, rope embedded in his neck, hacked off tail and maggots crawling over his body. Frozen in fear, it took Aurelia (his rescuer and Simi/Ash's too) three months to win his trust before I offered to find him a home in the UK. Neo came to me in early 2015 and I found a home for him soon afterwards in Wales. He was returned six months later after nipping a jogger. He came back to me a terrified wreck once more. Two more attempts at adoption failed so despite Tony's misgivings, we decided he was staying put.

- Crossbreed. Male. Spaniel mix.

- Personality: Some might say he's a bit of a split one but as the years have gone by, I've learnt to read him pretty well. Fiercely loyal to those he trusts and loves to be praised. Very intelligent. One of the most emotional moments I had once we arrived in France was seeing him run free, without his muzzle, for the first time – the joy was written all over his face. If you want a protection dog, Neo's your guy (even if he can only reach knees).

- Likes: Toys and will never destroy them. Being praised. Running fence lines and barking at nothing. Dancing with Mum. Sofas. Just being free to do what dogs do.

- Dislikes: People reaching out to touch him, sudden movements by strangers, drunk people, children who shriek. Puppies, entire male dogs, French Bulldogs, Arry being annoying, being disturbed by the other woofers when asleep (usually Arry again), nail clippers, derriere being brushed, water unless it’s in a bowl.

Gunshots, tractors, motorbikes.

Aragorn (Arry)

- Born 16th August 2015 somewhere near Rye, East Sussex. Came to me as a foster in February 2016 having already had two homes. Although we had home offers, it was Tony's suggestion that we keep as Macgyver (our then German Shepherd) was getting older and in T's words, "he's not going to cost anything". Ha. Arry turned out to have allergies to virtually everything, had never been trained and never socialised with other dogs (the first time he met Evee, he jumped onto a floor chest in terror). Macgyver died suddenly 6 months later and Arry became my grief outlet. He's almost died 3 times, put me in hospital and cost a fortune in veterinary bills. He's totally Mamma's boy.

- German Shepherd. Male.

- Personality: 'Why sit in first gear when warp speed is so much more exciting?' Tony once joked that God had given me almost ten years with the perfect dog so he felt I needed a challenge. Arry has zero spacial awareness, is easily distracted, above normal prey drive and will chase anything that moves. Loves everyone and will greet with tail slaps and monster kisses. Adores his Mum and knows when she needs his nose boop. Arry is the ultimate 'high octane' dog.

- Likes: Alice. Alice is his girl, his love, his other half (see below). From the day they met, hearts went boom. Water, any kind. Rivers, swimming pools, hosepipes, puddles, seas - if there's water, he has to be a part of it. Rocks (or cailloux) are an obsession whatever size, he'll pick them up and put them in his 'sanctuary'. Adores all small dogs especially terriers and French Bulldogs. All humans especially children and people who like working outside. Hiking anywhere for as long as he can. Chasing rabbits, cats, pigeons.

- Dislikes: Dogs bigger than him or mouthy ones. Scared of Springer Spaniels, Weimaraners and Dachshunds – and Evee. Mum being sad.

Alice

• Born 9th May 2017 in Somerset, England. I bought Alice as a 25th wedding anniversary present for Tony as he had always loved the breed thanks to Dame Carmen Callil's 'Effie'. Tony chose her name after picking her out of the litter; named after Alice in Wonderland although Clare, her breeder, oddly enough had already given her the same moniker.

• Border terrier. Female. (with exceptionally cute feet she says)

• Personality: If there was ever a dog born to be part of our lives, it's Alice. A tiny piece of dynamite that became the core of the pack. Despite the size of his nose when they first met, Alice became Arry's soulmate on sight and nothing has changed. If Evee is the matron of the pack, Alice is the psychiatric nurse. But if there was ever a dog born to be a mother, Alice is. She loves, nurtures and takes care of every member of the pack. Alice has the most positive outlook on life and sees every day as an opportunity to get food out of someone.

- Likes: Arry, bossing Arry, cleaning Arry, anything Arry. Relocating shoes. Food. Destroying toys in under 5 minutes. Running the hills. Sleep. All dogs.

- Dislikes: Getting up before 9 a.m unless we are going running. Lack of food or someone else getting food and not being part of it. Deep water.

ACKNOWLEDGEMENTS

This book would never have been written without the support of those who read The Widow plus Woofers every Sunday and to WordPress for publishing it. To be honest the idea for the book belongs to all of you who 'bullied' me endlessly into putting pen to paper especially the late, great, indomitable Dame Carmen Callil. And my editor, Sally Dixon whose eye for detail and a belief in the way I 'speak' that bought my story to life. Ally @allymuirart.com, I will be forever grateful to you for bringing Maxime's drawings to life and designing my somewhat unique cover!

To those who stood by me through the years, both then and now, I will never have the words to say how much I love and appreciate you. My bestie Rene, who I subjected to my driving with 9 woofers and a Peace Lily across France – you are the best friend a girl could have. Natalie and Iain, you were always there when I messed up or needed a helping hand, it was because of you Danny Boy found the perfect home – for that there are no words. Thank you, Dickie for being so patient with me and for being there when I needed a hug. To Doctor Tom, I don't know how I would have stayed upright without you and to all the paramedics and police, thank you for well just doing what you do above and beyond the call of duty. To the unsung nurse who broke the rules so we could get Tony to the car – thank you. And Phil, darling Phil, what would have done without you – you are and always will be my shoulder even if you kept a secret. Serena, I will miss our weekly catch ups over a glass or two – thank you for your ears when I just wanted to vent and never judging. To Tessa, you never fail to make me laugh, the Who loves you. Sophy and Sean – how would I have got down that hill without you – when you need hands, you were there. And to you Steven, for your devotion to the Woofers and your steadfastness in keeping me moving forward. To Judy-Ann and Mark, thank you for getting me out of sticky situations and to Vanya and Susan for so many years of dogged devotion. To Lareine, thank you for being the kookiest girlfriend and if ever I feel the need to dance, I'll think of you. Michelle and Trevor, what can I say? Thank you seems inadequate considering how much you did for both Callum and I myself in those early days – I hope all the vases came back intact. Darling Alex, what did you get me into? But it made a great chapter. And to your Tom for taking Callum under your wing – you gave Callum the confidence he needed. Of course, we never would have had the most superior of wheels without Chris Jones at

Motorhome Depot and for selling the Mothership afterwards – you're the best.

Here's to family. And mothers, whether you are lucky to have one like mine and are just as lucky to have an incredible mother-in-law. Mumo, without your advice, grammatical corrections and patience, this opus would never have been finished – I'm sure you could recite the whole thing in your sleep by now. I love you. Jenny, your words of wisdom and constant support for not just the book but the blog too have kept my fingers on the keyboard. To Frannie, who will always be my 'sis', we may not have been born with the same blood but I couldn't imagine going through the past few years without my 'big but slightly shorter' sister. Steve, I know you'll cringe but thanks for listening to me. And so, to my siblings and their corresponding spawn, well what can I say? Simon, Moth and Bong – I know I drive you nuts but I know I have and always will have your backs. Bong, I know we argue more than we should but your support and hugs in those most painful of days are imprinted in my heart. And to Louis, Callum, Maxime, Katie, Elisse, Jojo and Tigger – remember those halcyon days at L'Horte.

My dearest, darling Callum. The day you were born, my heart was lost. You were then and always will be the most important person in my life. Your ability to overcome all that has been thrown at you but stay standing plus having a mother who messages you every five minutes deserves a medal. I am so proud of you.

And you'd think I'd forgotten the woofers? Never. To Aragorn (Arry), Simi, Danny Boy, Neo, Mo Cridhe, Coco 'George Clooney' Loco, Alice, Yogi Bear, Evee and Gizmo, thank the Gods you can't read otherwise this would go to your heads. *Je t'adore* you smelly, disobedient, noisy, greedy, hairy lot of mix of wonderful personalities. And no that doesn't mean an extra piece of liver treat. Oh, and to the 'other one', you're not in this book but you are the most precious of beings.

This book was never written as a self-help book or a guide to how to get through 'Widow 101' DipEd. BSc. PHD. This is just my journey. But for all those that have loved and lost I have a little bit of advice – write it down. And talk to the stars…

Biography

Born in 1966, Sophi Stewart had a charmed childhood growing up in South-East Asia and Africa with her brothers and sister. Despite having a varied career from restaurant management, a brief modelling stint and several years as a fitness trainer trainer (where she met Tony Stewart), it wasn't until after a chance conversation with her vet that she decided to go to University and study Applied Animal Behaviour. She became a trainee with, and ultimately the owner of, Dog Hollow, a business she was to stay in for over 20 years. Sophi's weekly blog "the widow plus woofers" was launched in January 2019 and is still penned every Sunday.

These days Sophi still lives in South-West France next-door to her Mum and only a few minutes away from her beloved L'Horte. And of course, she still has her woofers, well most of them. Now retired from dog training, she has become quite the country girl, preferring to spend her days in boots and jeans out in all weathers unless she's running up the hills with the younger woofers. A passionate gardener, environmentalist and with an addiction to second-hand furniture stores, Sophi has embraced village life and its community. She is currently writing the 'what happened next' sequel to "Stop the car, I want to get out!" which she hopes will be published later in 2023.